INTERNATIONAL CENTRE FOR MECHANICAL SCIENCES

COURSES AND LECTURES - No. 273

DYNAMICS OF ROTORS

STABILITY AND SYSTEM IDENTIFICATION

EDITED BY

O. MAHRENHOLTZ

TECHNISCHE UNIVERSITÄT HAMBURG - HARBURG

SPRINGER-VERLAG WIEN GMBH

This volume contains 286 illustrations.

Originally published by Springer-Verlag Wien New York in 1984

ISBN 978-3-211-81846-6 ISBN 978-3-7091-2812-1 (eBook)
DOI 10.1007/978-3-7091-2812-1

PREFACE

The knowledge of the dynamic behaviour of turbo-machinery is of great importance in power production engineering and in adjacent fields. The safe and efficient operation of power plants calls for a proper understanding of the mechanical behaviour of turbine rotors.

The development of computers and computer techniques and instrumentation and vibration monitoring has improved the tools for handling rotor vibrations and stability.

It was the aim of the course DYNAMICS OF ROTORS, held at the International Centre for Mechanical Sciences (CISM), Udine, in October 1980 to present modern trends and new scientific and engineering results. In general, basic knowledge of the state-of-the-art in rotor dynamics was required from the audience and is hence required from the reader.

It has taken some time to get the lecture notes refined and printed. Although the editor has tried to coordinate the contributions,some overlapping was unavoidable since the authors presented their subjects independently. This, on the other hand, has the advantage to focus on some phenomena from different points of view.

After an introduction to modal analysis in rotor dynamics the dynamic behaviour is widely treated with particular respect to the influence of journal bearings and to rotor instability. New calculation methods in balancing rotors are discussed and so are measurement and identification techniques. Special topics such as crack problems, blade vibrations and torsional vibrations conclude the book.

The lecturers and the editor hope that this book may serve as a source of information for engineers and scientists working in the field of rotor dynamics.

O. Mahrenholtz

CONTENTS

PART I

MODAL ANALYSIS IN ROTOR DYNAMICS

CHAPTER 1.1

MODAL ANALYSIS IN ROTOR DYNAMICS

R. Nordmann

1 Introduction

Many investigations in linear rotor dynamics deal with the problems of

- natural vibrations
- unbalance vibrations
- transient vibrations.

Powerful approximation methods, e.g. the finite element method are available for calculation of the above problems. In most cases a fine partitioning of the rotor model is necessary and this leads to large linear differential equation systems for the unknown displacements (displacement method).
With such large systems the calculation is very time consuming, especially in the case of transient vibrations (short circuit, blade break etc.). Therefore methods are needed, which allow a reduction and possibly a decoupling of the equations.
A standard technique for calculating the response of nonrotating elastic

systems with symmetric matrices and proportional damping is "modal analysis". The idea is to reduce a system of simultaneous ordinary differential equations to a set of independent ordinary differential equations. The successful application of the method requires the solution of an eigenvalue problem associated with the given system. The eigenvectors or natural modes possess the orthogonality property, which permits the formulation of an expansion theorem for the response. The expansion in terms of the system natural modes leads to a set of independent ordinary differential equations of the same form as that describing the behaviour of a single degree of freedom system.

In rotor dynamics the classical modal analysis fails to uncouple the nonconservative equations. The system matrices have skewsymmetric (gyroscopic effects, internal damping) and nonsymmetric (journal bearings) parts.
An expansion in terms of natural modes of a corresponding conservative system permits a reduction of the number of equations, but they still stay coupled.

On the other hand an expansion with eigenvectors of the nonconservative system leads to an uncoupling as well as a reduction of the equations of motion. Both methods are described in this lecture.

2 Modal Analysis for Nonrotating Structures. The methods of calculation for a linear one degree of freedom system (Fig. 1) are well developed in mechanical vibrations. The solution for the displacement $\tilde{u}(t)$ can be readily obtained for different exciting forces $\tilde{F}(t)$.

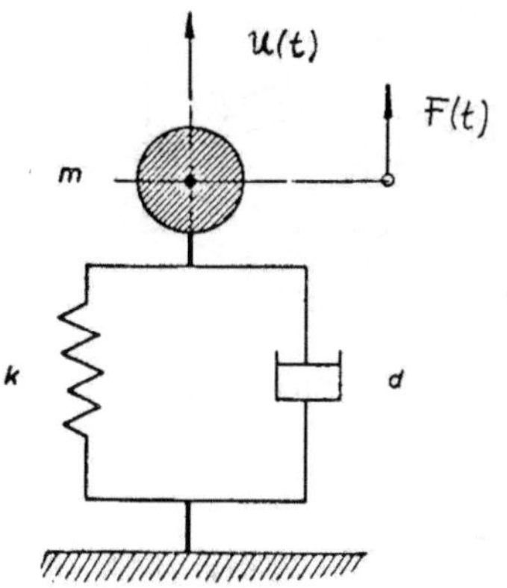

Fig. 1 SDOF-system

Real elastomechanic structures usually cannot be modeled by a single degree of freedom system. Those systems have a continuous mass and stiffness distribution and their number of degrees of freedom is infinite. They can be represented approximately by a finite number of coordinates resulting in a set of N coupled ordinary differential equations of motion of the system.

The calculation for large systems is much easier if the simultaneous equations can be transformed to a set of independent equations, each one describing the motion of a one degree of freedom system. This is possible by "modal analysis". For explanation we consider the elastic system in Fig. 2, a steel foundation for a power unit.

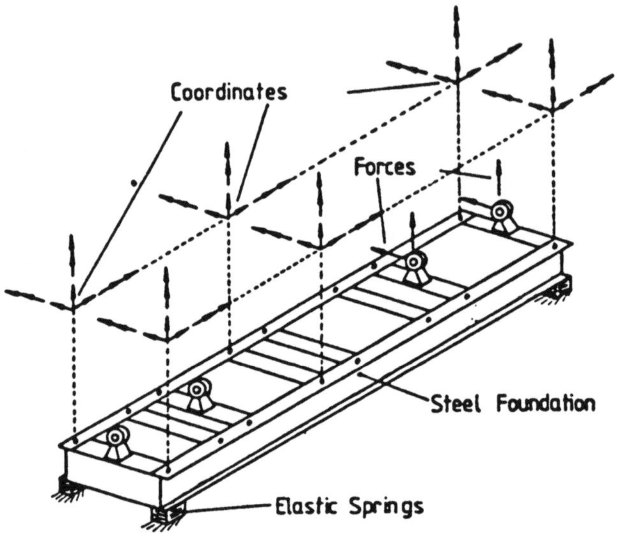

Fig. 2 Steel foundation for a power unit

The equations of motion, derived by means of the finite element method, are

$$\underline{M}\,\ddot{\underline{u}} + \underline{C}\,\dot{\underline{u}} + \underline{K}\,\underline{u} = \underline{F}(t) \tag{1}$$

with symmetric matrices

$\underline{M}$ mass matrix (order NxN)

$\underline{C}$ damping matrix (order NxN)

$\underline{K}$ stiffness matrix (order NxN)

and

$\tilde{\underline{F}}$ vector of exciting forces.

The problem is finding the unknown displacements $\tilde{\underline{u}}(t)$. Because the equations are coupled, they have to be solved simultaneously, which is very time consuming.

A simplification - especially in the case of transient vibrations - is possible, if a suitable transformation for the coordinates

$$\tilde{\underline{u}}(t) = \underline{\phi}\ \tilde{\underline{q}}(t) \tag{2}$$

with

$\tilde{\underline{u}}(t)$ displacement vector (order N)

$\underline{\Phi}$ transformation matrix (order NxN)

$\tilde{\underline{q}}(t)$ generalized coordinates (order N)

can be found, which leads to uncoupled equations.

The modal matrix $\underline{\Phi}$ of the undamped system represents such a linear transformation matrix, provided that the damping matrix $\underline{C}$ is a linear combination of the mass and stiffness matrix. The modal matrix consists of the N natural modes

$$\underline{\Phi} = \left\{\phi_1, \phi_2, \phi_3, \ldots \phi_N\right\} \tag{3}$$

Investigations of the dynamic behaviour of structures usually start with the computation of the N natural frequencies ω_j as well as the corresponding natural modes (eigenvectors) for the undamped system

$$(\underline{K} - \omega_j^2\ \underline{M})\ \underline{\Phi}_j = 0 \tag{4}$$

In this way natural frequencies and modes are usually present before

calculating forced vibrations and they can be employed for the transformation. Fig. 3 shows for example the three lowest natural modes $\underline{\phi}_j$ of the steel foundation with the corresponding natural frequencies.
In the case of symmetric matrices $\underline{K}$ and $\underline{M}$ the natural modes possess the following orthogonality properties

$$\underline{\phi}_j \underline{M}\, \underline{\phi}_k \cdot \begin{cases} 0 & \text{for } j \neq k \\ M_j & \text{for } j = k \end{cases} \qquad \underline{\phi}_j \underline{K}\, \underline{\phi}_k \cdot \begin{cases} 0 & \text{for } j \neq k \\ K_j & \text{for } j = k \end{cases} \tag{5}$$

$$\omega_j^2 = K_j / M_j \tag{6}$$

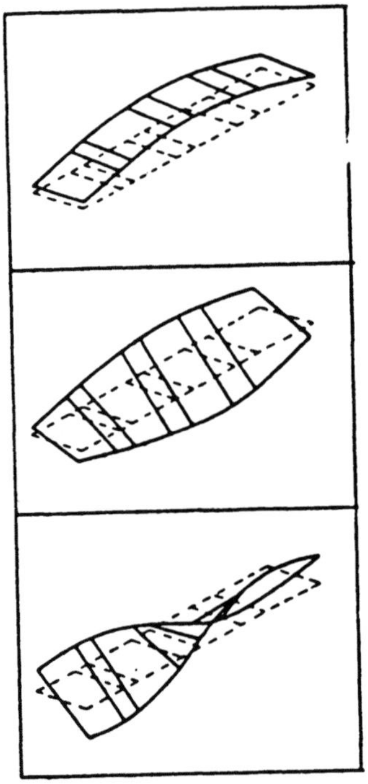

first bending mode with 21.8 Hz

bending and torsional mode with 3o.4 Hz

torsional mode with 65.o Hz

We can also write

$$\underline{\Phi}^T \underline{M}\, \underline{\Phi} = \text{diag}\,\{M_j\} \tag{5a}$$

$$\underline{\Phi}^T \underline{K}\, \underline{\Phi} = \text{diag}\,\{K_j\}$$

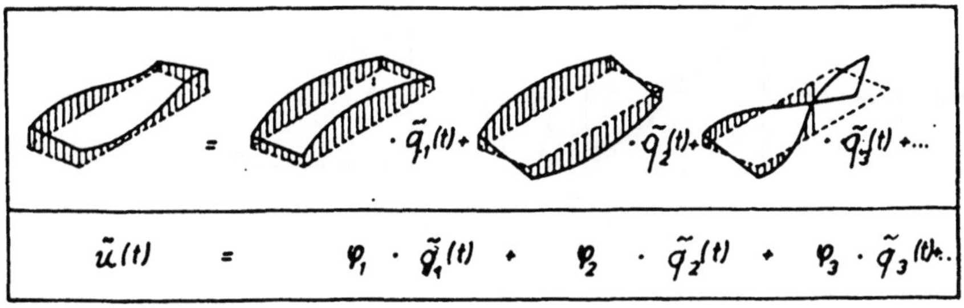

Fig. 4 Expansion in terms of natural modes

Expressing the system response by an expansion in terms of the natural modes (Fig. 4)

$$\tilde{\underline{u}}(t) = \Sigma \; \underline{\phi}_j \, \tilde{q}_j = \underline{\Phi} \, \tilde{\underline{q}} \tag{7}$$

and substituting of eq. (1.3-7) in the equations of motion we obtain after premultiplying with $\underline{\Phi}^T$

$$\underline{\Phi}^T \underline{M} \, \underline{\Phi} \, \ddot{\tilde{\underline{q}}} + \underline{\Phi}^T \underline{C} \, \underline{\Phi} \, \dot{\tilde{\underline{q}}} + \underline{\Phi}^T \underline{K} \, \underline{\Phi} \, \tilde{\underline{q}} = \underline{\Phi}^T \tilde{\underline{F}}$$

$$\mathrm{diag}\,\{M_j\} \, \ddot{\tilde{\underline{q}}} + \underline{\Phi}^T \underline{C} \, \underline{\Phi} \, \dot{\tilde{\underline{q}}} + \mathrm{diag}\,\{K_j\} \, \tilde{\underline{q}} = \underline{\Phi}^T \tilde{\underline{F}} \tag{8}$$

The matrices $\underline{\Phi}^T \underline{M} \, \underline{\Phi}$ and $\underline{\Phi}^T \underline{K} \, \underline{\Phi}$ are diagonal, which is usually not the case for the transformed damping matrix $\underline{\Phi}^T \underline{C} \, \underline{\Phi}$. $\underline{\Phi}^T \tilde{\underline{F}}$ is the vector of generalized forces acting upon the system.

If the damping matrix $\underline{C}$ is a linear combination of the mass and stiffness matrices

$$\underline{C} = \alpha \, \underline{M} + \beta \, \underline{K} \, , \tag{9}$$

$\underline{\Phi}^T \underline{C}\, \underline{\Phi}$ is also diagonal:

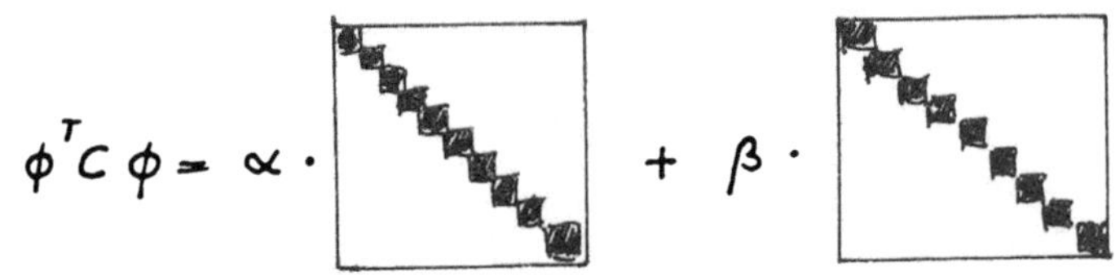

$$\underline{\Phi}^T \underline{C}\, \underline{\Phi} = \alpha \text{ diag } \{M_j\} + \beta \text{ diag } \{K_j\} = \text{diag } \{C_j\} \tag{10}$$

Now we have N decoupled equations

$$M_j \ddot{\tilde{q}}_j + \{\alpha M_j + \beta K_j\} \dot{\tilde{q}}_j + K_j \tilde{q}_j = \Phi_j^T \tilde{F} \tag{11}$$

which can be written in other form

$$\ddot{\tilde{q}}_j + \{\alpha + \beta \omega_j^2\} \dot{\tilde{q}}_j + \omega_j^2 q_j = \frac{1}{M_j} \Phi_j^T \tilde{F} \tag{11a}$$

With the above mentioned methods of calculation for a single degree of freedom system the unknown generalized coordinates $\tilde{q}_j$ can be calculated. The system response is composed of the natural modes $\underline{\Phi}_j$ with the generalized coordinates q_j as factors (eq. 7).
This method of obtaining the response of a system by using the modal matrix $\underline{\Phi}$ as the transformation matrix is called "modal analysis".

3 Equations of Motion of Rotating Machinery. Contrary to nonrotating structures the dynamic behaviour of rotating machines (turborotors etc.) is influenced by additional effects. Of great importance are selfexciting and damping effects (nonconservative effects), caused by the oil film forces of journal bearings, forces in sealings and ex-

ternal and internal damping forces. Furthermore gyroscopic effects have to be considered in certain circumstances.

For calculation of vibrations a mechanical model of the rotor is needed. Fig. 6 shows as an example a model of a turborotor. It consists of a flexible shaft with many beam-elements. Each beam-element is characterized by constant parameters: bending stiffness, mass per unit length, moments of inertia, damping coefficients (internal and external) and mass eccentricity. The shaft is running in journal bearings with angular of velocity Ω.

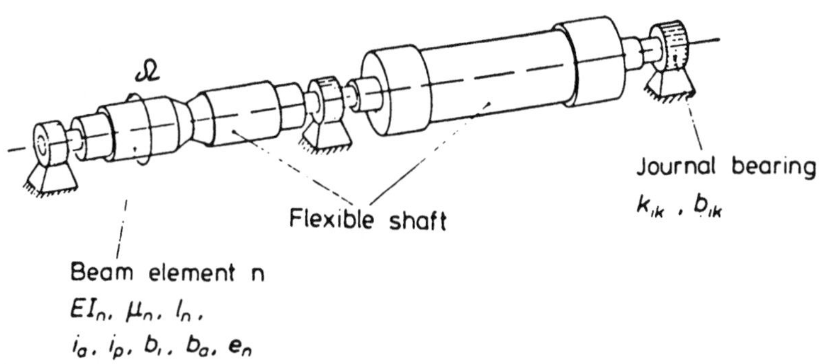

Fig. 6 Mechanical model of a turborotor

It is well-known from lubrication theory, that the dynamic forces acting from the oil film to the shaft journal depend on the displacements and the velocities of the journal (linear theory)

$$\begin{bmatrix} \tilde{F}_x \\ \tilde{F}_y \end{bmatrix}_B = - \begin{bmatrix} k_{xx} & k_{xy} \\ k_{yx} & k_{yy} \end{bmatrix}_B \cdot \begin{bmatrix} \tilde{u} \\ \tilde{v} \end{bmatrix}_B - \begin{bmatrix} c_{xx} & c_{xy} \\ c_{yx} & c_{yy} \end{bmatrix}_B \cdot \begin{bmatrix} \dot{\tilde{u}} \\ \dot{\tilde{v}} \end{bmatrix}_B \tag{12}$$

$$\tilde{\underline{F}}_B = - \underline{K}_B \quad \tilde{\underline{u}}_B - \underline{C}_B \quad \dot{\tilde{\underline{u}}}_B$$

k_{xx}, k_{xy}, k_{yx}, k_{yy} are the stiffness coefficients, c_{xx}, c_{xy}, c_{yx}, c_{yy} the damping coefficients of the oil film. They are functions of the running speed Ω. The matrices $\underline{K}_B$ and $\underline{C}_B$ are nonsymmetric, which is of great importance for further methods of calculations.
Exciting forces are caused by unbalance of the rotor. Other exciting forces may occur, caused by impacts to the shaft, a blade break, a short circuit etc.

Different possibilities exist, to describe the motion of the above mentioned rotor with continuous mass and stiffness distribution. It is obvious to take as mathematical model the partial differential equations together with the appropriate boundary conditions. Only in simple cases it is possible to find analytical solutions.
For practical calculations usually a discrete mathematical model is employed with a finite number of coordinates. The system behaviour is described by ordinary differential equations. Working with matrices is very useful in such cases.
The equations of motion mostly will be formulated with the principle of virtual work. In the expression of virtual work the unknown deflection-functions are substituted by assumed deflection shapes with free parameters. In the finite element method local trial functions are applied to each beam element. Free parameters are the deflections and angles at the boundaries of the elements. Finally we obtain a discrete mathematical model, which consists of ordinary differential equations for the unknown displacements at the boundaries of the elements (displacement method).
The equations can be written with the system matrices $\underline{M}$, $\underline{C}$, $\underline{K}$, the displacement vector $\underline{\tilde{u}}$ and the vector of external forces $\underline{\tilde{F}}$ (Fig. 7).

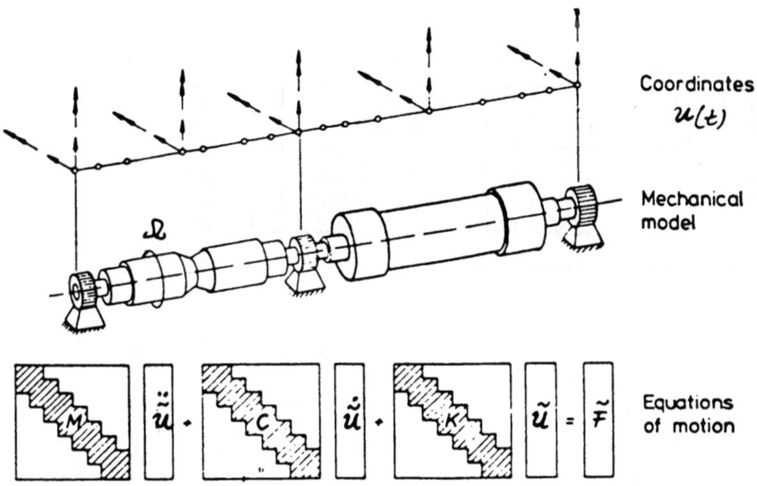

Fig. 7 Equations of motion

The matrices have bandstructure with an overlapping of the element matrices. An investigation of the matrices $\underline{M}$, $\underline{C}$, $\underline{K}$ for the described rotating structure shows differences opposite to the matrices for nonrotating structures (Fig. 8).

The mass matrix here is also symmetric and describes the translatory and rotatory inertia behaviour of the rotor.

Damping and stiffness matrices contain besides the symmetric terms also skewsymmetric and nonsymmetric terms.

Skewsymmetry is caused by gyroscopic effects (damping matrix) and by internal damping (stiffness matrix).

Nonsymmetry, caused by the oil film, appears in both stiffness and damping matrices. Therefore $\underline{K}$ and $\underline{C}$ are nonsymmetric in general with speed dependent elements.

There is also a coupling in the two planes (horizontal and vertical) caused by the mentioned effects.

The special nature of the matrices (nonsymmetry, skewsymmetry) for rotating systems requires other solution methods, especially in modal analysis.

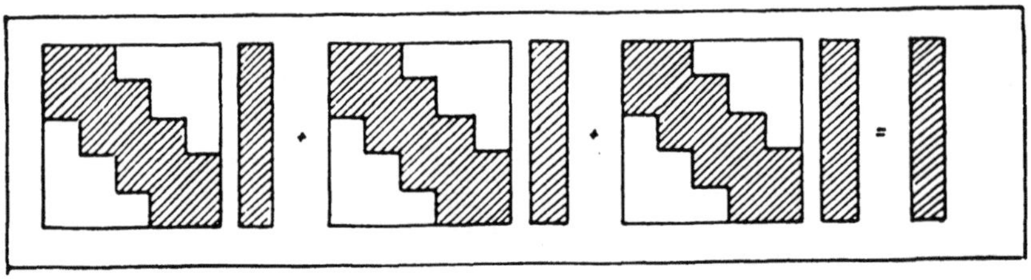

$$\underline{M}\,\ddot{\tilde{\underline{u}}} + \underline{C}(\Omega)\,\dot{\tilde{\underline{u}}} + \underline{K}(\Omega)\,\tilde{\underline{u}} = \tilde{\underline{F}}$$

symmetric	symmetric	symmetric	external
translatory and	external damping	beam bending	forces
rotatory inertia	internal damping		
	skewsymmetric	skewsymmetric	
	gyroscopic effects	internal damping	
	nonsymmetric	nonsymmetric	
	journal bearings	journal bearings	

Fig. 8 System matrices for rotating structures

4 Modal Analysis for Rotating Structures. In chapter 2 we found out, that the equations of motion for nonrotating structures can be decoupled, if the system matrices are symmetric and the damping matrix is a linear combination of mass and stiffness matrices.

For rotating structures the symmetry of the matrices normally is not given (chapter 3) . The question is, in what way forced vibrations in rotor dynamics can be treated by a modal analysis.

Two different ways are described in this chapter.

Modal analysis in rotor dynamics with natural modes of a conservative system. We start with the equations of motion for a rotor described in chapter 3

$$\underline{M}\,\ddot{\underline{\tilde{u}}} + \underline{C}\,\dot{\underline{\tilde{u}}} + \underline{K}\,\underline{\tilde{u}} = \underline{\tilde{F}}\,(t) \tag{13}$$

with nonsymmetric and Ω-dependent matrices $\underline{K}\,(\Omega)$ and $\underline{C}\,(\Omega)$.
Similar to the consideration in chapter 2 the basic idea is calculating the natural modes of a simple corresponding system and to employ them in the transformation matrix $\underline{\Phi}$.
The system matrices $\underline{M}$, $\underline{C}$, $\underline{K}$ always can be subdivided into a symmetric and a skewsymmetric part (Fig. 9). The symmetric parts of mass and stiffness matrices as well as the skewsymmetric part of the damping matrix are conservative. On the other hand the symmetric part of the damping matrix and the skewsymmetric part of the stiffness matrix are nonconservative. Disregarding the damping matrix $\underline{C} = \underline{C}_S + \underline{C}_A$, we take only the symmetric part of the stiffness matrix $\underline{K}_s$ and the symmetric mass matrix $\underline{M}_s$ and calculate eigenvalues and eigenvectors of this special conservative problem

$$(\underline{K}_s - \omega_j^2\,\underline{M}_s)\,\underline{\Phi}_{js} = 0 \tag{14}$$

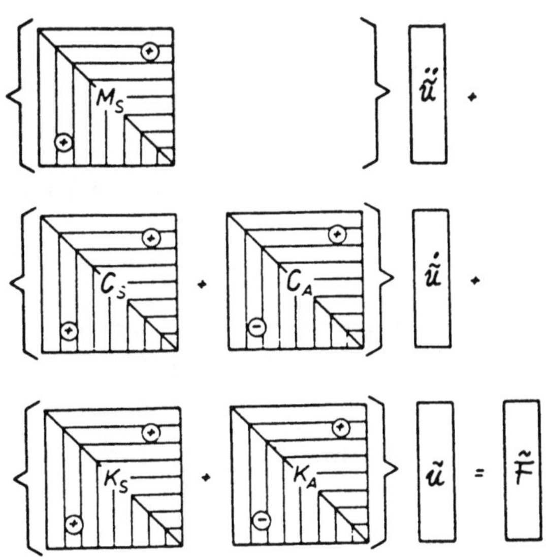

Fig. 9 Symmetric and skewsymmetric matrices

The eigenvalue routines are well developed for this case.
The natural modes $\underline{\Phi}_{js}$, respectively the modal matrix $\underline{\Phi}_s$ possess again the orthogonality properties

$$\underline{\Phi}_s^T \underline{K}_s \underline{\Phi}_s = \text{diag}\,\{K_j\} \; ; \; \underline{\Phi}_s^T \underline{M}_s \underline{\Phi}_s = \text{diag}\,\{M_j\} \tag{15}$$

Calculating the forced vibrations for the original nonconservative system (eq. 13), we express the system response by an expansion in terms of the natural modes

$$\tilde{\underline{u}}(t) = \sum_j^N \underline{\Phi}_{js}\, \tilde{q}_j(t) = \underline{\Phi}_s \tilde{\underline{q}} \tag{16}$$

Substituting eq. 16 in eq. 13 and premultiplying with $\underline{\Phi}_s^T$, we obtain

$$\underline{\Phi}_s^T \underline{M}_s \underline{\Phi}_s \ddot{\tilde{\underline{q}}} + \underline{\Phi}_s^T \underline{C} \underline{\Phi}_s \dot{\tilde{\underline{q}}} + \underline{\Phi}_s^T \{\underline{K}_s + \underline{K}_A\} \tilde{\underline{q}} = \underline{\Phi}_s^T \tilde{\underline{F}} \tag{17}$$

or in regard to orthogonality relationships

$$\text{diag}\{M_j\} \ddot{\tilde{\underline{q}}} + \underline{\Phi}_s^T \underline{C} \underline{\Phi}_s \dot{\tilde{\underline{q}}} + \{\text{diag}\,(K_j) + \underline{\Phi}_s^T \underline{K}_A \underline{\Phi}_s\} \tilde{\underline{q}} = \underline{\Phi}_s^T \tilde{\underline{F}} \tag{17a}$$

Fig. 10 shows, that in the transformed system only the matrices $\underline{\Phi}_s^T \underline{M}_s \underline{\Phi}_s$ and $\underline{\Phi}_s^T \underline{K}_s \underline{\Phi}_s$ are diagonal. The remaining matrices are filled and therefore

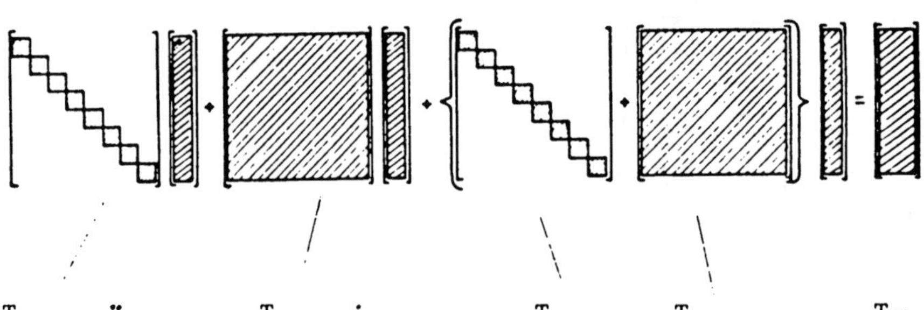

$$\underline{\Phi}_s^T \underline{M}_s \underline{\Phi}_s \ddot{\tilde{\underline{q}}} + \underline{\Phi}_s^T \underline{C} \underline{\Phi}_s \dot{\tilde{\underline{q}}} + (\underline{\Phi}_s^T \underline{K}_s \underline{\Phi}_s + \underline{\Phi}_s^T \underline{K}_A \underline{\Phi}_s) \tilde{\underline{q}} = \underline{\Phi}_s^T \tilde{\underline{F}}$$

Fig. 10 Coupling of the transformed system

a decoupling of the equations is not possible in this case. However the number of equations can be reduced by taking only some of the natural modes in the transformation matrix $\underline{\Phi}_s$.

Modal analysis in rotor dynamics with natural modes of the non-conservative system. In the last chapter we found, that the classical modal analysis with natural modes of a conservative system fails to uncouple the equations of motion in rotor dynamics. However working with eigenvalues and natural modes of the nonconservative system leads to the desired decoupling.

Eigenvalues and natural modes of the nonconservative system. The first step in modal analysis is always the determination of eigenvalues and natural modes. Therefore at first we give some remarks about eigenvalues and eigenvectors (natural modes) in rotor dynamics.
The case, in which the force vector $\bar{\underline{F}} = o$, is called free vibrations and is characterized by the homogeneous equation

$$\underline{M}\,\ddot{\breve{\underline{u}}} + \underline{C}\,\dot{\breve{\underline{u}}} + \underline{K}\,\breve{\underline{u}} = \underline{O} \tag{18}$$

The solution of eq. 1.3-18 is of the form

$$\breve{\underline{u}}(t) = \underline{\Phi}\, e^{\lambda t} \tag{19}$$

Substitution yields the quadratic eigenvalue problem

$$\{\lambda^2 \underline{M} + \lambda\, \underline{C} + \underline{K}\}\, \underline{\Phi} = \underline{O} \tag{20}$$

with 2N eigenvalues λ_j and corresponding natural modes $\underline{\Phi}_j$, if the order of the matrices is N.
The eigenvalues as well as the eigenvectors mainly occur in conjugate complex pairs (real eigenvalues and eigenvectors are not considered)

Eigenvalues: $\lambda_j = \alpha_j + i\omega_j \qquad \lambda_j = \alpha_j - i\omega_j$ (21)

Eigenvectors: $\underline{\Phi}_j = \underline{s}_j + i\underline{t}_j \qquad \underline{\Phi}_j = \underline{s}_j - i\underline{t}_j$ (22)

The part of the solution, which belongs to such a conjugate complex pair, can be written as

$$\underline{\breve{u}}_j(t) = B_j e^{\alpha_j t} \{\underline{s}_j \sin(\omega_j t + \gamma_j) + \underline{t}_j \cos(\omega_j t + \gamma_j)\} \tag{23}$$

ω_j is the circular frequency and α_j the damping constant (decay constant). The damping constant α_j (real part of λ_j) determines, whether the solution $\underline{\breve{u}}_j(t)$ decreases ($\alpha_j < o$) or increases ($\alpha_j > o$) Fig. 11 shows the plane motion of one point of the shaft for the three cases $\alpha_j < o$, $\alpha_j = o$ (stability limit) and $\alpha_j > o$.

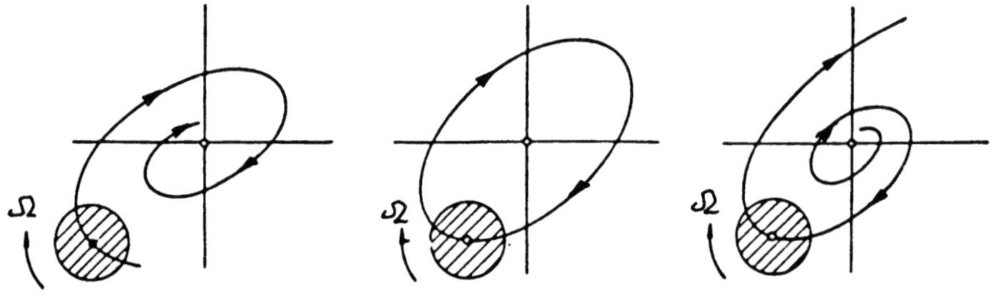

$\alpha_j < o$ stable $\qquad \alpha_j = o$ stability limit $\qquad \alpha_j > o$ unstable

Fig. 11 Plane motion of a point of the shaft

The constants B_j and the phase angle γ_j depend on the initial conditions.

Normally all of the conjugate complex pairs contribute to the solution of the natural vibrations. Discussing the natural modes we assume, that the initial conditions are chosen in a way, that only the above part $\underline{\breve{u}}_j(t)$ contributes to the solution. Then the systems vibrates with the

circular frequency ω_j and the corresponding natural mode.
To explain the natural mode is not so easy as in the case of conservative systems. The expression in parentheses { } of eq. 23 can be defined as natural mode, representing a time-dependent curve in space.
Fig. 12 shows the modal shapes for two different points of time t_1 and t_2. There is no constant modal shape, proportions and relative phasing in general vary from point to point at the shaft.

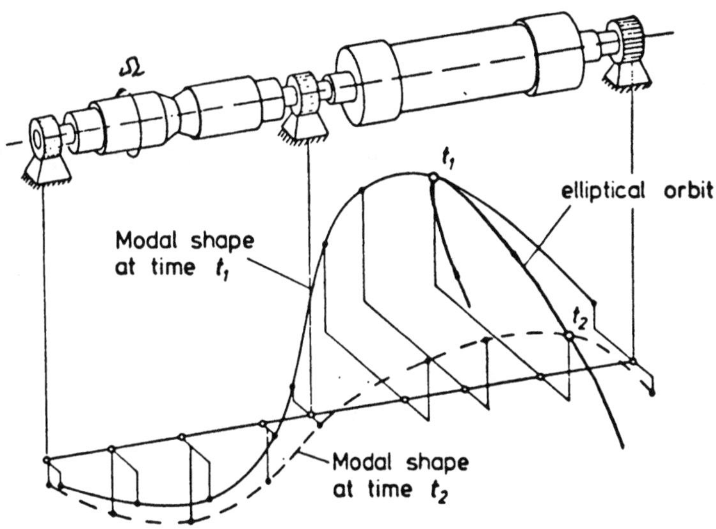

Fig. 12 Natural mode of a nonconservative rotor

Considering only one point of the shaft, the plane motion is an elliptical orbit (Fig. 12).

Orthogonality properties of left and right eigenvectors. The equations of motion of a N-degree of freedom system can be converted to 2N first order differential equations with the dependent variables $\tilde{\underline{r}}$ constituting a 2N dimensional state vector

$$\underbrace{\begin{bmatrix} \underline{M} & \underline{O} \\ O & -\underline{K} \end{bmatrix}}_{\underline{A}} \cdot \underbrace{\begin{bmatrix} \dot{\underline{u}} \\ \underline{u} \end{bmatrix}}_{\underline{r}} - \underbrace{\begin{bmatrix} \underline{O} & \underline{M} \\ \underline{M} & \underline{C} \end{bmatrix}}_{\underline{B}} \cdot \underbrace{\begin{bmatrix} \ddot{\underline{u}} \\ \dot{\underline{u}} \end{bmatrix}}_{\dot{\underline{r}}} = \underbrace{\begin{bmatrix} 0 \\ 0 \end{bmatrix}}_{\underline{O}} \tag{24}$$

The matrices $\underline{A}$ and $\underline{B}$ are real and nonsymmetric.

The corresponding eigenvalue problem

$$\{\underline{A} - \lambda_i \underline{B}\}\, \underline{r}_i = \underline{O} \tag{25}$$

has the same eigenvalues λ_i as the problem (20) and the right eigenvectors

$$\underline{r}_i = \begin{bmatrix} \lambda_i \underline{\Phi}_i \\ \underline{\Phi}_i \end{bmatrix} \tag{26}$$

obtain the eigenvectors $\underline{\Phi}_i$ of (20).
If can be shown, that the transposed eigenvalue problem

$$\{\underline{A}^T - \lambda_j \underline{B}^T\}\, \underline{l}_j = \underline{O} \tag{27}$$

also has the eigenvalues λ_j but other eigenvectors - the left eigenvectors - consisting of

$$\underline{l}_j = \begin{bmatrix} \lambda_j \psi_j \\ \psi_j \end{bmatrix} \tag{28}$$

If we premultiply two different eigenvalue equations (25) and (27) with $\underline{l}_j^T$ and $\underline{r}_i^T$, transpose the second and substract the second from the first, we obtain

$$\underline{l}_j^T \underline{A} \underline{r}_i \quad - \quad \lambda_i \underline{l}_j^T \underline{B} \underline{r}_i = \underline{0} \tag{29}$$

$$\underline{r}_i^T \underline{A}^T \underline{l}_j \quad - \quad \lambda_j \underline{r}_i^T \underline{B}^T \underline{l}_j = \underline{0} \tag{30}$$

$$(\lambda_j - \lambda_i)\, \underline{l}_j^T \underline{B} \underline{r}_i = \underline{0} \tag{31}$$

From these last equations we find the bi-orthogonality properties

$$\underline{l}_j^T \underline{B} \underline{r}_i \;\begin{cases} 0 & \text{for } j \neq i \\ d_j & \text{for } j = i \end{cases} \qquad \underline{l}_j^T \underline{A} \underline{r}_i \;\begin{cases} 0 & \text{for } j \neq i \\ c_j & \text{for } j = i;\; c_j = \lambda_j\, d_j \end{cases} \tag{32}$$

If we represent the left and right eigenvectors in the modal matrices $\underline{R}$ and $\underline{L}$

$$R = \{r_1, r_2, \ldots\ldots r_{2N}\} \quad ; \quad L = \{l_1, l_2, \ldots\ldots l_{2N}\}$$

the orthogonality properties can be written in the form

$$\underline{L}^T \underline{A} \underline{R} = \text{diag}\,(c_j) \quad ; \quad \underline{L}^T \underline{B} \underline{R} = \text{diag}\,(d_j) \tag{33}$$

Normalizing the eigenvectors to get $d_j = 1$ for all eigenvectors is possible.

Decoupling of equations of motion. The equations of motion for the forced vibrations may be developed in first order form

$$\underbrace{\begin{bmatrix} M & O \\ O & -K \end{bmatrix}}_{\underline{A}} \underbrace{\begin{bmatrix} \dot{\tilde{u}} \\ \tilde{u} \end{bmatrix}}_{\tilde{\underline{r}}} - \underbrace{\begin{bmatrix} O & M \\ M & C \end{bmatrix}}_{\underline{B}} \underbrace{\begin{bmatrix} \ddot{\tilde{u}} \\ \dot{\tilde{u}} \end{bmatrix}}_{\dot{\tilde{\underline{r}}}} = \underbrace{\begin{bmatrix} O \\ -\tilde{F} \end{bmatrix}}_{\tilde{\underline{f}}} \tag{34}$$

Again we express the unknown system response by an expansion in terms of the right eigenvectors of the nonconservative system

$$\underline{\tilde{r}}(t) = \sum_j^{2N} \underline{r}_j \tilde{q}_j = \underline{R}\,\mathbf{q} \tag{35}$$

with the generalized coordinates $\tilde{q}_j$. Substituting in eq. (34) and premultiplying with $\underline{L}^T$ we obtain

$$\underline{L}^T \underline{A}\,\underline{R}\,\underline{\tilde{q}} - \underline{L}^T \underline{B}\,\underline{R}\,\underline{\dot{\tilde{q}}} = \underline{L}^T \underline{\tilde{f}} \tag{36}$$

The matrices $\underline{L}^T \underline{A}\,\underline{R}$ and $\underline{L}^T \underline{B}\,\underline{R}$ are diagonal.

We normalize them to get

$$\underline{L}^T \underline{A}\,\underline{R} = \mathrm{diag}\,(\lambda_j) = \underline{\Lambda}$$

$$\underline{L}^T \underline{B}\,\underline{R} = \mathrm{diag}\,(1) = \underline{I} \tag{37}$$

The system of ordinary differential equations (34) is reduced to a set of independent complex equations of first order for the generalized coordinates

$$\lambda_j \tilde{q}_j - \dot{\tilde{q}}_j = \underline{l}_j^T \underline{\tilde{f}} = - \underline{\psi}_j^T \underline{\tilde{F}} \tag{38}$$

The generalized forces are represented by the left eigenvectors $\underline{\psi}_j$ and the force vector $\underline{F}$.
The decoupled equations can be solved easily for actual load cases (harmonic excitation, impulse etc.). The general solution for the generalized coordinates is given by the combination of the complementary solution (homogeneous equation) and the particular solution.

5 Example-Shaft in Two Journal Bearings. Applying the above described modal analysis we investigate a simple symmetric shaft, running in two equal journal bearings with angular velocity Ω (Fig. 13).

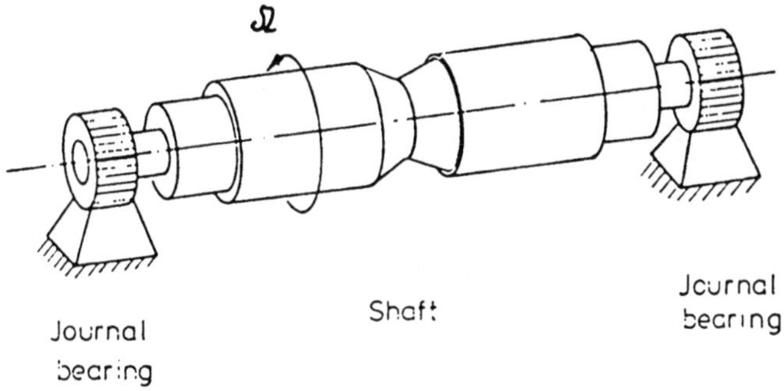

Fig. 13 Symmetric shaft in two equal journal bearings

Fig. 14 shows the five lowest natural frequencies ω_j and the corresponding damping constant α_j versus the running speed Ω. If all of the damping constants α_j are negative, the natural vibrations decrease. When the stability limit is reached one of the damping constants become zero (this is the case for α_1 in Fig. 14). Above the stability threshold speed ($\alpha_1 > o$) the natural vibrations increase with time.

From the plotted natural modes it can be explained, that the stiffness of the shaft is high compared with the bearings.
The intersections of the natural frequency curves ω_j with the straight line $\omega_j = \Omega$ indicate resonances of the system. Whether the individual resonances with their corresponding natural modes are well excited or not, depends on the distribution of the exciting forces.

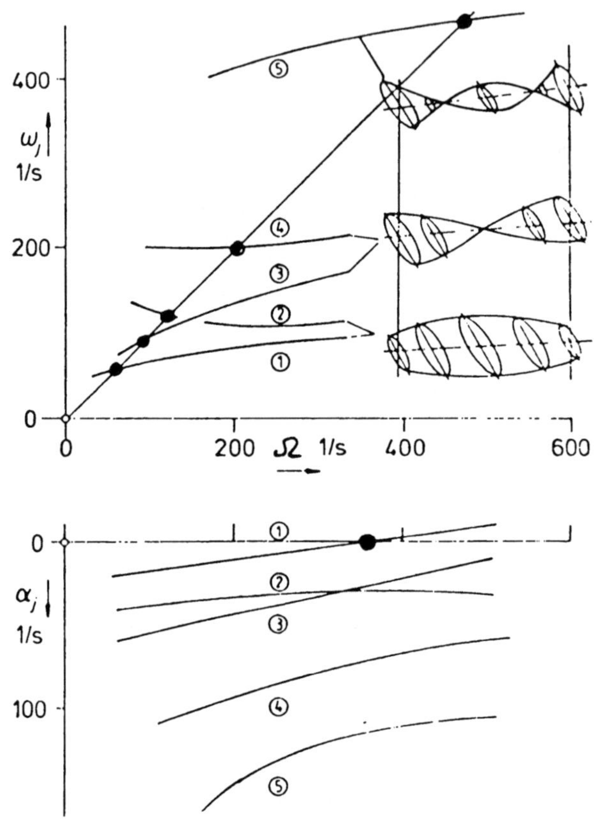

Fig. 14 Eigenvalues and natural modes

In Fig. 15 forced unbalance vibrations are investigated in the case of uniform distributed mass eccentricity.

If the rotor is running with the angular velocity Ω, the system is excited by harmonic unbalance forces with exciter frequency Ω. The steady state response is harmonic too and the plane motion of one point of the shaft is an elliptical orbit (see Fig. 12).

In the upper diagram of Fig. 15 (eigenfrequencies versus speed) the intersections are indicating resonances.

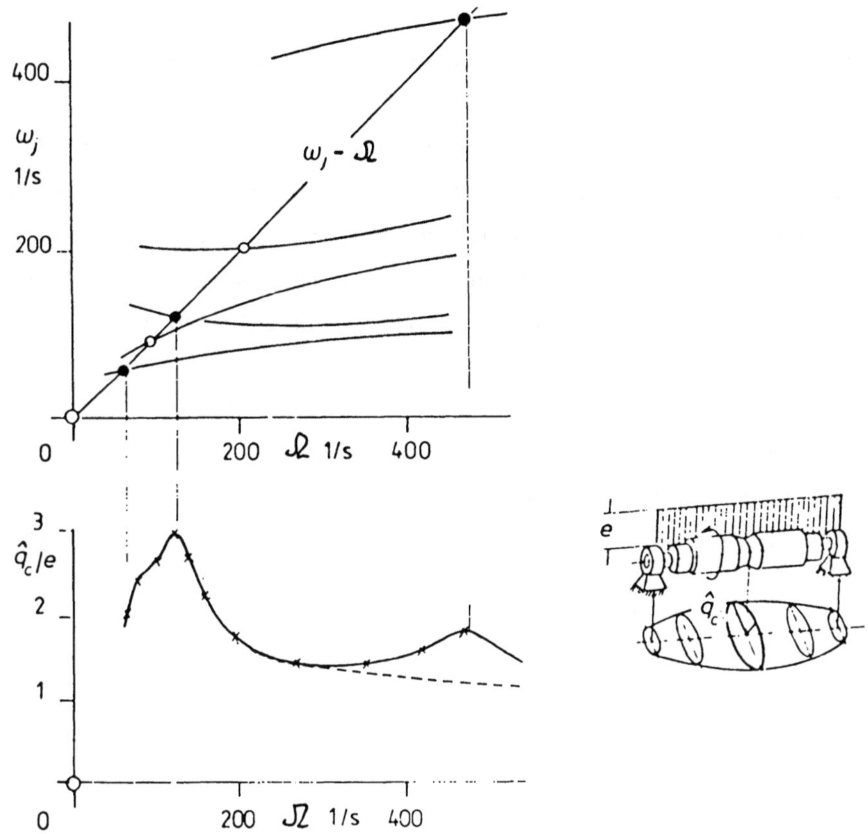

Fig. 15 Steady state response caused by unbalance forces

In the lower diagram the nondimensional major axis of the ellipse (center of the shaft) is plotted versus the running speed.

As a result of the symmetrical distributed exciting forces only the symmetrical mode shapes are excited. The solid line is the exact unbalance response of the model calculated with a complex Gauss algorithm. The solution found by "modal analysis" is plotted with crosses. Only the six lowest natural modes - of the nonconservative system - were employed and there is a good agreement with the exact solution. Working only with two eigenvectors the response is still good in the lower frequency range (dotted line in Fig. 15). The represented calculation of unbalance vibrations with modal analysis is convenient, if the complex eigenvalues and eigenvectors - all depending on the speed Ω - are present for the individual running speeds Ω. Otherwise it is too time consuming and the direct calculation solving a linear algebraic equation system is more effective.

On the other hand the application of modal analysis is convenient calculating disturbing cases during operation with constant running speed Ω. In such cases there is no variation in the system matrices (left hand side of equations of motion), only the exciting forces are changed. The eigenvalues and eigenvectors need to be calculated only for the one running speed Ω.

Fig. 16 shows the transient vibrations at the center of the shaft before and after a simulated blade break, which is equivalent with a local change in mass eccentricity. The running speed Ω is constant (Ω = 314 1/s) The complementary as well as the particular solution have to be taken into account.

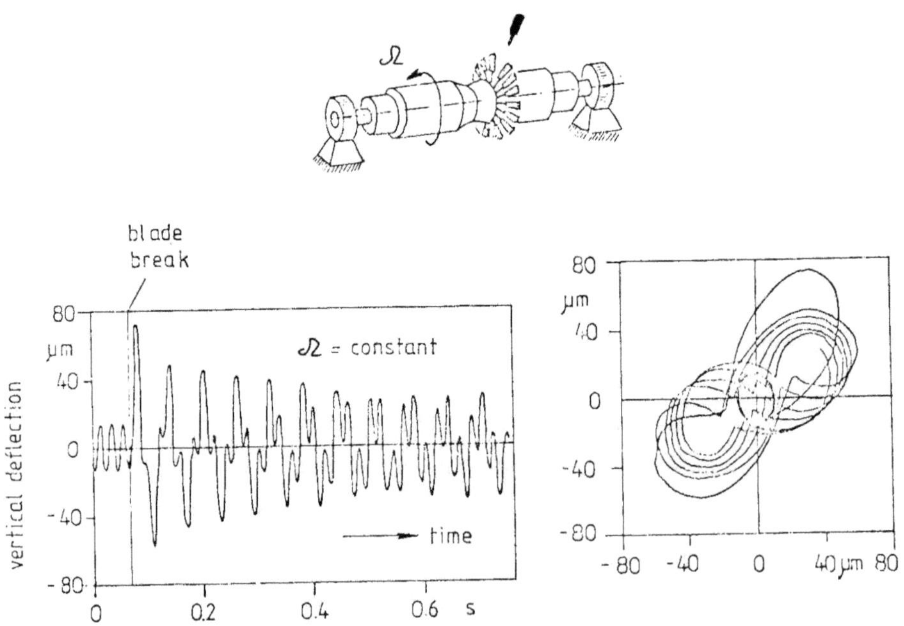

Fig. 16 Simulation of a blade break

The left plot shows the deflections in vertical direction, the right plot the plane motion. The amplitude magnification after the break is about five times higher than in the steady state solution before the break.

Besides the operating frequency Ω especially the first eigenfrequency ω_1 with a corresponding low damping (Fig. 14) determines the behaviour of the response. Good results were obtained by modal analysis calculation working again with the first six eigenvectors.

[illegible]

PART II

DYNAMIC BEHAVIOUR OF ROTORS

CHAPTER 2.1

THEORY OF BEARING STATIC AND DYNAMIC PROPERTIES

Z.A. Parszewski

1 Hydrodynamic Forces.

Bearing dynamics characteristics is necessary for rotor-bearing system dynamic analysis. The characteristics should make possible obtaining the forces acting on the journal at any moment of its vibrations. Discussed here will be hydrodynamic journal bearings, applied widely in machines. These bearings at sufficient oil supply, produce oil film and generate oil pressure and separate fully rotating journal from the sleeve.

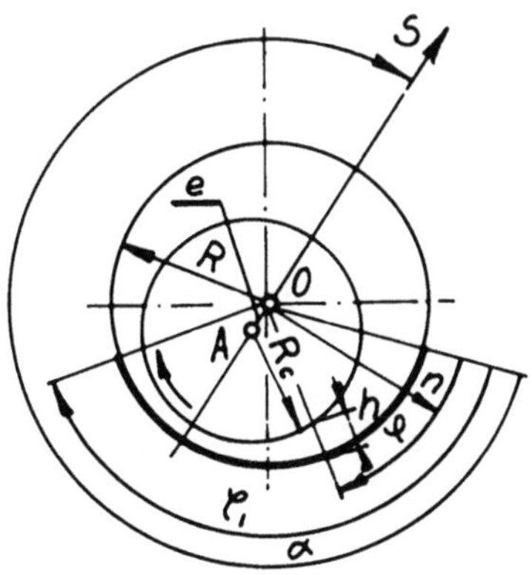

Fig. 1

The oil film can be produced at proper radial clearance in the bearing corresponding to the radii difference of the sleeve and the journal $r = R-R_c$

The pressure distribution on the sleeve can be obtained from the equations of oil flow in the bearing clearance.

The Navier-Stokes fluid flow equations together with the continuity equations, simplified for the bearing narrow clearance and integrated across the oil film thickness reduce, for laminar flow and corresponding Reynolds numbers, to the Reynolds equation. That equation generalised for the unsteady state of journal plane vibrations in the sleeve [1], [2], [3], has the form

$$\frac{\partial}{\partial x}\left(\frac{h^3}{\eta}\frac{\partial p}{\partial x}\right) + \frac{\partial}{\partial z}\left(\frac{h^3}{\eta}\frac{\partial p}{\partial z_1}\right) = 6\frac{\partial}{\partial x}(\omega Rh) + 12\frac{\partial h}{\partial t} \qquad (1)$$

Introduction of nondimensional co-ordinates

$$\phi = \frac{x}{R}\ ; \qquad z = 2\frac{z_1}{R}\ ; \qquad H = \frac{h}{r} \qquad (2)$$

and the pressure function

$$u = H^{3/2}\,\frac{p\left(\frac{r}{R}\right)^2}{\eta\,\omega} \qquad (3)$$

and denotation of the derivatives by corresponding indices gives for the equation (1) the following nondimensional form

$$u_{\phi\phi} + \frac{D^2}{L^2}u_{zz} + d(\phi,c,\alpha)u = b(\phi,c,\alpha,C_t,A) \qquad (4)$$

where

$$d(\phi,c,\alpha) = -\frac{3}{4}H^{-2}\left[H_\phi^2 + 2H(1-H)\right]$$

$$b(\phi,c,\alpha,C_t,A) = 6H^{-\frac{2}{3}}K \qquad (5)$$

and

$$K = - Ac \sin (\phi - \alpha) + 2C_t \cos(\phi - \alpha) \tag{6}$$

Equation (4) can be programmed for computer after previous replacement of the derivatives by corresponding ratios of finite differences in the whole area of the developed sleeve.

End conditions to be fulfilled are

$$(u)_{\phi=o} = p_z \;; \quad (u)_{\phi=\phi_1} = p_w \;; \quad (u)_{z=\pm 1} = p_o \tag{7}$$

and real area of pressure generation on the sleeve is to be found. This last is found directly in the computation process of successive approximations if the conditions (7) are completed by the requirement [1]

$$u \geq 0 \tag{8}$$

for incompressible fluid as oil is. Pressure distribution on a considered sleeve is computed in this way for journal positions described by the eccentricity ratio $c = \frac{e}{R}$ and the angle α. The journal rotates at the same time with speed ω and vibrates with a plane motion having the velocity of the following radial and circumferential components

$$C_t = \frac{1}{\omega} \frac{\partial c}{\partial t} \;; \qquad A = 1-2\alpha_t \tag{9}$$

In the above conditions on the circumference $\phi = o$ and $\phi = \phi_1$ give the beginning and the end edge of the sleeve, $z = \pm 1$ gives outer edges of the sleeve and p_z, p_w - correspondingly denote the inlet and outlet oil pressure, p_o - is the external (atmospheric pressure).

Components of the hydrodynamic force in the radial s and perpendicular

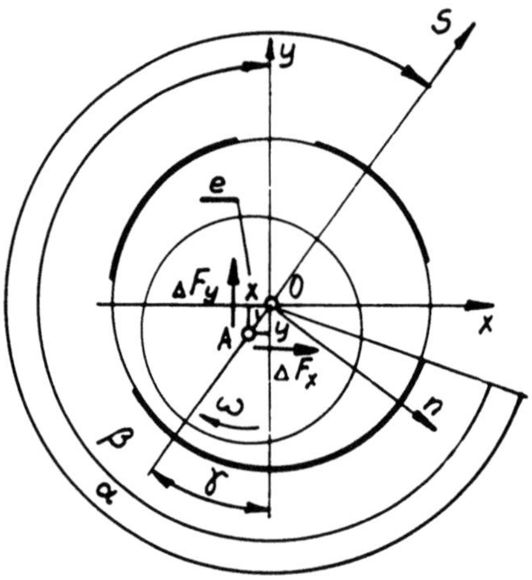

Fig. 2

to it direction n (Fig. 1) , generated by the described journal motion are hence given by the integrals

$$F_{sp} = -\tfrac{1}{2}\int_0^1\int_0^{\phi_{1p}} \frac{p\left(\frac{r}{R}\right)^2}{\eta\omega}\cos(\phi-\alpha_p)\,d\phi dz$$

(10)

$$F_{np} = -\tfrac{1}{2}\int_0^1\int_0^{\phi_{1p}} \frac{p\left(\frac{r}{R}\right)^2}{\eta\omega}\sin(\phi-\alpha_p)\,d\phi dz$$

The index p denotes the considered sleeve.

2 Static Characteristics

For stability or antivibratory reasons modern journal bearings are often built up from a few partial sleeves, concentric or eccentric. Fig. (2) shows a three lobe bearing having three independent oil films, hence three stationary partial sleeves.

The components of hydrodynamic forces of a bearing of n partial sleeves with independent supply and outflow of oil, (hence of independent n oil films) are obtained as geometric sums of those forces on all the partial sleeves

$$F_s = F_s(e,\alpha,C_t,A) = \sum_{p=1}^{n} F_{sp}(e,\alpha,C_t,A)$$

(11)

$$F_n = F_n(e,\alpha,C_t,A) = \sum_{p=1}^{n} F_{np}(e,\alpha,C_t,A)$$

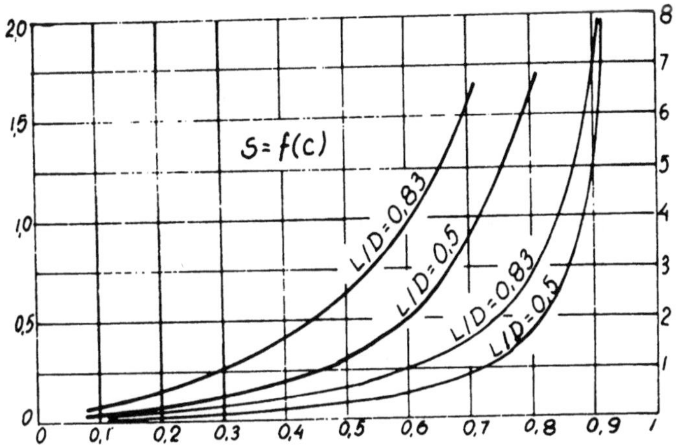

Fig. 3

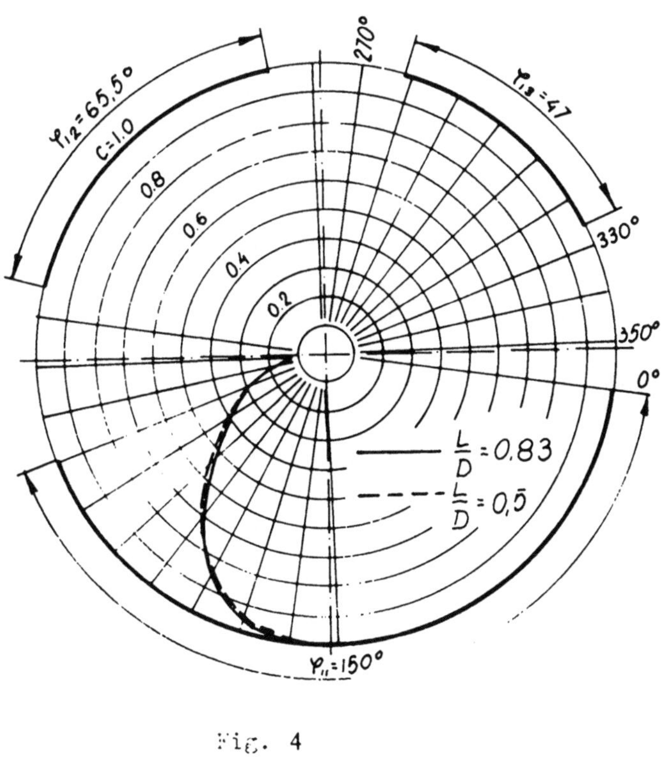

Fig. 4

The components F_y, F_x of these forces in the direction opposite to the load and normal to it (Fig. 2) are

$$F_y = F_s \cos(\alpha-\beta) - F_n \sin(\alpha-\beta)$$

$$F_x = F_s \sin(\alpha-\beta) + F_n \cos(\alpha-\beta) \qquad (12)$$

For steady state are $C_t = 0$, $A = 1$ and the component perpendicular to the load vanishes

$$F_x = 0$$

Corresponding eccentricity ratios c and angles α give the locus of the equilibrium positions of the journal centre in the bearing

$$c = c(\alpha) \qquad (13)$$

The component F_y opposite to the load gives then the nondimensional load capacity (Sommerfield's number) of the bearing

$$S = S(c) \tag{14}$$

The bearing nondimensional load capacity (14) , defined for a given direction as nondimensional load (corresponding to the load P_o) at an equilibrium position and at a given speed ω, is given by the formula

$$S = \frac{P_o\left(\frac{r}{R}\right)^2}{2RL\eta\omega} \tag{15}$$

Relations (14) and (13) give hence the bearing static characteristics. Computed results [4] are given in Figs. 3 and 4 for a three lobe bearing with the partial sleeve arcs of

$$\phi_{11} = 150^o; \quad \phi_{12} = 65.5^o; \quad \phi_{13} = 47^o$$

and $\frac{L}{2R} = \frac{L}{D} = 0.83$ and 0.5

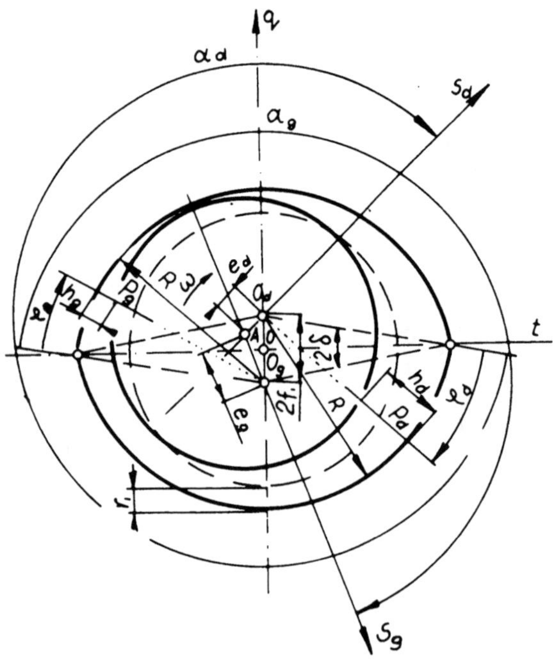

Fig. 5

Corresponding curves [5] for a bearing with eccentric sleeves - lemon shaped- (Fig. 5) are given in the Fig. 6 and 7 a,b,c, for $\frac{L}{D}$ = 0.851 and three values of the sleeve eccentricity parameter $f = f_1/(f_1 + r_1)$ Both types of bearings are used in generators of Dolmel-Wroclaw production.

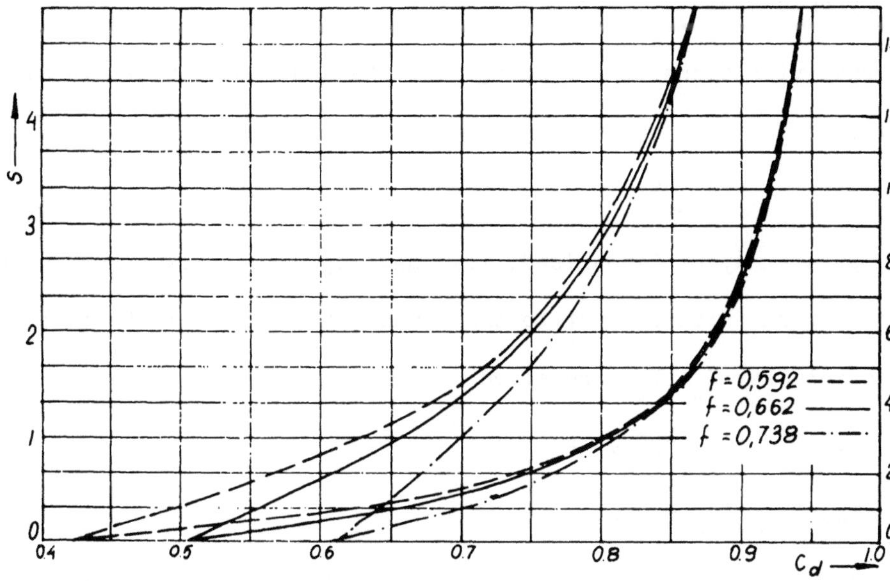

Fig. 6

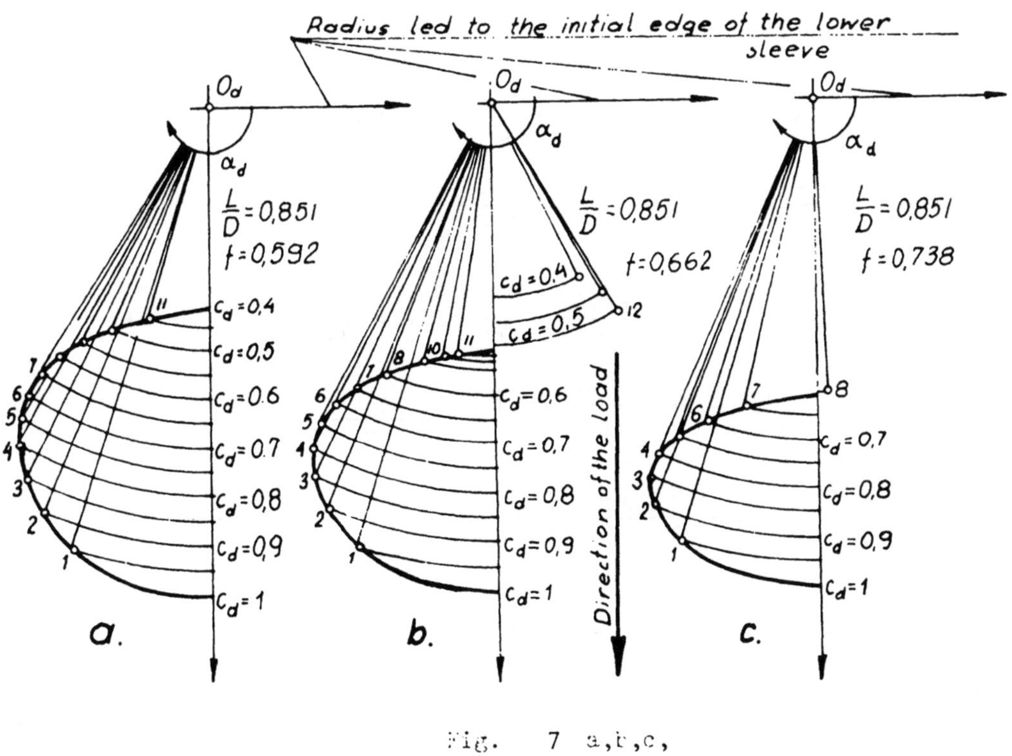

Fig. 7 a,b,c,

3 Dynamic Characteristics

Journal vibrations in a bearing generate additional hydrodynamic forces [6]

$$\Delta F_y = F_y - S \qquad \Delta F_x = F_x \tag{16}$$

These functions may be linearised, i.e. developed in a power series and only linear terms with respect to the component displacements X, Y and vibration velocity X_t, Y_t, preserved

$$\begin{aligned} \Delta F_x &= -(K_{xx}X + K_{xy}Y + C_{xx}X_t + C_{xy}Y_t) \\ \Delta F_y &= -(K_{yx}X + K_{yy}Y + C_{yx}X_t + C_{yy}Y_t) \end{aligned} \tag{17}$$

All displacements and velocities are here nondimensional. Nondimensional stiffness K and damping C coefficients of the bearing oil film are hence expressed in the form

$$K_{xx} = \frac{\partial F_s}{\partial c}\sin^2\gamma + \frac{\partial F_s}{\partial \alpha}\frac{\sin\gamma\cos\gamma}{c} + \frac{F_s}{c} + \frac{\partial F_n}{\partial c}\sin\gamma\cos\gamma + $$

$$+ \frac{\partial F_n}{\partial \alpha}\frac{\cos^2\gamma}{c} - C_{xx}\frac{\cos^2\gamma}{c}c_t + C_{xy}\left(\frac{1-A}{2} + \frac{\sin\gamma\cos\gamma}{c}c_t\right)$$

$$K_{xy} = \frac{\partial F_s}{\partial c}\sin\gamma\cos\gamma - \frac{\partial F_s}{\partial \alpha}\frac{\sin^2\gamma}{c} + \frac{\partial F_n}{\partial c}\cos^2\gamma - \frac{\partial F_n}{\partial \alpha}\frac{\sin\gamma\cos\gamma}{c} + $$

$$+ \frac{F_n}{c} - C_{xx}\left(\frac{1-A}{2} - \frac{\sin\gamma\cos\gamma}{c}c_t\right) - C_{xy}\frac{\sin^2\gamma}{c}c_t$$

$$K_{yx} = \frac{\partial F_s}{\partial c}\sin\gamma\cos\gamma + \frac{\partial F_s}{\partial \alpha}\frac{\cos^2\gamma}{c} - \frac{\partial F_n}{\partial c}\sin^2\gamma - \frac{\partial F_n}{\partial \alpha}\frac{\sin\gamma\cos\gamma}{c} + $$

$$- C_{yx}\frac{\cos^2\gamma}{c}c_t + C_{yy}\left(\frac{1-A}{2} + \frac{\sin\gamma\cos\gamma}{c}c_t\right)$$

$$K_{yy} = \frac{\partial F_s}{\partial c}\cos^2\gamma - \frac{\partial F_s}{\partial \alpha}\,\frac{\sin\gamma\cos\gamma}{c} - \frac{\partial F_n}{\partial c}\sin\gamma\cos\gamma + \frac{\partial F_n}{\partial \alpha}\,\frac{\sin^2\gamma}{c} +$$

$$- C_{yx}\left[\frac{1-A}{2} - \frac{\sin\gamma\cos\gamma}{c}\,C_t\right] - C_{yy}\frac{\sin^2\gamma}{c}C_t \qquad (18)$$

$$C_{xx} = \frac{\partial F_s}{\partial C_t}\sin^2\gamma - 2\frac{\partial F_s}{\partial A}\,\frac{\sin\gamma\cos\gamma}{c} + \frac{\partial F_n}{\partial C_t}\sin\gamma\cos\gamma - 2\frac{\partial F_n}{\partial A}\,\frac{\cos^2\gamma}{c}$$

$$C_{xy} = \frac{\partial F_s}{\partial C_t}\sin\gamma\cos\gamma + 2\frac{\partial F_s}{\partial A}\,\frac{\sin^2\gamma}{c} + \frac{\partial F_n}{\partial C_t}\cos^2\gamma + 2\frac{\partial F_n}{\partial A}\,\frac{\sin\gamma\cos\gamma}{c}$$

$$C_{yx} = \frac{\partial F_s}{\partial C_t}\sin\gamma\cos\gamma - 2\frac{\partial F_s}{\partial A}\,\frac{\cos^2\gamma}{c} - \frac{\partial F_n}{\partial C_t}\sin^2\gamma + 2\frac{\partial F_n}{\partial A}\,\frac{\sin\gamma\cos\gamma}{c}$$

$$Cyy = \frac{\partial F_s}{\partial C_t}\cos^2\gamma + 2\frac{\partial F_s}{\partial A}\,\frac{\sin\gamma\cos\gamma}{c} - \frac{\partial F_s}{\partial C_t}\sin\gamma\cos\gamma - 2\frac{\partial F_s}{\partial A}\,\frac{\sin^2\gamma}{c}$$

where $y = \alpha - \beta$

For journal vibrations round its equilibrium position, with small velocity is

$$K_{xx} = \frac{\partial F_s}{\partial c}\sin^2\gamma + \frac{\partial F_s}{\partial \alpha}\,\frac{\sin\gamma\cos\gamma}{c} + \frac{F_s}{c} + \frac{\partial F_n}{\partial c}\sin\gamma\cos\gamma + \frac{\partial F_n}{\partial \alpha}\,\frac{\cos^2\gamma}{c}$$

$$K_{xy} = -\frac{\partial F_s}{\partial c}\sin\gamma\cos\gamma - \frac{\partial F_s}{\partial \alpha}\,\frac{\sin^2\gamma}{c} + \frac{\partial F_n}{\partial c}\cos^2\gamma - \frac{\partial F_n}{\partial \alpha}\,\frac{\sin\gamma\cos\gamma}{c} + \frac{F_n}{c}$$

$$K_{yx} = \frac{\partial F_s}{\partial c}\sin\gamma\cos\gamma + \frac{\partial F_s}{\partial \alpha}\,\frac{\cos^2\gamma}{c} - \frac{\partial F_n}{\partial c}\sin^2\gamma - \frac{\partial F_n}{\partial \alpha}\,\frac{\sin\gamma\cos\gamma}{c}$$

$$K_{yy} = \frac{\partial F_s}{\partial c}\cos^2\gamma - \frac{\partial F_s}{\partial \alpha}\,\frac{\sin\gamma\cos\gamma}{c} - \frac{\partial F_n}{\partial c}\sin\gamma\cos\gamma + \frac{\partial F_n}{\partial \alpha}\,\frac{\sin^2\gamma}{c}$$

$$C_{xx} = \frac{\partial F_s}{\partial C_t}\sin^2\gamma - 2\frac{\partial F_s}{\partial A}\,\frac{\sin\gamma\cos\gamma}{c} + \frac{\partial F_n}{\partial C_t}\sin\gamma\cos\gamma - 2\frac{\partial F_n}{\partial A}\,\frac{\cos^2\gamma}{c}$$

$$C_{xy} = \frac{\partial F_s}{\partial C_t} \sin\gamma\cos\gamma + 2\frac{\partial F_s}{\partial A}\frac{\sin^2\gamma}{c} + \frac{\partial F_n}{\partial C_t}\cos^2\gamma + \frac{\partial F_n}{\partial A}\frac{\sin\gamma\cos\gamma}{c}$$

$$C_{yx} = \frac{\partial F_s}{\partial C_t} \sin\gamma\cos\gamma - 2\frac{\partial F_s}{\partial A}\frac{\cos^2\gamma}{c} - \frac{\partial F_n}{\partial C_t}\sin^2\gamma + 2\frac{\partial F_n}{\partial A}\frac{\sin\gamma\cos\gamma}{c}$$

$$C_{yy} = \frac{\partial F_s}{\partial C_t} \cos^2\gamma + 2\frac{\partial F_s}{\partial A}\frac{\sin\gamma\cos\gamma}{c} - \frac{\partial F_n}{\partial C_t}\sin\gamma\cos\gamma - 2\frac{\partial F_n}{\partial A}\frac{\sin^2\gamma}{c} \tag{19}$$

The stiffness matrix

$$[K] = \begin{bmatrix} K_{xx} & K_{xy} \\ K_{yx} & K_{yy} \end{bmatrix} \tag{20}$$

and the damping matrix

$$[C] = \begin{bmatrix} C_{xx} & C_{xy} \\ C_{yx} & C_{yy} \end{bmatrix} \tag{21}$$

as functions of parameters of the equilibrium state (i.e. of the eccentricity ratio c) give hence together the bearing dynamic characteristics.

The flow diagram for computation of the elements of these matrices for any bearing is given in Fig. 8.

Computation results for the same two types of bearings, as described in the previous point are presented graphically in the Figs. 9 and 10 for the three lobe bearing [7] and in Figs. 11 and 12 for the eccentric (lemon) bearing [8].

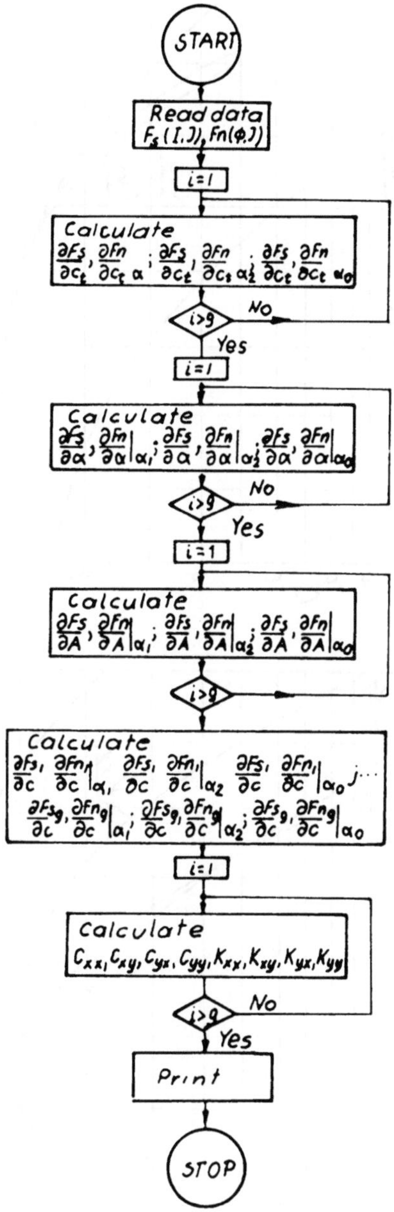
START
Read data
Calculate
Calculate
Calculate
Calculate
Calculate
No
Yes
Print
STOP

Fig. 8

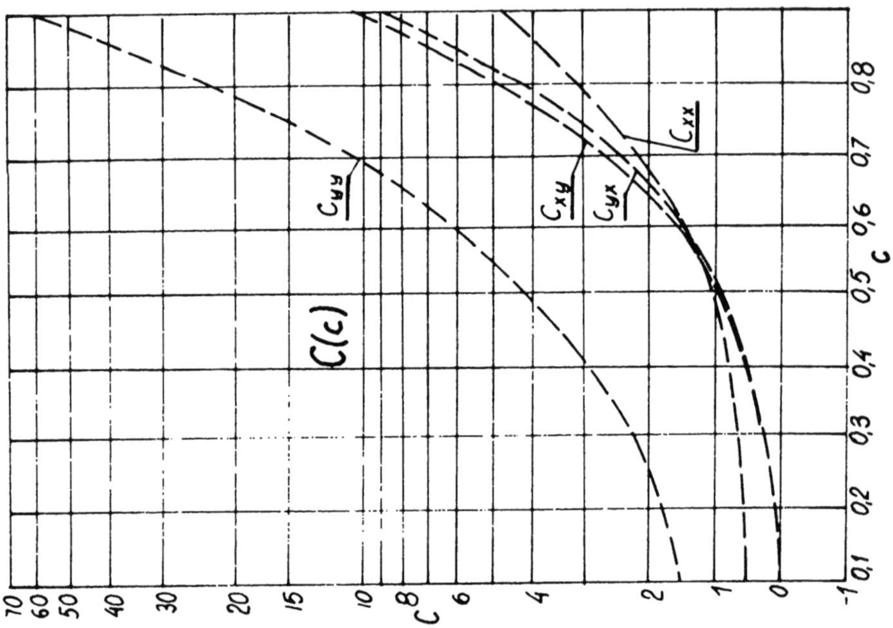

Fig. 10

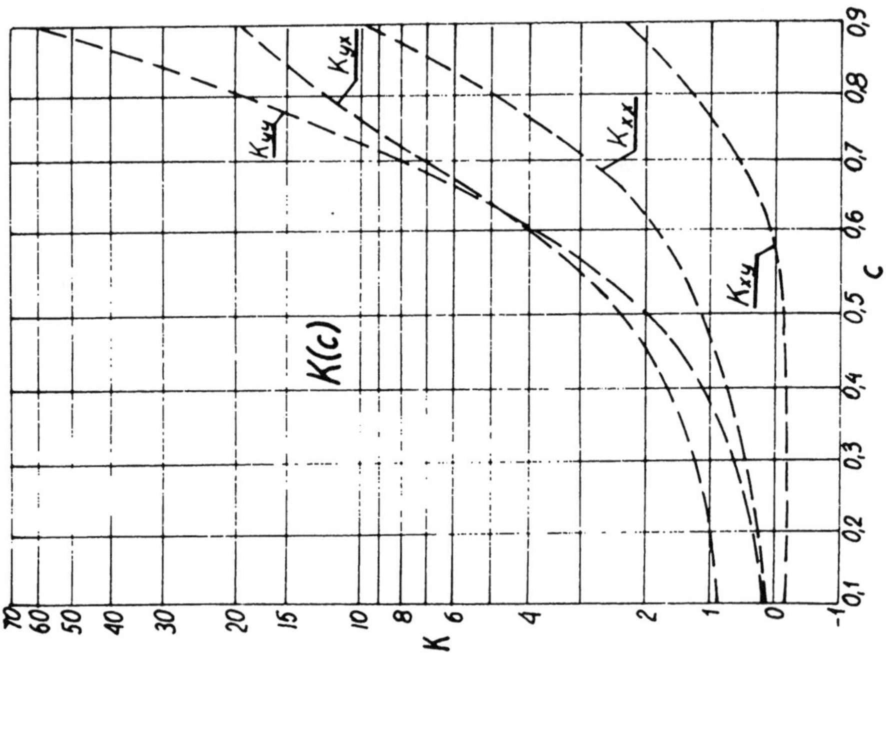

Fig. 9

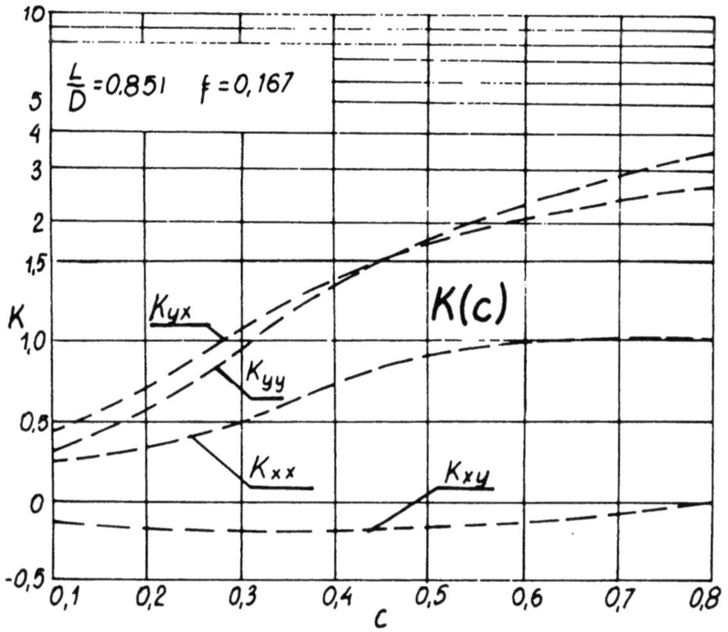

L/D = 0.851 f = 0,167
K(c)
K
K_{yx}
K_{yy}
K_{xx}
K_{xy}
10
5
4
3
2
1,5
1,0
0,5
0
-0,5
0,1
0,2
0,3
0,4
0,5
0,6
0,7
0,8
c

Fig. 11

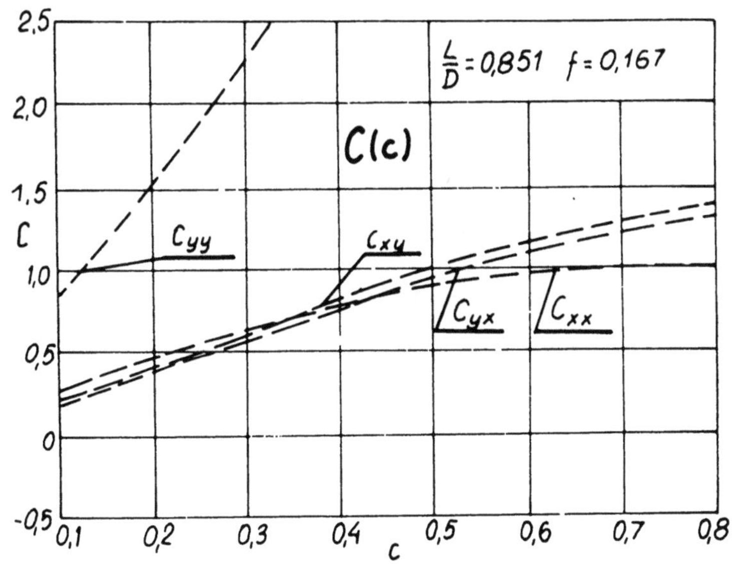

L/D = 0,851 f = 0,167
C(c)
C
C_{yy}
C_{xy}
C_{yx}
C_{xx}
2,5
2,0
1,5
1,0
0,5
0
-0,5
0,1
0,2
0,3
0,4
0,5
0,6
0,7
0,8
c

Fig. 12

References

(1) Parszewski, Z.A., Roszkowski, M., Gluza, J., Pressure Distribution and Hydrodynamic Forces of Journal Bearing Partial Sleeves. Arch.Bud.Masz, 14, 3, Warszawa 1967, 455-523

(2) Korovcinskij, M.V., Teorieticeckije asnovy roboty podszipnikov skolzenija, Masgiz, Moskow 1959.

(3) Pinkus, O., Sternlicht, B., Theory of Hydrodynamic Lubrication, McGraw Hill, New York 1961.

(4) Parszewski, Z., Roszkowski, M., Load Capacity of Multi-Sleeve Journal Bearing, Arch. Bud. Masz, 14, 4, Warszawa 1967, 577-593.

(5) Parszewski, Z., Roszkowski, M., Pressure Distribution and Load Capacity of Journal Bearing with Eccentric Sleeves, Arch.Bud.Masz, 16, 2, Warszawa 1969, 155-171.

(6) Roszkowski, M., Dynamics of Plane Motion of a Journal in a Partial Slide Bearing, Arch. Bud. Masz., 18, 1, Warszawa 1971, 33-75.

(7) Hien, N.H., Roszkowski, M., Stiffness and Damping Matrices of Multi-Sleeve Journal Bearings, Proc. V Congr. TMM, Montreal 1979, 1364-1367.

(8) Olszewski, W., Characterystyka sztywnosci i tlumienia łozysk z panewkami ekscentrycznymi o wspolzaleznych filmach olejowych, D.Eng.Sc. thesis, P.L. Lodz 1978.

(9) Parszewski, Z.A., Drgania i Dynamika Maszyn, WNT, Warszawa 1982, 420p.

CHAPTER 2.2

STIFFNESS AND DAMPING CHARACTERISTICS OF TILTING-PAD JOURNAL BEARINGS

H. Springer

1. INTRODUCTION

Lateral vibrations of large turborotors of steamturbines or hydroelectric power plants are highly influenced by the stiffness and damping characteristics of the bearings. Large turborotors are usually guided in hydrodynamic journal bearings. Tilting-pad journal bearings are used particularly for horizontal or vertical shafts of high diameters. If the amplitude of the shaft vibration is small compared with the clearance of the bearing, the characteristics of the hydrodynamic oil film can be described, corresponding to a linear theory, by unsymmetrical stiffness and damping matrices of dimension (2x2), containing four stiffness and four damping coefficients, respectively. When assuming a rigid and immovable bearing house then the stiffness and damping coefficients of the oil film depend only upon the static bearing load, the clearance and speed of the shaft and the specifications of the lubricant, see Ref.1,2. They are not influenced by the frequency of lateral shaft oscillations. However, when the bearing house is flexible or movable - for example, when a tilting-pad bearing with flexibly pivoted pads is used - then an essential influence of lateral shaft vibrations upon the stiffness and damping coefficients could occur [3,4,5]. This holds even for small amplitudes of vibrations where a linear theory can be applied.

In this paper journal bearings of vertical turborotors for hydroelectric power plants are investigated [6]. Figure 1. shows a schematic diagram of a vertical turborotor guided by four tilting-pad journal bearings and carried by one thrust bearing which is balancing the weight of the rotor and axial forces of the runner. Under transient operating conditions - for example a pump or a turbine start up - the turborotor can be exposed to severe hydraulically excited vibrations. Then very high radial forces are acting upon the guide bearings causing radial and tangential displacements of the flexibly supported pads. When the flexibility of the support of the pads is of same order of magnitude as the flexibility of the oil film one can not neglect the movability of the pads. In other words, all degrees of freedom of the pads have to be taken into account if one is to determine correct stiffness and damping coefficients of such a guide bearing. Let us assume a small lateral harmonic excitation force F_x acting upon the shaft with an angular frequency ω_e, see Figure 2. Then

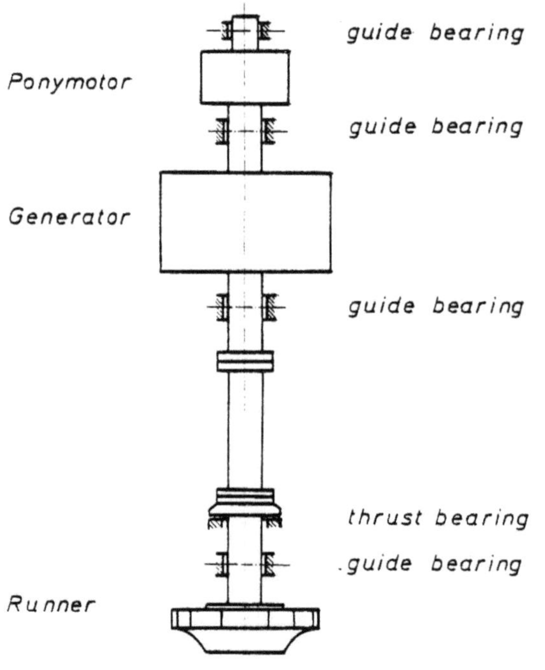

Figure 1. Schematic diagram of a vertical turborotor for a hydroelectric power plant

the shaft as well as all pads of the bearing will oscillate with the same frequency ω_e. For small amplitudes of vibrations we can establish a set of linear differential equations of motion for the bearing system. Now the degrees of freedom of all pads can be eliminated from the equations of motion and a condensed dynamic stiffness matrix, describing the motion of the shaft center only, is set up. The resulting dynamic stiffness matrix is of dimension (2x2) corresponding to the fact that two degrees of freedom for the movement of the shaft center remain after the condensation.

The objective of this paper is the calculation of condensed dynamic stiffness matrices and flexibility matrices for large tilting-pad bearings that guide a vertical turborotor. Socalled local bearing systems

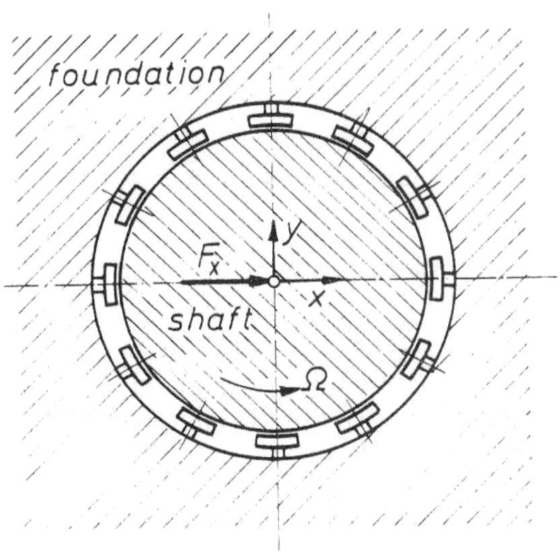

Figure 2. Schematic cross-section of a tilting-pad bearing with 12 pads. Radial excitation force F_x.

are assumed in this investigation, i.e., interactions between different bearings through the foundation are neglected.

2. HYDRODYNAMIC OIL FILM PRESSURE OF A FLEXIBLY SUPPORTED PAD

Figure 3 shows a detailed diagram of a pivoted pad for the radial guide bearing as drawn in Figure 2. The support of the pad is assumed to have a linear viscoelastic characteristic. Flexibility as well as movability of the foundation, carrying the pads, are not considered. The symbols used in Figure 3 have the following meaning [7,8,9]:

$x - y$		coordinate system for the entire bearing
$\bar{x} - \bar{y}$		coordinate system for one pad (called subsystem)
h_o	+)	radial clearance of the bearing (for $\Omega = 0$)
$h(\bar{\varphi},t)$		thickness function of the oil film of a pad
$u,v,\bar{u},\bar{v}$		displacements of the shaft center in the x-y and $\bar{x}$-$\bar{y}$ system, resp.
$\bar{u}_s,\bar{v}_s,\alpha_s$		radial, tangential and angular displacements of the pad, resp.
r		radius of the shaft
R_s	+)	radius of the curvature of the pad
Λ_s	+)	thickness of the pad at the supporting point
Φ		angular position of the pad with respect to the x-axis
m_s	+)	mass of a pad
I_s	+)	moment of inertia of a pad with respect to the supporting point
$2\varphi_o$	+)	sector angle of a pad
k_r,k_t,k_α	+)	stiffness- } coefficients of visco-elastic supporting in radial, tangential and angular direction, respectively
c_r,c_t,c_α	+)	damping- }
$K_{\bar{x}},K_{\bar{y}}$		oil film forces acting upon the pad
Ω		angular velocity of the shaft, assumed to be constant
t		time

+) indicated parameters are assumed to be equal for all pads of the bearing

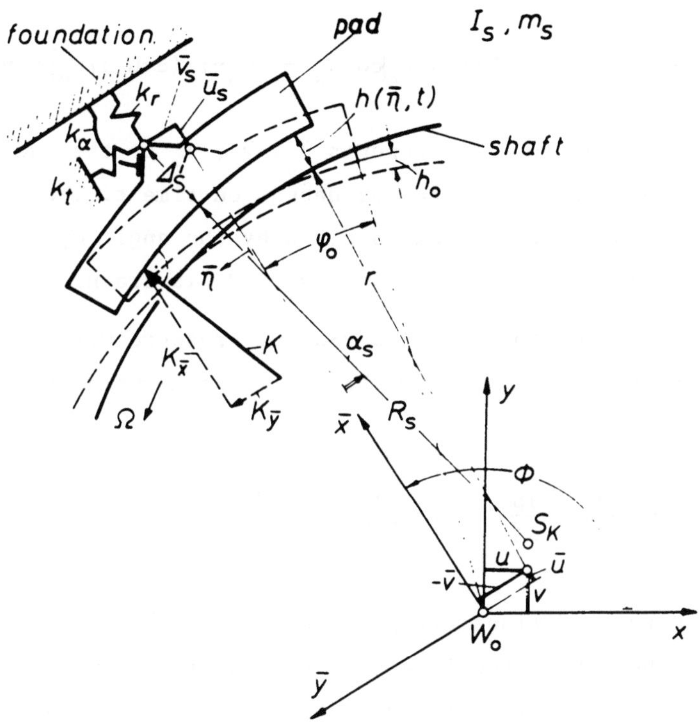

Figure 3. Flexibly supported pad of a tilting-pad journal bearing

In the following we consider a guide bearing with $n \geqslant 3$ equally shaped pads which are flexibly supported to the foundation and symmetrically mounted along the circumference of the shaft, see Figure 2. Using References 7,8,9 the thickness function of the oil film for an out of center pivoted pad is given by

$$H(\bar{\eta},t) = h(\bar{\eta},t)/h_o = C_o + C_1(t)\cos(2\varphi_o\bar{\eta}/L) + S_1(t)\sin(2\varphi_o\bar{\eta}/L) \quad (2.1)$$

where the coefficients

$$C_o = (R_s - r)/h_o \quad (2.2a)$$

$$C_1 = \left[\alpha_s(R_s+\Delta_s)/h_o + (\bar{v}-\bar{v}_s)/h_o\right]\sin\varphi_s - \left[C_o - 1 + (\bar{u}-\bar{u}_s)/h_o\right]\cos\varphi_s \quad (2.2b)$$

$$S_1 = -\left[\alpha_s(R_s+\Delta_s)/h_o + (\bar{v}-\bar{v}_s)/h_o\right]\cos\varphi_s - \left[C_o - 1 + (\bar{u}-\bar{u}_s)/h_o\right]\sin\varphi_s \quad (2.2c)$$

depend on time t since the displacements $\bar{u}_s$, $\bar{v}_s$, α_s of the pad and the displacements $\bar{u}$, $\bar{v}$ of the shaft center depend on time t. The eccentricity of the supporting point of the pad is given by the angle φ_s, see Ref.7. In Figure 3 a centrally pivoted pad with $\varphi_s = 0$, is drawn.

The distribution of the non-steady-state pressure within the oil film of a pad can be evaluated from the wellknown Reynolds differential equation [2],

$$\frac{\partial}{\partial\bar{\eta}}\left(\frac{h^3}{\mu}\frac{\partial p}{\partial\bar{\eta}}\right) + \frac{\partial}{\partial\bar{\zeta}}\left(\frac{h^3}{\mu}\frac{\partial p}{\partial\bar{\zeta}}\right) = 6r\Omega\frac{\partial h}{\partial\bar{\eta}} + 12\frac{\partial h}{\partial t} \quad (2.3)$$

The simple boundary conditions are $p(\bar{\eta},\bar{\zeta},t) = 0$ at $(\bar{\eta} = \pm L/2,\bar{\zeta},t)$ and $(\bar{\eta},\bar{\zeta} = \pm B/2,t)$, where L is the length of the pad in circumferential direction, B is the width in axial direction and μ is the dynamic viscosity of the lubricant.

An approximate solution of Equation (2.3) was given by the author [7], applying the method of Galerkin [10] to the variational principle corresponding to Equation (2.3). By using that method the non-steady-state pressure distribution $p(\bar{\eta},\bar{\zeta},t)$ within the oil film can be approximately determined in the form

$$p(\bar{\eta},\bar{\zeta},t) = \frac{\mu L r\Omega}{h_o^2} f(2\bar{\eta}/L,t)g(2\bar{\zeta}/B) \quad (2.4)$$

where

$$f(\eta,t) = \sum_{\nu=1}^{N}\{a_\nu(t)\left[1-T_{2\nu}(\eta)\right] + b_\nu(t)\left[T_1(\eta)-T_{2\nu+1}(\eta)\right]\} \quad (2.5)$$

is a Chebyshev-expansion - T_ν being the Chebyshev-polynomial [11] of order

$\nu \leq 2N+1$ - which describes the pressure distribution in circumferential direction, and

$$g(\zeta) = 1 - |\zeta|^m \tag{2.6}$$

is a parabolic function of order $m \geq 2$ which determines the distribution of pressure in the axial direction. The calculation of the vector

$$\underline{z}(t) = \{a_1, a_2, \ldots a_N, b_1, b_2, \ldots b_N\}^T \tag{2.7}$$

containing the Chebyshev coefficients of the above series and the calculation of the exponent m are shown in Ref.7,8. The solution as given by Equation (2.4) to (2.6) is very useful for numerical computation since the Chebyshev-series has a very high rate of convergence. Hence, a lot of computer time can be saved in comparison with other numerical methods established in the literature. If cavitation occurs within the oil film of a pad then modified boundary conditions have to be applied [2,4,7]. The components of the oil film forces acting upon the pad, see Figure 3, can be calculated in matrix notation as follows [7,8],

$$\{K_{\bar{x}}, K_{\bar{y}}\}^T = \frac{2m}{m+1} B\mu r\Omega(\varphi_o/\psi)^2 \underline{\Lambda}_s \underline{z} \tag{2.8}$$

where $\underline{\Lambda}_s$ is a (2x2N) dimensioned geometrical coefficient matrix which can be evaluated from Ref.7,8. In Equation (2.8) φ_o is the sector angle, $\psi = h_o/r$ is the dimensionless clearance, and B is the width of a pad. The moment of the oil film forces with respect to the supporting point of the pad is given in the form

$$M_\alpha = -K_{\bar{y}}(R_s + \Delta_s) \tag{2.9}$$

3. DYNAMIC STIFFNESS AND FLEXIBILITY MATRIX OF THE TILTING-PAD BEARING

A guide bearing with $n \geq 3$ equally shaped pads is considered. The pads are assumed to be symmetrically mounted along the circumference of the shaft, see Figure 2. Then, the angular positions of the pads are given by $\Phi_j = 2\pi(j-1)/n$ where $j = 1,2,3...n$ indicates the number of a pad. The angular speed Ω of the shaft is assumed to be constant.

In the general case an external static load is acting upon the shaft in the radial direction. For that given load there exists an equilibrium position for the shaft center at (u_o,v_o) and for the pads at $(\bar{u}_{so},\bar{v}_{so},\alpha_{so})_j$ where $j=1,2,...n$. The equilibrium position of the bearing system can be calculated by iteration using Equations (2.8) and (2.9). The iteration has to be continued to the end that the nonlinear force exerted by the oil film upon each pad is in balance with the linear force exerted by the visco-elastic supporting of the pad,and further,the resulting nonlinear oil film force acting upon the shaft is in equilibrium with the given external static load. The nonlinear calculation is carried out through a computer program.

In this paper the simple case of a vertical shaft is assumed and there is no external static load acting in the radial direction of the bearing system. Hence, for a symmetrical bearing the equilibrium position of the shaft is given by $u_o=\bar{u}_o=o$, $v_o=\bar{v}_o=o$,and the equilibrium position of each pad is given by $(\bar{u}_{so})_j = \bar{u}_{so}$, $(\bar{v}_{so})_j = \bar{v}_{so}$, $(\alpha_{so})_j = \alpha_{so}$ as drawn in Figure 4. The values of $\bar{u}_{so}$, $\bar{v}_{so}$, α_{so} are found by a nonlinear numerical calculation.

If the equilibrium state of the bearing is disturbed by external perturbation forces acting upon the shaft then the bearing system is excited to vibrations, see Figure 2. Assuming small amplitudes of oscillations compared with the clearance of the bearing a linear calculation is made possible. Then, in general,the bearing system is completely characterized by its dynamic stiffness matrix which is of dimension (2x2) and contains stiffness and damping of the oil film as well as inertia effects and viscoelastic supporting effects of the pads. Inertia effects of the rotor itself are not included into the bearing dynamics. Hence, the mass of the shaft is assumed to be zero for this investigation. In order to calculate the dynamic stiffness matrix of the whole bearing a subsystem is considered which contains the massless shaft and one pad only, being in the above mentioned equilibrium state, see Figure 4.

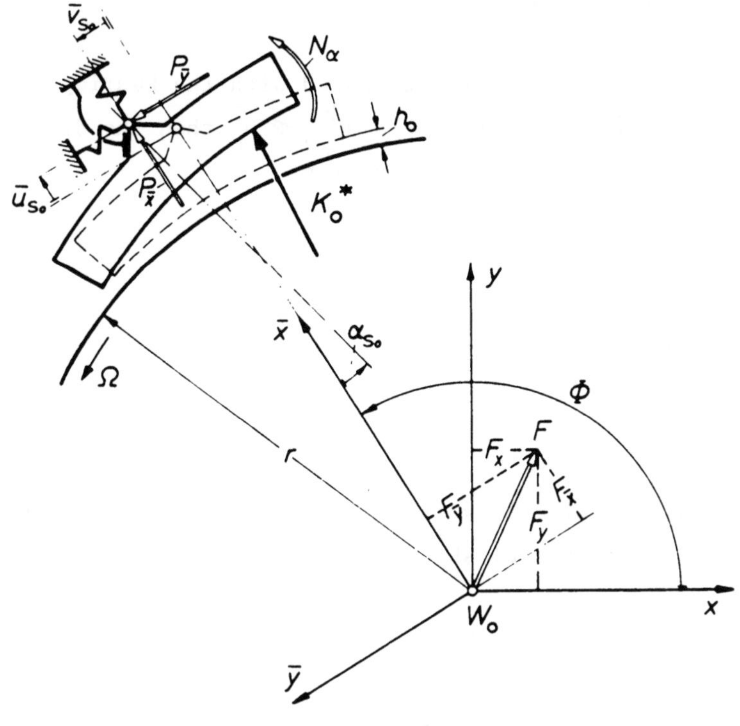

Figure 4. Zero-load state of the vertical shaft at $u_o = v_o = 0$ and of the pad at $\bar{u}_{so}$, $\bar{v}_{so}$, α_{so}. External perturpation forces $F_{\bar{x}}$, $F_{\bar{y}}$, $P_{\bar{x}}$, $P_{\bar{y}}$, N_α.

Now, the equilibrium state of the subsystem is disturbed by a small harmonic perturbation force (using complex notation with $i = \sqrt{-1}$)

$$\underline{F}_j(t) = \{F_{\bar{x}}, F_{\bar{y}}, P_{\bar{x}}, P_{\bar{y}}, N_\alpha\}_j^T = \hat{\underline{F}}_j e^{i\omega_e t} \qquad (3.1)$$

where j indicates the number of the subsystem to be considered and

$\omega_e = 2\pi f_e$ stands for the angular frequency of excitation. The components of the exciting force vector $\underline{F}_j$ are indicated in Figure 4. The number of components of $\underline{F}_j$ is equal to the number of degrees of freedom of the subsystem. After transient vibrations have died out the bearing subsystem is excited to steady-state oscillations as given by the displacement vector

$$\underline{z}_j(t) = \{\bar{u}, \bar{v}, \bar{u}_s, \bar{v}_s, \alpha_s\}_j^T \tag{3.2}$$

If the magnitude of $\underline{z}_j(t)$ is small compared with the bearing clearance h_o, i.e., $\max\left[\bar{u}^2 + \bar{v}^2 + \bar{u}_s^2 + \bar{v}_s^2 + (r\alpha_s)^2\right]_j \ll h_o^2$ then the displacement vector is approximately a harmonic function

$$\underline{z}_j(t) = \hat{\underline{z}}_j \, e^{i\omega_e t} \tag{3.3}$$

Then a linear relation

$$\hat{F}_j = \underline{S}_j^!(i\omega_e)\hat{\underline{z}}_j \tag{3.4}$$

between the amplitudes of the exciting forces $\hat{\underline{F}}_j$ and the amplitudes of displacements $\hat{\underline{z}}_j$ can be established. The complex valued matrix

$$\underline{S}_j^!(i\omega_e) = -\omega_e^2\underline{M}_j + i\omega_e\underline{C}_j + \underline{K}_j \tag{3.5}$$

is called the dynamic stiffness matrix of the bearing subsystem j, with $\underline{M}_j$, $\underline{C}_j$ and $\underline{K}_j$ being the mass, damping and stiffness matrix of the subsystem, respectively [9]. $\underline{S}_j^!$ is of dimension (5x5). Considering the equilibrium conditions for the massless oil film of a pad and using Equation (2.2) the dynamic stiffness matrix can be written in the form of Equation (3.6). In that Equation $\omega_e = 2\pi f_e$ is the angular frequency of external excitation. Furthermore, the complex valued matrix

$$\underline{s}_j = \begin{bmatrix} s_{\bar{x}\bar{x}} & s_{\bar{x}\alpha} \\ s_{\alpha\bar{x}} & s_{\alpha\alpha} \end{bmatrix}_j = \begin{bmatrix} k_{\bar{x}\bar{x}}+i\omega_e c_{\bar{x}\bar{x}} & k_{\bar{x}\alpha}+i\omega_e c_{\bar{x}\alpha} \\ k_{\alpha\bar{x}}+i\omega_e c_{\alpha\bar{x}} & k_{\alpha\alpha}+i\omega_e c_{\alpha\alpha} \end{bmatrix}_j$$

represents the constant stiffness and damping coefficients of the oil film of the subsystem. It can be shown that four independent stiffness coefficients $k_{\mu\nu}$ only and four independent damping coefficients $c_{\mu\nu}$ only exist for a pad[9].

$$\underline{S}_j'(i\omega_e) = \left[\begin{array}{cc|ccc} \multicolumn{2}{c|}{\underline{A}_j} & \multicolumn{3}{c}{\underline{H}_j} \\ s_{\bar{x}\bar{x}} & \dfrac{-s_{\bar{x}\alpha}}{r_s} & -s_{\bar{x}\bar{x}} & \dfrac{s_{\bar{x}\alpha}}{r_s} & -s_{\bar{x}\alpha} \\ \dfrac{-s_{\alpha\bar{x}}}{r_s} & \dfrac{s_{\alpha\alpha}}{r_s^2} - \dfrac{K_o^*}{r_s} & \dfrac{s_{\alpha\bar{x}}}{r_s} & \dfrac{-s_{\alpha\alpha}}{r_s^2} + \dfrac{K_o^*}{r_s} & \dfrac{s_{\alpha\alpha}}{r_s} \\ \hline -s_{\bar{x}\bar{x}} & \dfrac{s_{\alpha\bar{x}}}{r_s} & \begin{array}{c} s_{\bar{x}\bar{x}}+t_{\bar{x}} \\ -m_s\omega_e^2 \end{array} & \dfrac{-s_{\bar{x}\alpha}}{r_s} & s_{\bar{x}\alpha} \\ \dfrac{s_{\alpha\bar{x}}}{r_s} & \dfrac{-s_{\alpha\alpha}}{r_s^2} + \dfrac{K_o^*}{r_s} & \dfrac{-s_{\alpha\bar{x}}}{r_s} & \begin{array}{c} \dfrac{s_{\alpha\alpha}}{r_s^2} - \dfrac{K_o^*}{r_s} \\ +t_{\bar{y}} - m_s\omega_e^2 \end{array} & \dfrac{s_{\alpha\alpha}}{r_s} \\ -s_{\alpha\bar{x}} & \dfrac{s_{\alpha\alpha}}{r_s} & s_{\alpha\bar{x}} & \dfrac{-s_{\alpha\alpha}}{r_s} & \begin{array}{c} s_{\alpha\alpha}+t_{\alpha} \\ -I_s\omega_e^2 \end{array} \\ \multicolumn{2}{c|}{\underline{V}_j} & \multicolumn{3}{c}{\underline{B}_j} \end{array}\right]_j \qquad (3.6)$$

The term K_o^*/r_s, where $r_s = R_s + \Delta_s$, depends upon the oil film force $K_o^* \geqslant 0$ which is acting between a pad and the shaft in the zero-load equilibrium state of the bearing system. K_o^* is a socalled "reactive force" which, in general, does not vanish. As can be seen from the diagonal of the matrix Equation (3.6) the term $-K_o^*/r_s$ yields a destabilizing effect but in general K_o^* is very small compared with the term $k_{\alpha\alpha}/r_s$ and no instability appears.

Furthermore, in Equation (3.6) $\{t_{\bar{x}}, t_{\bar{y}}, t_\alpha\}_j = \{k_r, k_t, k_\alpha\}_j + i\omega_e\{c_r, c_t, c_\alpha\}_j$ represents the stiffness and damping coefficients of the visco-elastic supporting of a pad, see Figure 3.

Inertia effects of the pad are described by the mass m_s and the moment of inertia I_s, see Figure 3. $\underline{A}_j$, $\underline{H}_j$, $\underline{V}_j$, $\underline{B}_j$ are submatrices to be needed later in Equation (3.8).

Under practical operating conditions of a rotor bearing system the pads are not excited by external perturbation forces. Hence, the components of the amplitude force vector in Equation (3.1) reduce to $\hat{F}_{\bar{x}j} \neq 0$, $\hat{F}_{\bar{y}j} \neq 0$, $\hat{P}_{\bar{x}j} = \hat{P}_{\bar{y}j} = 0$, $\hat{N}_{\alpha j} = 0$. Using this, it is possible to eliminate the amplitudes of the displacements of the pad as $(\hat{\bar{u}}_s, \hat{\bar{v}}_s, \hat{\alpha}_s)j$ from Equation (3.4) and a reduced system

$$\left\{\hat{F}_{\bar{x}}, \hat{F}_{\bar{y}}\right\}_j^T = \underline{S}_j(i\omega_e)\left\{\hat{\bar{u}}, \hat{\bar{v}}\right\}_j^T \tag{3.7}$$

is obtained which no more contains the displacements of the pad but only the displacements of the shaft. Excitation forces are acting upon the massless shaft and correspond to the subsystem j. The new matrix $\underline{S}_j(i\omega_e)$ is called the condensed dynamic stiffness matrix of the subsystem j and is of dimension (2x2). Using the submatrices as indicated in Equation (3.6) yields the condensation

$$\underline{S}_j(i\omega_e) = \underline{A}_j - \underline{H}_j\underline{B}_j^{-1}\underline{V}_j \tag{3.8}$$

which can be calculated by using numerical means. In order to determine the dynamic stiffness matrix of the whole bearing system, including the shaft and all pads, the force and displacement vector in Equation (3.7) have to be transformed from the local $\bar{x}$ - $\bar{y}$ system of a pad into the bearing system x - y, see Figure 3. This is carried out by the orthogonal transformation matrix

$$\underline{T}_j = \begin{bmatrix} \cos\Phi & \sin\Phi \\ -\sin\Phi & \cos\Phi \end{bmatrix}_j \qquad (3.9)$$

After superposing the forces of all subsystems j = 1,2,..n the resultant dynamic stiffness matrix

$$\underline{S}(i\omega_e) = \begin{bmatrix} s_{xx} & s_{xy} \\ s_{yx} & s_{yy} \end{bmatrix} = \sum_{j=1}^{n} \underline{T}_j^{-1}\, \underline{S}_j(i\omega_e)\, \underline{T}_j \qquad (3.10)$$

is obtained and the wellknown relation

$$\left\{\hat{F}_x, \hat{F}_y\right\}^T = \underline{S}(i\omega_e)\,\left\{\hat{u},\hat{v}\right\}^T \qquad (3.11)$$

holds. Equation (3.11) determines the force vector in terms of the displacement vector for a harmonicaly excited massless shaft guided in a tilting-pad journal bearing. The inverse of $\underline{S}$

$$\underline{\alpha}(i\omega_e) = \underline{S}^{-1}(i\omega_e) \qquad (3.12)$$

is called the dynamic flexibility or receptance matrix. The frequency response of the bearing system is clearly characterized by drawing the elements of $\underline{\alpha}(i\omega_e)$ in a complex plane (Nyquist-plot). Furthermore,

$$\underline{K}(\omega_e) = \mathrm{Re}\left[\underline{S}(i\omega_e)\right] = \begin{bmatrix} k_{xx} & k_{xy} \\ k_{yx} & k_{yy} \end{bmatrix} \qquad (3.13)$$

and

$$\underline{C}(\omega_e) = \frac{1}{\omega_e}\,\mathrm{Im}\,\left[\underline{S}(i\omega_e)\right] = \begin{bmatrix} c_{xx} & c_{xy} \\ c_{yx} & c_{yy} \end{bmatrix} \qquad (3.14)$$

respectively, are the stiffness and damping matrices of the tilting-pad bearing system. In general they depend upon the frequency ω_e of the shaft excitation.

4. NUMERICAL EXAMPLE

A tilting-pad bearing for a large vertical turborotor is considered, see Figure 2. From Figure 3 the following specifications are assumed: Number of pads n = 12, centrally pivoted. Length and width of equally shaped pads L = 0.25 m and B = 0.25 m, respectively. Thickness of a pad Δ_s = 0.085 m, radius of curvature R_s = 0.6515 m and sector angle $2\varphi_o$ = 22°. Mass and moment of inertia of a pad m_s = 50 kg and I_s = 0.26 kgm^2, re-

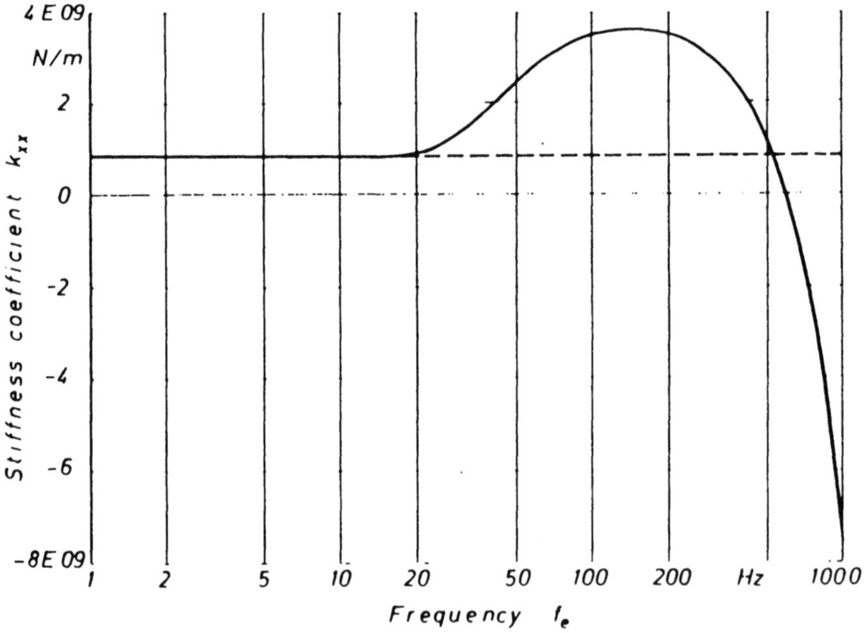

Figure 5. Stiffness coefficient as a function of the frequency of lateral shaft excitation. Specifications as assumed above.
---- value of the stiffness for zero frequency

spectively. Radius of the shaft $r = 0.650$ m. Radial clearance of the bearing $h_o = 200$ μm. Speed of the shaft $n_s = 375$ rpm. Stiffness and damping of the support of the pad $k_r = 7\times10^8$ N/m, $k_t = \infty$, $k_\alpha = 10^5$ Nm and $c_r = 7.5\times10^4$ Ns/m, $c_\alpha = 100$ Nsm, respectively. Dynamic viscosity of the lubricant $\mu = 0.023$ Ns/m^2. Using the above specifications, the equilibrium state of the bearing system is evaluated to $\bar{u}_{so} = 0.1038\, h_o$, $\bar{v}_{so} = 0$, $\alpha_{so} = 1.1078\, h_o/r$ for each pad and $u_o = 0$, $v_o = 0$ for the shaft.

From the above, four stiffness and four damping coefficients for the oil film of each pad can be calculated and yields the matrix $\underline{s}_j$, see Equation (3.6). The reactive force, acting between the shaft and each pad at the equilibrium state, is evaluated to $K_o^* = 14520$ N. As a result of the symmetry of the bearing system we obtain $s_{xx} = s_{yy}$ and $s_{xy} = -s_{yx}$. Neglecting the moment of inertia of a pad and setting k_α and c_α to zero

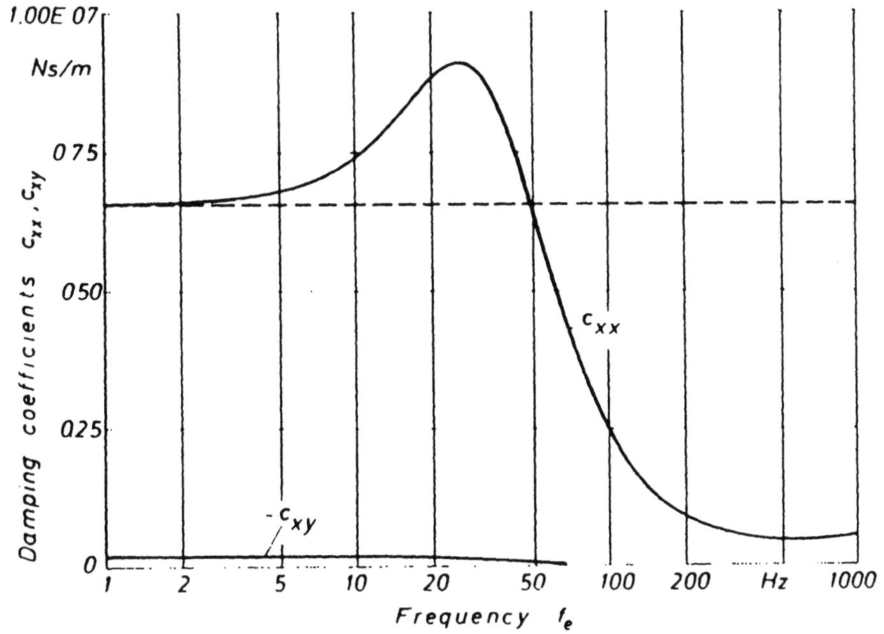

Figure 6. Damping coefficients of a tilting-pad bearing as a function of the frequency of shaft excitation.
---- damping value c_{xx} for zero frequency

an isotropic bearing system results, for which the cross-coupling terms $s_{xy} = -s_{yx}$ vanish.

Figure 5 shows the coefficient of stiffness k_{xx} of the bearing as a function of the frequency f_e of shaft excitation. For the example considered, the cross-coupling term k_{xy} is negligible small compared with the value of k_{xx}. It can be seen from the diagram that for values of the frequency less than about 20 Hz the stiffness remains nearly constant but changes rapidly when the frequency is exceeding the value of 20 Hz. When the frequency goes to infinity, the stiffness goes to negative infinity. This results from the inertia effects of the pads. The value of the stiffness for zero frequency (static case) is drawn in the diagram as a broken line.

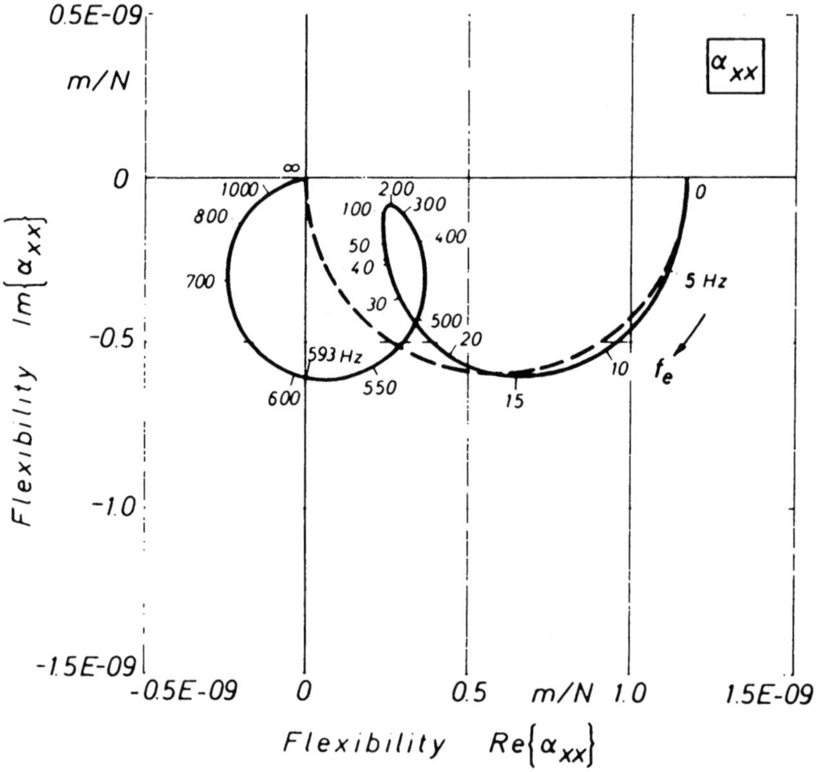

Figure 7. Nyquist plot of the dynamic flexibility of a tilting-pad bearing.
—— exact solution
---- approximation for constant stiffness and damping

Figure 6 shows the coefficients of damping c_{xx} and c_{xy} in terms of frequency f_e. Obviously the magnitude of c_{xy} is negligible compared with the magnitude of c_{xx}. Only for a small frequency range up to 3 Hz the damping values of the bearing can be assumed as nearly constant. If the frequency increases, then the dynamic stiffness of the oil film grows and the resulting damping coefficient of the bearing decreases to a value which can be determined from the damping coefficient of the support of the pads only.

Figure 7 shows a Nyquist-plot of the element α_{xx} of the flexibility or receptance matrix of the bearing system considered. The solid line represents the correct solution, the broken line is an approximation where stiffness and damping coefficients of the bearing are assumed to remain at their static values. There is an acceptable agreement between the

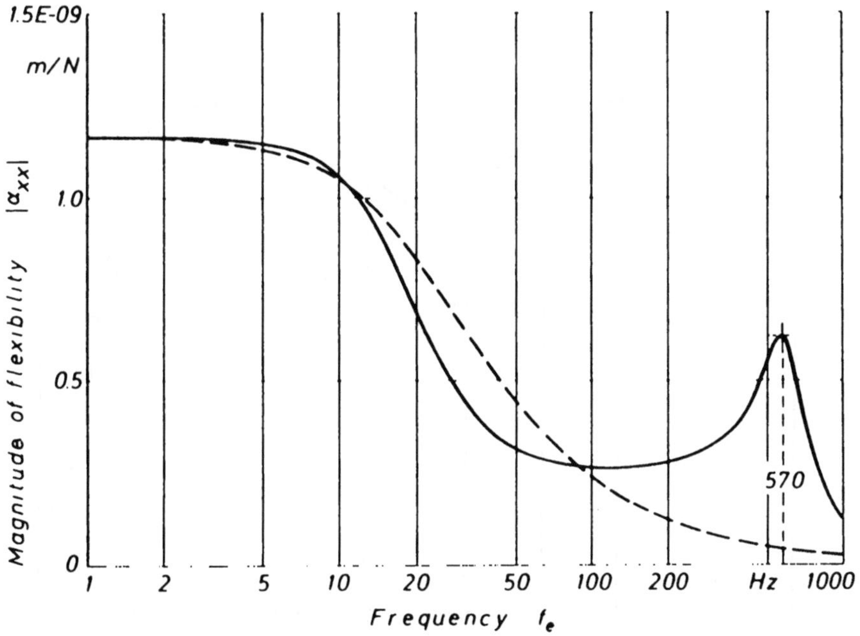

Figure 8. Frequency response for the magnitude of the dynamic flexibility
—— exact solution
---- approximation for constant stiffness and damping

approximation and the exact solution only in a small frequency range up to 5 Hz. The first circle in the diagram (solid line up to f = 200 Hz) depends primarily upon the characteristics of the oil film and on the stiffness and damping of the supports of the pads. The second circle is mainly determined by the mass of the pads and the stiffness and damping of the supports. Hence, the influence of inertia effects on the bearing flexibility grows with increasing values of the frequency. Figure 8 shows the frequency response for the magnitude of the point flexibility α_{xx}. The broken line represents an approximation where constant values for stiffness and damping coefficients are assumed. At f = 570 Hz a resonance occurs corresponding to the mass-spring system of the elastically mounted pad.

The frequency response of the phase angle ε_{xx} is drawn in Figure 9. For low values of the frequency up to 5 Hz a good agreement between the exact solution and the approximation is obtained. For high values of the

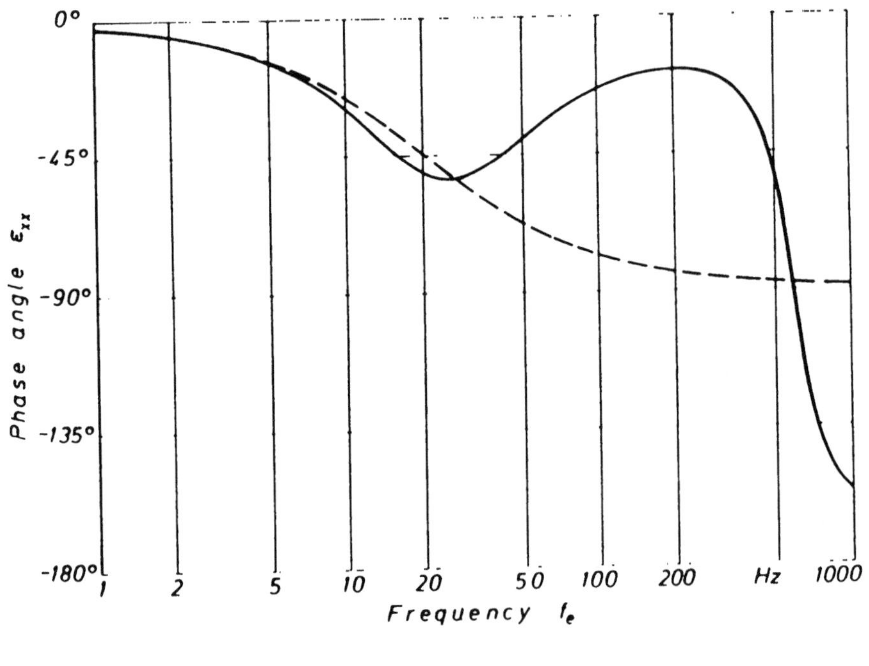

Figure 9. Frequency response for the phase angle of the dynamic flexibility
—— exact solution
---- approximation for constant stiffness and damping

frequency the phase angle ε_{xx} approaches -180° according to inertia effects of the pads. The coefficient of the dynamic cross-flexibility $\alpha_{xy} = -\alpha_{yx}$ is drawn as a Nyquist plot in Figure 10. The magnitude of α_{xy} is very small compared with the magnitude of α_{xx} and decreases for increasing values of the frequency.

5. CONCLUSION

A linear theory for small amplitudes of oscillations is developed which makes it possible to calculate dynamic stiffness and flexibility matrices of tilting-pad journal bearings. The investigation shows that stiffness and damping coefficients of a bearing system with movable pads

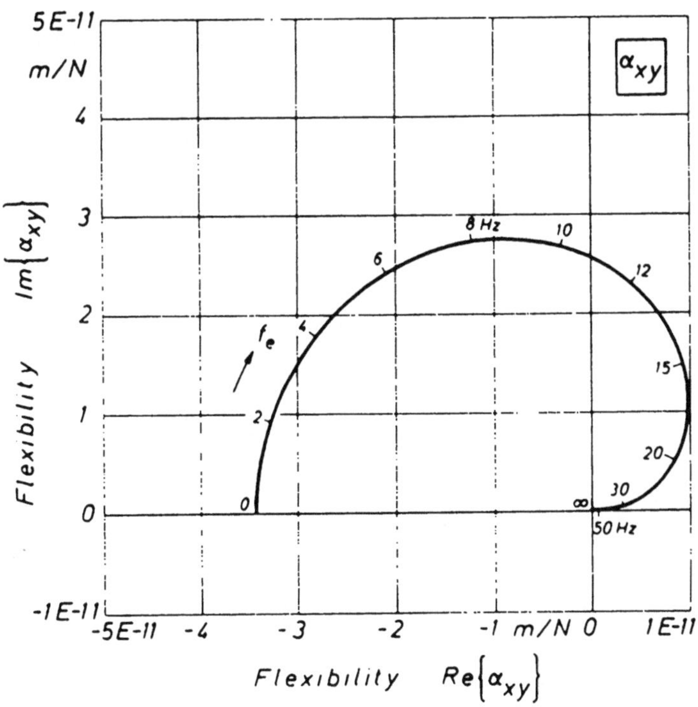

Figure 10. Nyquist plot of the cross-flexibility of a tilitng-pad bearing

depend not only on the Sommerfeldnumber of the bearing but on the frequency of lateral shaft oscillations also. This result has to be taken into account if one is performing an eigenfrequency analysis or calculating critical speeds of a turborotor system.

6. REFERENCES

1. Glienicke, J., Feder- und Dämpfungskonstanten von Gleitlagern für Turbomaschinen und deren Einfluß auf das Schwingungsverhalten des einfachen Rotors. *Dissertation, TU Karlsruhe*, 1966.
2. Varga, Z.E., Wellenbewegung, Reibung und Öldurchsatz beim segmentierten Radialgleitlager von beliebiger Spaltform unter konstanter und zeitlich veränderlicher Belastung. *Dissertation, ETH Zürich*, 1971.
3. Lund, J.W., Spring and damping coefficients for the tilting-pad journal bearing. *ASLE Transactions 7*, 342, 1964.
4. Klumpp, R., Ein Beitrag zur Theorie von Kippsegmentlagern. *Dissertation, TU Karlsruhe*, 1975.
5. Malcher, L., Die Federungs- und Dämpfungseigenschaften von Gleitlagern für Turbomaschinen (Experimentelle Untersuchungen). *Dissertation, TU Karlsruhe*, 1975.
6. Giersig, K., Rodund II pumped storage scheme commissioned. *Water Power & Dam Construction*, 5, 21, 1977.
7. Springer, H., Zur Berechnung hydrodynamischer Lager mit Hilfe von Tschebyscheff-Polynomen. *Forschg.Ing.Wesen*, 44, 126, 1978.
8. Springer H., Nichtlineare Schwingungen schwerer Rotoren mit vertikaler Welle und Kippsegmentradiallagern. *Forschg.Ing.Wesen*, 45, 119, 1979.
9. Springer, H., Dynamische Eigenschaften von Gleitlagern mit beweglichen Segmenten. *VDI-Bericht 381*, 177, 1980.
10. Finlayson, B.A., *The method of weighted residuals and variational principles*. Academic Press, New York, 1972.
11. Clenshaw, C.W., *Chebyshev series for mathematical functions. Her Majesty's stationary office*, London, 1962.

CHAPTER 2.3

PRINCIPLES OF ROTOR SYSTEM INSTABILITY

Z.A. Parszewski

1 Stability Threshold

Equations of free transverse vibrations of a rotating shaft [1], [2], [3] describing small vibrations of the shaft around its equilibrium position at the speed ω, are of the form

$$\begin{aligned} \ddot{x} + \frac{1}{A\rho}\frac{\partial^2}{\partial z^2}\left(EI\frac{\partial^2 x}{\partial z^2}\right) + 2h_e\dot{x} + 2h_i(\dot{x}+\omega y) &= 0 \\ \ddot{y} + \frac{1}{A\rho}\frac{\partial^2}{\partial z^2}\left(EI\frac{\partial^2 y}{\partial z^2}\right) + 2h_e\dot{y} + 2h_i(\dot{y}-\omega x) &= 0 \end{aligned} \tag{1}$$

External h_e and internal h_i damping coefficients, reduced to the reference unit mass $(A\rho)_o$ of the shaft are

$$h_e = \frac{c_e}{2(A\rho)_o} \; ; \quad h_i = \frac{c_i}{2(A\rho)_o} \tag{2}$$

c_e and c_i are equivalent viscous coefficients of external and internal damping and $A\rho$ and EI are respectively mass of unit length and stiffness of the shaft, both variable along the axial coordinate z.

The above is a set of partial differential equations, linear and homogenous. Its solution can be represented as a series in functions $f_x(z)$ and $f_y(z)$ corresponding to the characteristic (eigen) functions describing the principal modes of vibrations of the shaft (in rotating coordinates).

In the case of rotary symmetry of the shaft and support these functions are identical with the characteristic functions and are the same for each axial plane

$$f_x = f_y = f$$

hence

$$x = \sum_{j=1}^{\infty} f_j(z)x_j(t); \qquad y = \sum_{j=1}^{\infty} f_j(z)y_j(t) \tag{3}$$

The characteristic functions are defined by the equation

$$\alpha_j^2 f_j = \frac{1}{A\rho}\frac{d^2}{dz^2}\left(El\frac{d^2 f_j}{dz^2}\right) \tag{4}$$

Equations (1) give hence for each x_j and y_j

$$\begin{aligned} \ddot{x}_j + \alpha_j{}^2 x_j + 2h_{ej}\dot{x}_j + 2h_{ij}(\dot{x}_j + \omega y_j) &= 0 \\ \ddot{y}_j + \alpha_j{}^2 y_j + 2h_{ej}\dot{y}_j + 2h_{ij}(\dot{y}_j - \omega x_j) &= 0 \end{aligned} \tag{5}$$

The solution for free vibration can be sought in the form

$$x_j = X_j e^{s_j t} \qquad y_j = Y_j e^{s_j t} \tag{6}$$

The following set of algebraic linear on homogenous equations is obtained for the constants X_j and Y_j

$$\begin{aligned} [s_j^2 + 2(h_{ej} + h_{ij})s_j + \alpha_j^2]X_j + 2h_{ij}\omega Y_j &= 0 \\ -2h_{ij}\omega X_j + [s_j^2 + 2(h_{ej} + h_{ij})s_j + \alpha_j^2]Y_j &= 0 \end{aligned} \tag{7}$$

that gives for X_j and Y_j non zero solution and hence vibrations of the form (6) are possible only when the determinant vanishes giving the characteristic equation

$$[s_j^2 + 2(h_{ej} + h_{ij})s + \alpha_j^2]^2 + 4h_{ij}\,\omega^2 = 0$$

or (8)

$$s_j^2 + 4(h_{ej} + h_{ij})s_j^3 + 2[2(h_{ej} + h_{ij})^2 + \alpha_j^2]s_j^2 + 4(h_{ej} + h_{ij})\alpha_j^2 s_j + 4h_{ij}^2\omega^2 + \alpha_j^4 = 0$$

which has the form

$$a_4 s_j^4 + a_3 s_j^3 + a_2 s_j^2 + s_1 s_j + a_0 = 0 \tag{9}$$

The considered system is stable, hence the vibrations will die out if the real parts of all the roots of the characteristic equations (8) are negative. The necessary condition (of Hurwitz criterion) is fulfilled (all coefficients are positive). The sufficent condition for the 4th order equation (9) is

$$a_1 a_2 a_3 - a_0 a_3^2 - a_1^2 a_4 > 0 \tag{10}$$

Generally Hurwitz criterion requires that all the subdeterminants of the Hurwitz determinant built from the characteristic equation coefficients are positive. Criterion (10) gives for the equation (8) after re-arrangement

$$\frac{\omega}{\alpha_j} < \frac{h_{ej} + h_{ij}}{h_{ij}}$$

or (11)

$$\frac{\omega}{\alpha_j} < \frac{c_{ej} + c_{ij}}{c_{ij}}$$

The stability threshold corresponds to the stability limit of the lowest mode

$$\omega < \frac{c_{ej} + c_{ij}}{c_{ij}} \alpha_1 \tag{12}$$

that will remain unstable at all higher speeds at which higher modes will be becoming unstable.

The stability condition is hence

$$\frac{\omega}{\alpha_1} < 1 + \frac{c_{e1}}{c_{i1}} \tag{13}$$

Here α_1 is the first natural frequency of the system. The instability will not occur for speeds lower than the first critical speed of the shaft $\omega_{cr1} = \alpha_1$. Over this speed instability onsets the lower the higher is the ratio of internal to external damping. The above did not consider the mechanism of energy transfer to the vibration at instability. It is easy however to name its source - it is the driving power of the shaft. Besides the mentioned supply of energy to the vibrations there always exists its dissipation.

If the self excited vibration (instability) occurs in a linear system (when more energy is supplied than dissipated) they would grow gradually unlimited. Or, when the dissipation is higher than supply, vibration will gradually disappear.

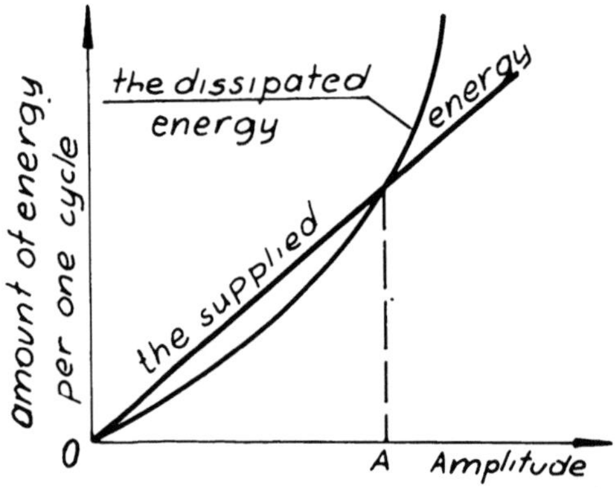

Fig. 1

The energy dissipated (supplied to) in each vibration cycle (Fig. 1) is a nonlinear function of vibration amplitude. Even for viscotic (linear) damping is

$$E = \pi c \alpha A^2$$

dissipated energy a square function of amplitude A. Limit cycle is hence attained with an amplitude (Fig. 1) corresponding to the balance of supplied and dissipated energy. This amplitude may not usually be allowed for the structure.

The mechanism of the energy transfer to the vibration relies on forces depending on vibration velocities in the described dynamic instability.

It is distinct from static instability depending on position forces (tending to increase the displacement of the system from its equilibrium position). The static instability can be hence described by negative stiffness (or natural frequency) whereas dynamic instability - by negative damping.

2 Free Vibration

In the complex notation

$$r = x + iy \tag{14}$$

the equations (3.2-5) give

$$\ddot{r}_j + 2(h_{ej} + h_{ij})\dot{r}_j + (\alpha_j^2 - 2ih_{ij}\omega)r_j = 0 \tag{15}$$

Partial solution of the form

$$r_j = e^{s_j t} \tag{16}$$

leads to the characteristic equation

$$s_j^2 + 2(h_{ej} + h_{ij})s_j + \alpha_j^2 - 2ih_{ij}\omega = 0 \tag{17}$$

hence

$$s_j = -(h_{ej} + h_{ij}) \pm \sqrt{(h_{ej} + h_{ij})^2 - \alpha_j^2 + 2ih_{ij}\omega}\ , \tag{18}$$

or

$$s_j = -\beta \pm i\lambda_j \tag{19}$$

where

$$\lambda_j = \sqrt{\tfrac{1}{2}\left[\alpha_j^2 - (h_{ej} + h_{ij})^2\right] + \tfrac{1}{2}\sqrt{\alpha_j^2 - (h_{ej} + h_{ij})^2 + 4h_{ej}\omega^2}}$$

$$\beta_j = h_{ej} + h_{ij} \mp \frac{h_{ij}\omega}{\lambda_j} \tag{20}$$

The general solution of the equation (15) representing the jth eigen-vector r_j of the shaft deflection in the stationary co-ordinate system is

$$r_j = D_j\, e^{-(h_{ij} + h_{ij} + \frac{h_{ij}}{\lambda_j}\omega)t}\, e^{-i\lambda_j t} + E_j\, e^{-(h_{ej} + h_{ij} - \frac{h_{ij}}{\lambda_j}\omega)t}\, e^{i\lambda_j t} \tag{21}$$

Here D_j and E_j are complex constants, λ_j is a real constant - circular frequency of damped vibrations (angular velocity of precession). The vibratory motion of any point of the shaft axis, described by the formula (21) gradually decreases only when the stability condition (11) is fulfilled. This formula represents two spiral motions. The first term in equation (21) describes a converging (decreasing) spiral motion of the shaft cross-section centres with orbiting circular velocity λ_j opposite to the shaft speed ω (opposite-retrograde precession). The second term corresponds to a spiral motion with angular velocity λ_j in ω direction (direct precession). This motion will be diverging (the deflection will be increasing) when the rotor speed ω exceeds the limit given by the condition (11)

Second term is always dominant (direct precession) as in case of stability it decreases slower than the first term and in case of instability it increases whereas the first term always decreases. The deflection of the shaft for both the terms of the formula (21) is according to (3) represented by the jth pair of the modes. When the shaft speed ω attains the stability limit of its jth mode (11)

$$\omega = \frac{h_{ej} + h_{ij}}{h_{ij}}\,\alpha_j \tag{22}$$

the formula (20) gives

$$\lambda_j = \alpha_j \tag{23}$$

The direct precession has then the angular velocity equal to the jth circular frequency of the shaft. The damping of this term (the second term in the formula (21)) vanishes. The first term damped effectively - vanishes gradually. Limit cycle is hence attained

$$r_j = E_j f_j(z)e^{i\alpha_j t}$$

with constant amplitude, at the limit shaft speed ω given by the formula (22) For higher ω values this component (second component in formula 21) increases in time.

General solution representing any small vibration of the shaft is represented in complex form as follows:

$$r = \sum_{j=1}^{\infty} f_j(z)r_j(t) \tag{24}$$

The vibrations gradually vanish and the shafts steady rotation and its corresponding equilibrium postion are dynamically stable, only when the condition (11) is fulfilled i.e. for speeds ω not exceeding the stability limit of the shaft's first mode

$$\omega < (1 + \frac{c_{e1}}{c_{i1}})\alpha_1 \tag{25}$$

It next remains unstable for all higher speeds.

The stability problem of higher modes is hence in this case usually immaterial as corresponding speeds can not be generally attained.

At the stability limit it is at the speed

$$\omega = \frac{h_{e1} + h_{i1}}{h_{i1}} \alpha_1 \tag{26}$$

the motion tends to the limit cycle in the first mode and direct precession (the other components are gradually damped out). At other speeds ω vibratory motion of the shaft is more compound. All modal components exist with modal deflection and precession angular velocities λ_j all depending on the shaft speed ω.

3 Parametric Instabilities

Introduction. Variation of some parameters of machine systems may be introducing dynamic phenomena, called parametric vibrations and parametric effects.

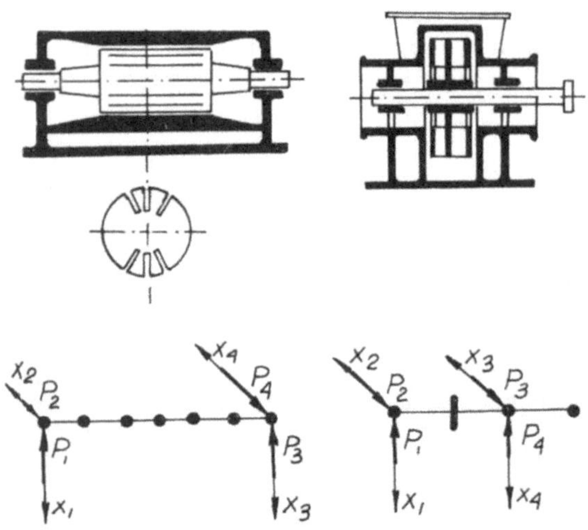

Fig. 2

Fig. 2 shows two examples of rotating machines with rotors of different principal stiffness in the first case and different principal moments of inertia in the second case. Time variation of corresponding system parameters (stiffness in the first case and moment of inertia in the second) will occur during rotor rotation.

Theory of parametric vibrations is well developed and hence the parametric effects are known qualitatively [1], [4], [5], [6]. It is hence known that instability of equilibrium position or steady state motion may occur in those systems at circular frequencies θ of variation of the parameters (circular frequency of parametric excitation) close to

$$\theta^{(\pm)}_{k,l,m} = \frac{\alpha_k \pm \alpha_l}{m} \tag{27}$$

here: k,l,m = 1, 2, 3, ...

α_k, α_l - any pair of the natural frequencies of the system for medium values of the variable parameters (zero coefficient of parametric excitation $\varepsilon = 0$).

The set of θ values for which instability occurs at k = 1 gives simple instability region, and that at k ≠ 1 gives compound instability region. The formula (27) gives the positions of instability regions but their number and width depend on the value of the parametric excitation coefficient ε (it is generally a nondimensional measure of the ratio of the parameter variation amplitude to its average value) and on the damping characteristics of the system.

In real machine systems instability regions may occur in the vicinity of $\theta^{(+)}_{k,l,m}$ for small m numbers (first order m = 1 and second order m = 2 instability regions) or may not occur at all. (Fig. 3)

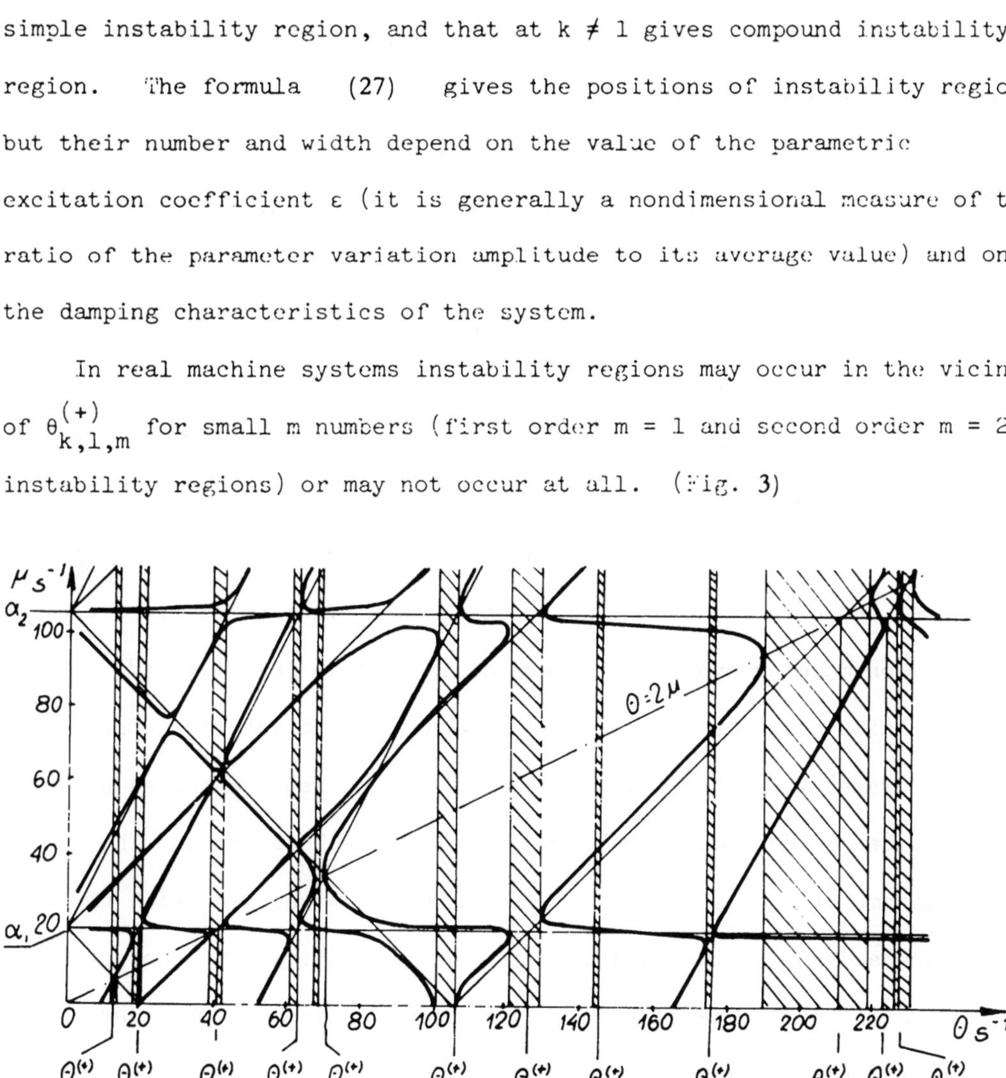

Fig. 3

Critical states corresponding to resonance are possible in stable regions, leading to high amplitude of steady vibrations. These states occur at circular frequencies θ given by the formula (28)

$$m\theta_{k,m}^{(\pm)} = \alpha_k \pm \nu\omega \; ; \quad m = 0, 1, 2, 3, \ldots, \tag{28}$$

here $\nu\omega$ is the circular frequency of the νth force harmonics. The quantities describing steady vibration (eg. its spectrum or marked trajectory) depend on: the exciting force amplitude, the value of parametric excitation coefficient and on damping.

Answers to the following questions are of importance for practical application:

1) Is the equilibrium stable for all working values of parametric excitation frequency.

2) What are the steady state vibration characteristics for working values of parametric excitation frequency.

Only quantitative analysis of a corresponding model of the system may answer those questions. Inclusion of the influence of supporting structure on machine dynamics is here of importance [6], [7], [8]. The model contains hence two dynamically interacting subsystems. One (rotor) being a parametric system, can be discretised and represented as having finite number of degrees of freedom (Fig. 8 bold lines). The second one nonparametric, contains the whole supporting structure (with the foundation and ground included), usually very complicated. Receptance (impedence) matrix is hence used to describe its dynamic behaviour. Its elements can then be measured (if not calculated).

Mathematical Model. Transverse vibration of parametric rotor interacting with supporting structure, can be represented by a set of linear differential equations with variable coefficients..

$$\mathbf{M}\ddot{\mathbf{x}} + \mathbf{C}\ddot{\mathbf{x}} + \mathbf{K}\ddot{\mathbf{x}} + (\boldsymbol{\gamma}_r + i^*\boldsymbol{\gamma}_u)\mathbf{x} + \varepsilon\left\{\sum_{j=1}^{\infty}[(\mathbf{m}_j^c \cos j\theta t + \mathbf{m}_j^s \sin j\theta t)\ddot{\mathbf{x}} + \right.$$

$$\left. + (\mathbf{c}_j^c \cos j\theta t + \mathbf{c}_j^s \sin j\theta t)\dot{\mathbf{x}} + (\mathbf{k}_j^c \cos j\theta t + \mathbf{k}_j^s \sin j\theta t)\mathbf{x}]\right\} = \mathbf{0} \tag{29}$$

here

θ - is circular frequency of parametric excitation

ε - is coefficient (or depth) of parametric excitation $\varepsilon = \frac{k_x - k_y}{k_x + k_y}$

M K C - are matrices of inertia, stiffness and of damping of the discrete parametric system

m k c - are matrices of parametric excitation

$\boldsymbol{\gamma}_r \boldsymbol{\gamma}_u$ - are real and imaginary parts of stiffness matrix of the supporting structure, along the connecting co-ordinates.

The solution of the set of equations (29) may be sought in the Floquet form

$$\mathbf{x} = e^{(\vartheta + i^*\mu)t} \sum_{i=0}^{\infty} (\mathbf{a}_i \sin i\theta t + \mathbf{b}_i \cos i\theta t). \tag{30}$$

here

$\vartheta + i^*\mu$ - is the Floquet's complex exponent.

The sum in equation (30) represents the Fourier series of a periodic function of the period

$$T = \frac{2\pi}{\theta}$$

At the stability limit (on the boundary of stability region) the real part ϑ of the Floquet's exponent is zero. Hence, for finding the stability limits, the partial solutions of the system (29) can be sought in the form

$$\mathbf{x} = e^{i*\mu t} \sum_{i=0}^{\infty} (\mathbf{a}_i \sin i\theta t + \mathbf{b}_i \cos i\theta t) \tag{31}$$

here

$\mathbf{a}_i$ and $\mathbf{b}_i$ are complex matrices.

Introduction of (31) into (29) gives for the constants $\mathbf{a}_i$, $\mathbf{b}_i$ an infinite set of linear algebraic equations. It has non zero solutions of the type (31) when its characteristic determinant is zero. The determinant has the following form

$$|\mathbf{W}_{01\ldots}(\theta,\mu)| =$$

$$\begin{array}{c|cccccc|}
 & 0 & 1 & 2 & 3 & 4 & 5 \\
0 & \mathbf{E}_{\gamma}(\beta_0^{(+)}) & \varepsilon\mathbf{e}_1(\beta_1^{(-)}) & \varepsilon\mathbf{f}_1(\beta_1^{(+)}) & \varepsilon\mathbf{e}_2(\beta_2^{(-)}) & \varepsilon\mathbf{f}_2(\beta_2^{(+)}) & \varepsilon\mathbf{e}_3(\beta_3^{(-)})\ldots \\
1 & \varepsilon\mathbf{e}_1(\beta_0^{(+)}) & \mathbf{E}_{\gamma}(\beta_1^{(-)}) & \varepsilon\mathbf{e}_2(\beta_1^{(+)}) & \varepsilon\mathbf{f}_1(\beta_2^{(-)}) & \varepsilon\mathbf{e}_3(\beta_2^{(+)}) & \varepsilon\mathbf{f}_2(\beta_3^{(-)})\ldots \\
2 & \varepsilon\mathbf{g}_1(\beta_0^{(+)}) & \varepsilon\mathbf{e}_2(\beta_1^{(-)}) & \mathbf{E}_{\gamma}(\beta_1^{(+)}) & \varepsilon\mathbf{e}_3(\beta_2^{(-)}) & \varepsilon\mathbf{f}_1(\beta_2^{(+)}) & \varepsilon\mathbf{e}_4(\beta_3^{(-)})\ldots \\
3 & \varepsilon\mathbf{e}_2(\beta_0^{(+)}) & \varepsilon\mathbf{g}_1(\beta_1^{(-)}) & \varepsilon\mathbf{e}_3(\beta_1^{(+)}) & \mathbf{E}_{\gamma}(\beta_2^{(-)}) & \varepsilon\mathbf{e}_4(\beta_2^{(+)}) & \varepsilon\mathbf{f}_1(\beta_3^{(-)})\ldots \\
4 & \varepsilon\mathbf{g}_2(\beta_0^{(+)}) & \varepsilon\mathbf{e}_3(\beta_1^{(-)}) & \varepsilon\mathbf{g}_1(\beta_1^{(+)}) & \varepsilon\mathbf{e}_4(\beta_2^{(-)}) & \mathbf{E}\gamma(\beta_2^{(+)}) & \varepsilon\mathbf{e}_5(\beta_3^{(-)})\ldots \\
5 & \varepsilon\mathbf{e}_3(\beta_0^{(+)}) & \varepsilon\mathbf{g}_2(\beta_1^{(-)}) & \varepsilon\mathbf{e}_4(\beta_1^{(+)}) & \varepsilon\mathbf{g}_1(\beta_2^{(-)}) & \varepsilon\mathbf{e}_5(\beta_2^{(+)}) & \mathbf{E}_{\gamma}(\beta_3^{(-)})\ \ldots \\
 & \vdots & \vdots & \vdots & \vdots & \vdots & \vdots
\end{array} = 0 \tag{32}$$

Where

$$\mathbf{E}_\gamma\left(\beta_i^{(\pm)}\right) =$$

$$= \begin{bmatrix} -\left(\beta_i^{(\pm)}\right)^2\mathbf{M} + \mathbf{K} + \boldsymbol{\gamma}_r\left(\beta_i^{(\pm)}\right) & -\beta_i^{(\pm)}\mathbf{C} - \boldsymbol{\gamma}_u\left(\beta_i^{(\pm)}\right) \\ +\beta_i^{(\pm)}\mathbf{C} + \boldsymbol{\gamma}_u\left(\beta_i^{(\pm)}\right) & -\left(\beta_i^{(\pm)}\right)^2\mathbf{M} + \mathbf{K} + \boldsymbol{\gamma}_r\left(\beta_i^{(\pm)}\right) \end{bmatrix}.$$

$$\mathbf{e}_j\left(\beta_i^{(\pm)}\right) = \begin{bmatrix} -\mathbf{w}_j\left(\beta_i^{(\pm)}\right) + \mathbf{v}_j\left(\beta_i^{(\pm)}\right) \\ +\mathbf{v}_j\left(\beta_i^{(\pm)}\right) + \mathbf{w}_j\left(\beta_i^{(\pm)}\right) \end{bmatrix},$$

$$\mathbf{f}_j\left(\beta_i^{(\pm)}\right) = \begin{bmatrix} +\mathbf{w}_j\left(\beta_i^{(\pm)}\right) - \mathbf{v}_j\left(\beta_i^{(\pm)}\right) \\ +\mathbf{v}_j\left(\beta_i^{(\pm)}\right) + \mathbf{w}_j\left(\beta_i^{(\pm)}\right) \end{bmatrix}, \tag{33}$$

$$\mathbf{g}_j\left(\beta_i^{(\pm)}\right) = \begin{bmatrix} +\mathbf{w}_j\left(\beta_i^{(\pm)}\right) + \mathbf{v}_j\left(\beta_i^{(\pm)}\right) \\ -\mathbf{v}_j\left(\beta_i^{(\pm)}\right) + \mathbf{w}_j\left(\beta_i^{(\pm)}\right) \end{bmatrix}.$$

and

$$\beta_i^{(\pm)} = i\theta \pm \mu,$$

$$\mathbf{w}_j\left(\beta_i^{(\pm)}\right) = \frac{1}{2}\left[-\left(\beta_i^{(\pm)}\right)^2\mathbf{m}_j^c - \beta_i^{(\pm)}\mathbf{c}_j^s + \mathbf{k}_j^c\right],$$

$$\mathbf{v}_j\left(\beta_i^{(\pm)}\right) = \frac{1}{2}\left[-\left(\beta_i^{(\pm)}\right)^2\mathbf{m}_j^s + \beta_i^{(\pm)}\mathbf{c}_j^c + \mathbf{k}_j^s\right]$$

$i = 0, 1, 2, \ldots \qquad j = 1, 2, \ldots$

Hence the problem of finding the boundaries of stability regions reduces to finding the zero points of the determinant $|\mathbf{W}_{01..}(\theta,\mu)|$.

The function $\mathbf{W} = |\mathbf{W}_{01..}(\theta,\mu)|$ is a continuous function of two variables and represents hence a surface over the coordinate plane μ,θ.

However, for real systems, the function $\mathbf{W} = |\mathbf{W}_{01..}(\theta,\mu)|$ is definite non-negative. Hence the zero points may have the character shown in the Fig. 3.

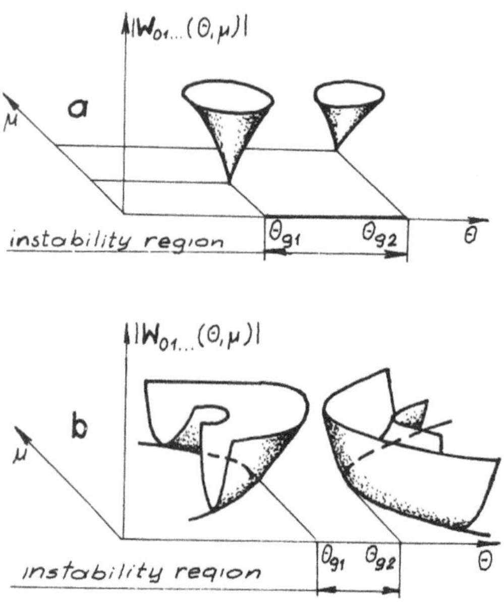

Fig. 4

The solutions of the a type may be expected for damped systems ($\mathbf{C} + \gamma_u \neq \mathbf{0}$), as in those cases nonincreasing and nondecreasing solutions are possible only on the stability regions boundaries hence for discrete values θ_{g1}, θ_{g2} (Fig. 4a).

Unlimited number of solutions of the type (31) may correspond to each θ value, for systems without damping ($\mathbf{C} + \gamma_u = \mathbf{0}$). Hence solutions giving continuous lines $\mu = \mu(\theta)$ may be expected, as in Fig. 4b.

Numerical Procedures.

Damped Systems ($\mathbf{C} + \gamma_u \neq \mathbf{0}$). The determinant is non-negative in these cases, hence its minimum as function of two variables θ, μ is sought and checked for zero value.

The minima were found with the application of the following numeric procedures.

Values of the determinant **W** are computed, with a step α, in the distance R from a point 0_o of the coordinate plane μ, θ (Fig. 5) until smallest value is found, at a point 0_1.

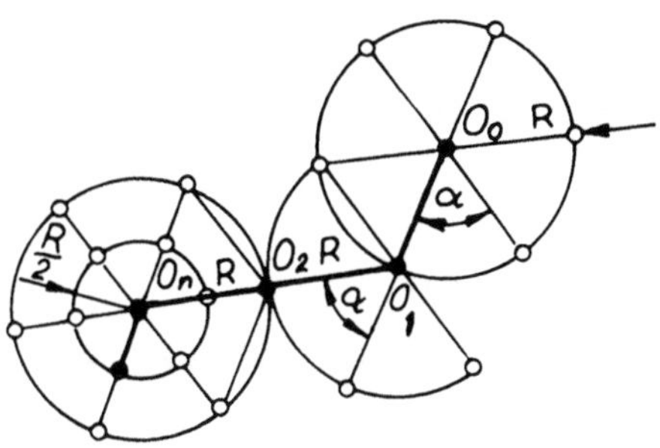

Fig. 5

The procedure is repeated for points in the distance R from 0_1, starting from the straight line 0_o0_1.

This process is followed until the determinant minimal value on a circle is higher than in its centre 0_n.

The whole procedure is next repeated from 0_n with the radius R halved. This is followed until the radius R attains a value smaller than assumed error ΔR.

The procedure was programmed and executed on a computer for the third order approximation of the determinant $\mathbf{W} = |\mathbf{W}_{012}(\mu, \theta)|$. The receptance functions $\gamma_r(\beta)$ and $\gamma_u(\beta)$ were read into the form of a table, which values were interpolated in the computing process.

The computation results, in the form of the determinant values along the computing trajectory were used for finding the determinant zero values (Fig. 8).

Flow Diagrams. The flow diagrams for computing the trajectory co-ordinates and the determinant $|\mathbf{W}_{01..}(\mu,\theta)|$ values are given in Fig. 6 for damped systems and in Fig. 7 for undamped systems.

T and P represent the θ and μ coordinates respectively, with TO, and PO being their initial values. R and DAL give steps and TD, TG, PD and PG define the considered area.

Instability Regions

Undamped Systems, $\mathbf{C} + \gamma_u = \mathbf{0}$. Instability regions, in such a system, are the sets of θ values for which the periodic solutions of the type (31) are lacking. These regions are shaded in the Fig. 3.

Damped Systems, $\mathbf{C} + \gamma_u \neq \mathbf{0}$ Instability regions are sets of values between the solutions of the type shown in Fig. 4a.

Those solutions are marked in the small bold circles in the Fig. 8.

The instability regions are shaded in this diagram. The continuous lines in Fig. 8 are the univalue curves (of the determinant $|\mathbf{W}_{01..}(\mu,\theta)|$) and the broken line gives the solution for no damping $\mathbf{C} + \gamma_u = \mathbf{0}$

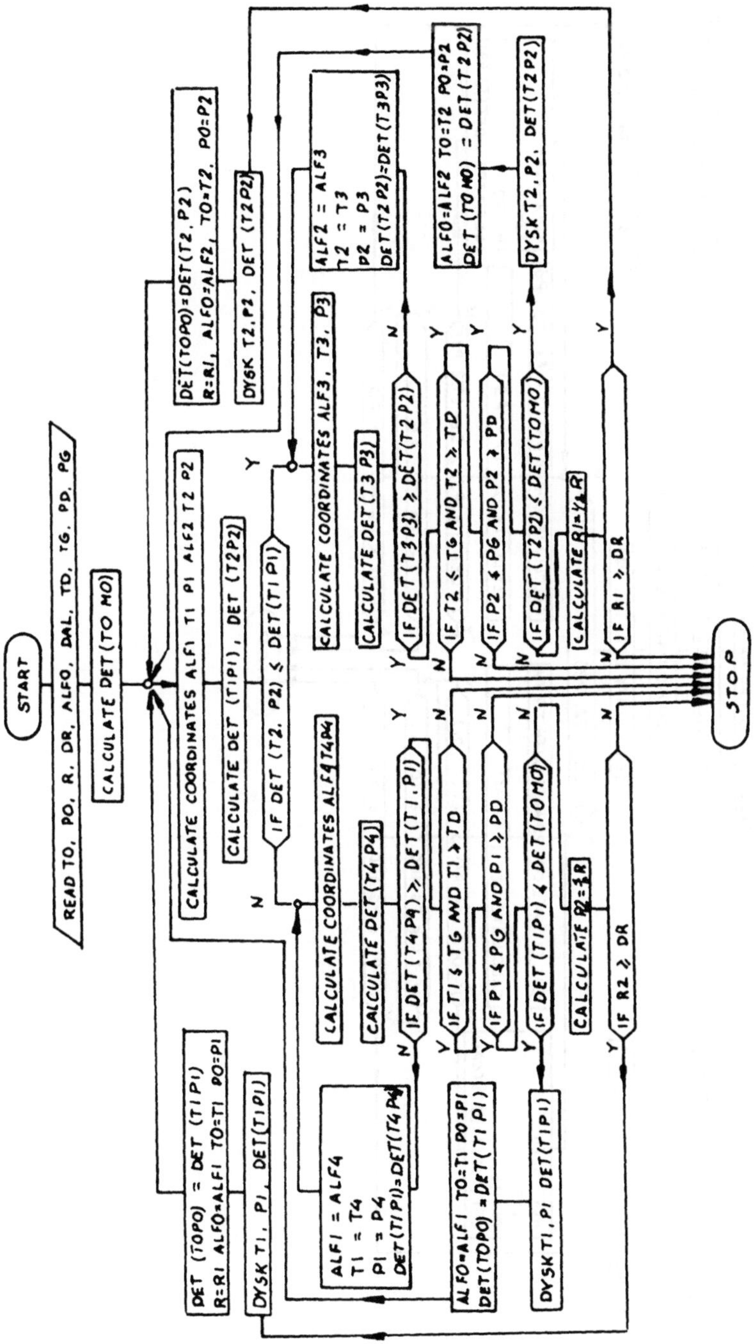
START
READ TO, PO, R, DR, ALFO, DAL, TD, TG, PD, PG
CALCULATE DET(TO MO)
CALCULATE COORDINATES ALF1 T1 P1 ALF2 T2 P2
CALCULATE DET (T1P1), DET (T2P2)
IF DET (T2, P2) ≤ DET(T1P1)
CALCULATE COORDINATES ALF3, T3, P3
CALCULATE DET(T3 P3)
IF DET(T3P3) ≥ DET(T2P2)
IF T2 ≤ TG AND T2 ≥ TD
IF P2 ≤ PG AND P2 ≥ PD
IF DET(T2P2) ≤ DET(TOMO)
CALCULATE R1=½R
IF R1 ≥ DR
CALCULATE COORDINATES ALF4T4P4
CALCULATE DET(T4P4)
IF DET(T4P4) ≥ DET(T1,P1)
IF T1 ≤ TG AND T1 ≥ TD
IF P1 ≤ PG AND P1 ≥ PD
IF DET(T1P1) ≤ DET(TOMO)
CALCULATE R2=3R
IF R2 ≥ DR
DET(TOPO)=DET(T2,P2)
R=R1, ALFO=ALF2, TO=T2, PO=P2
DYSK T2,P2, DET (T2P2)
ALF2 = ALF3
T2 = T3
P2 = P3
DET(T2P2)=DET(T3P3)
ALFO=ALF2 TO=T2 PO=P2
DET (TOMO) = DET (T2P2)
DYSK T2, P2, DET(T2P2)
DET (TOPO) = DET (T1P1)
R=R1 ALFO=ALF1 TO=T1 PO=P1
DYSK T1, P1, DET(T1P1)
ALF1 = ALF4
T1 = T4
P1 = P4
DET(T1P1)=DET(T4P4)
ALFO=ALF1 TO=T1 PO=P1
DET(TOPO) =DET(T1P1)
DYSK T1,P1 DET(T1P1)
STOP

Fig. 6

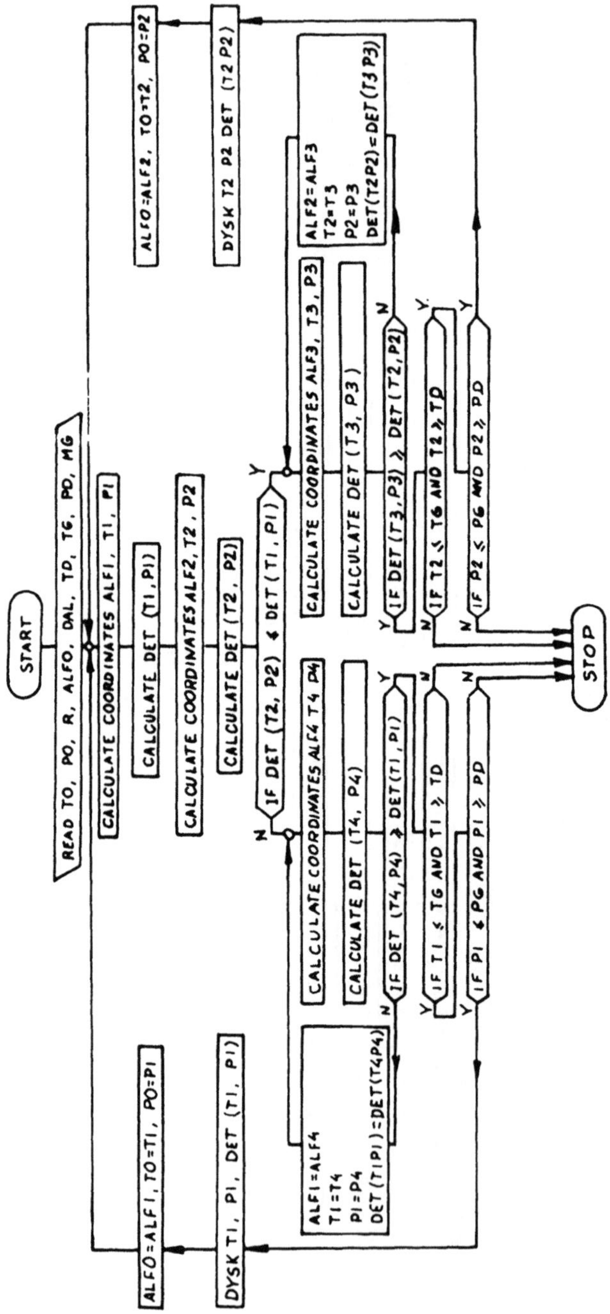
START
READ TO, PO, R, ALFO, DAL, TD, TG, PD, MG
CALCULATE COORDINATES ALF1, T1, P1
CALCULATE DET (T1, P1)
CALCULATE COORDINATES ALF2, T2, P2
CALCULATE DET (T2, P2)
IF DET (T2, P2) ≤ DET (T1, P1)
CALCULATE COORDINATES ALF3, T3, P3
CALCULATE DET (T3, P3)
IF DET (T3, P3) ≥ DET (T2, P2)
IF T2 ≤ TG AND T2 ≥ TD
IF P2 ≤ PG AND P2 ≥ PD
ALF2=ALF3
T2=T3
P2=P3
DET(T2P2)=DET(T3P3)
ALFO=ALF2, TO=T2, PO=P2
DYSK T2 P2 DET (T2 P2)
CALCULATE COORDINATES ALF4 T4 P4
CALCULATE DET (T4, P4)
IF DET (T4, P4) ≥ DET (T1, P1)
IF T1 ≤ TG AND T1 ≥ TD
IF P1 ≤ PG AND P1 ≥ PD
ALF1=ALF4
T1=T4
P1=P4
DET (T1P1)=DET(T4P4)
ALFO=ALF1, TO=T1, PO=P1
DYSK T1, P1, DET (T1, P1)
STOP

Fig. 7

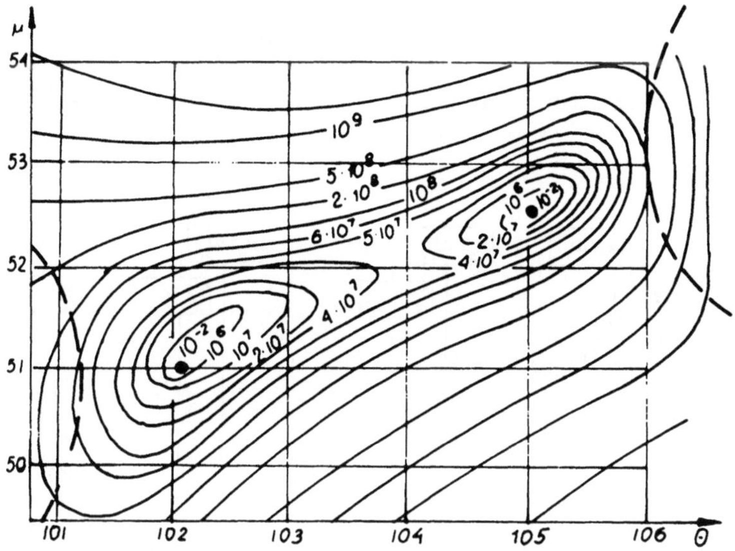

Fig. 8

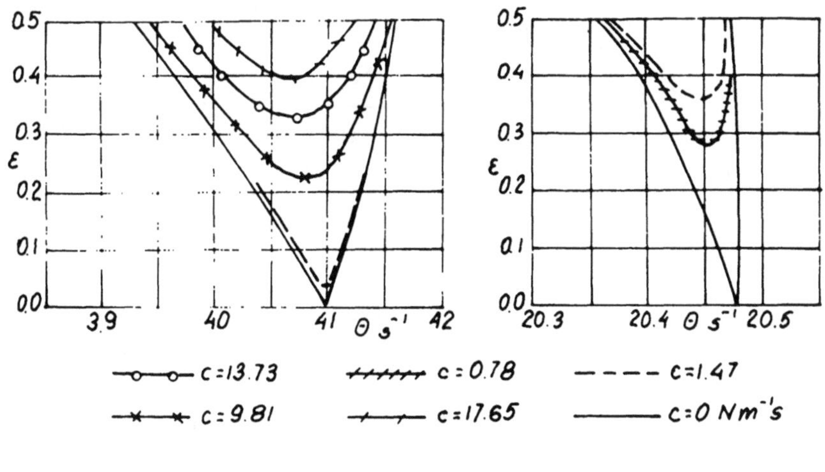

Fig. 9

Instability regions as functions of the coefficient ε value for various damping coefficients c are shown in Fig. 9.

First three instability regions for a rotor support system (an industrial ventilator) are shown in Fig. 10 together with the amplitude of its first four harmonics in the vertical x and horizontal y directions.

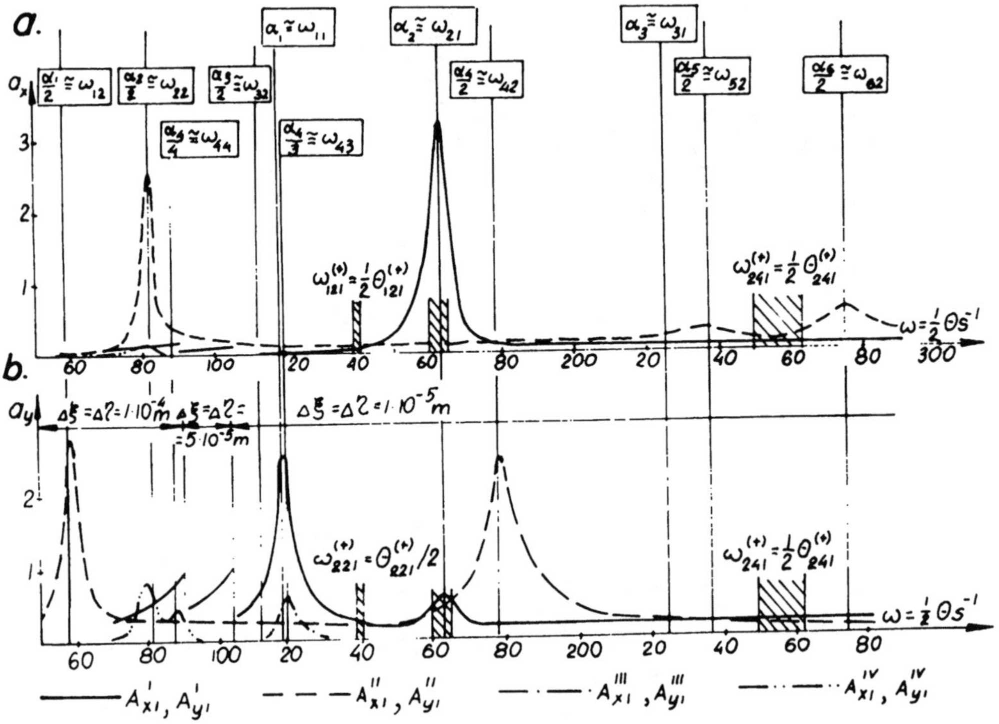

Fig. 10

The instability region corresponding to two adjacent natural frequencies as a function of the coefficient of parametric excitation is shown in Fig. 11c for nonconservative support with stiffness cross-coefficients varying with rotor speed (Fig. 11a) as is the case in hydrodynamic journal bearings. Corresponding vibration-time diagrams are given in Fig. 12 A - J.

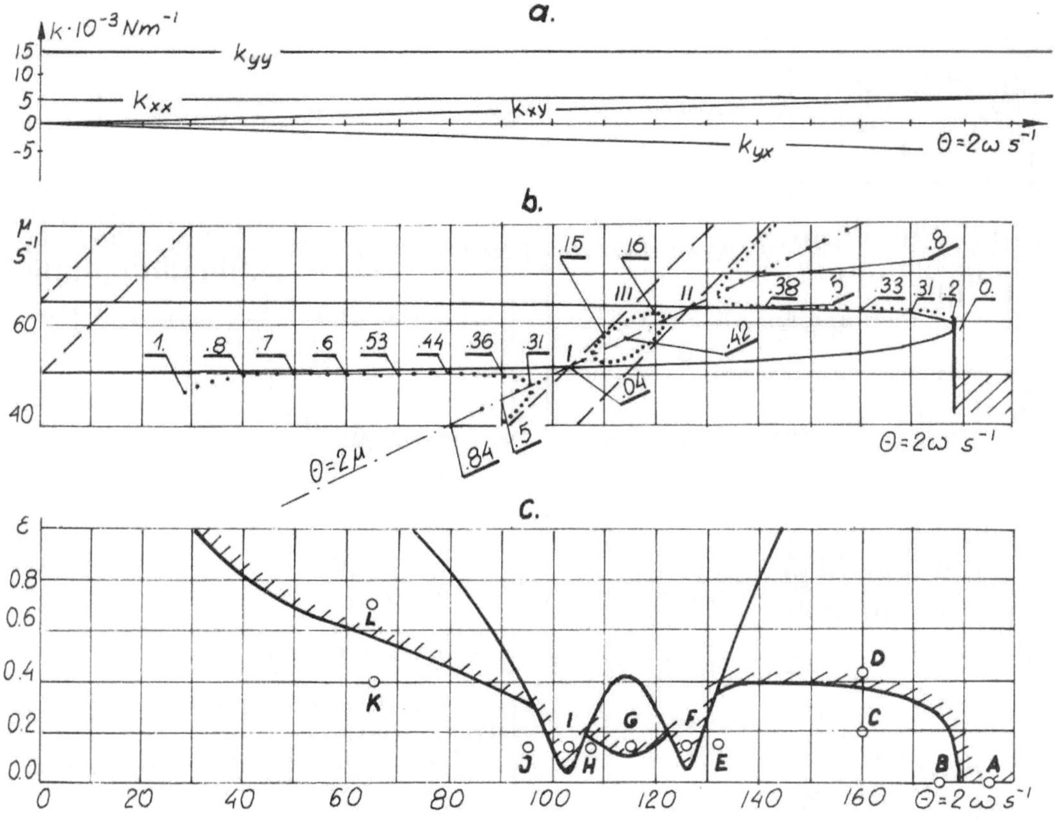
a.
k·10⁻³ Nm⁻¹
kyy
Kxx
Kxy
kyx
Θ=2ω s⁻¹
b.
μ s⁻¹
Θ=2μ
Θ=2ω s⁻¹
c.
ε
0 20 40 60 80 100 120 140 160 Θ=2ω s⁻¹

Fig. 11

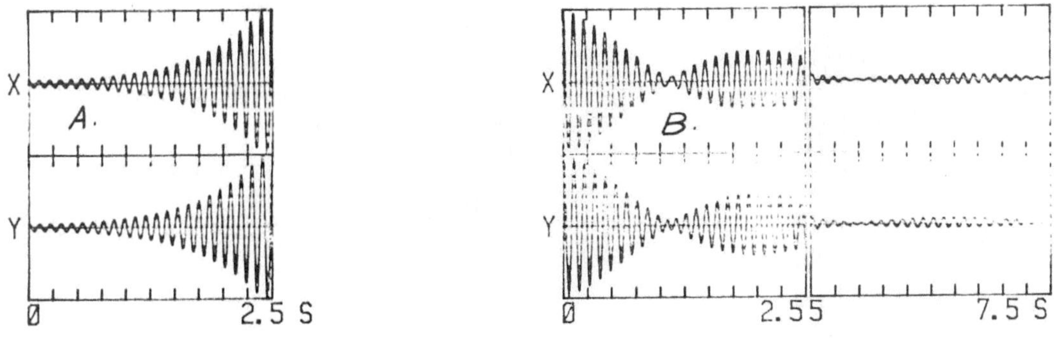
X
Y
A.
B.
Ø
2.5 S
2.55
7.5 S

Fig. 12

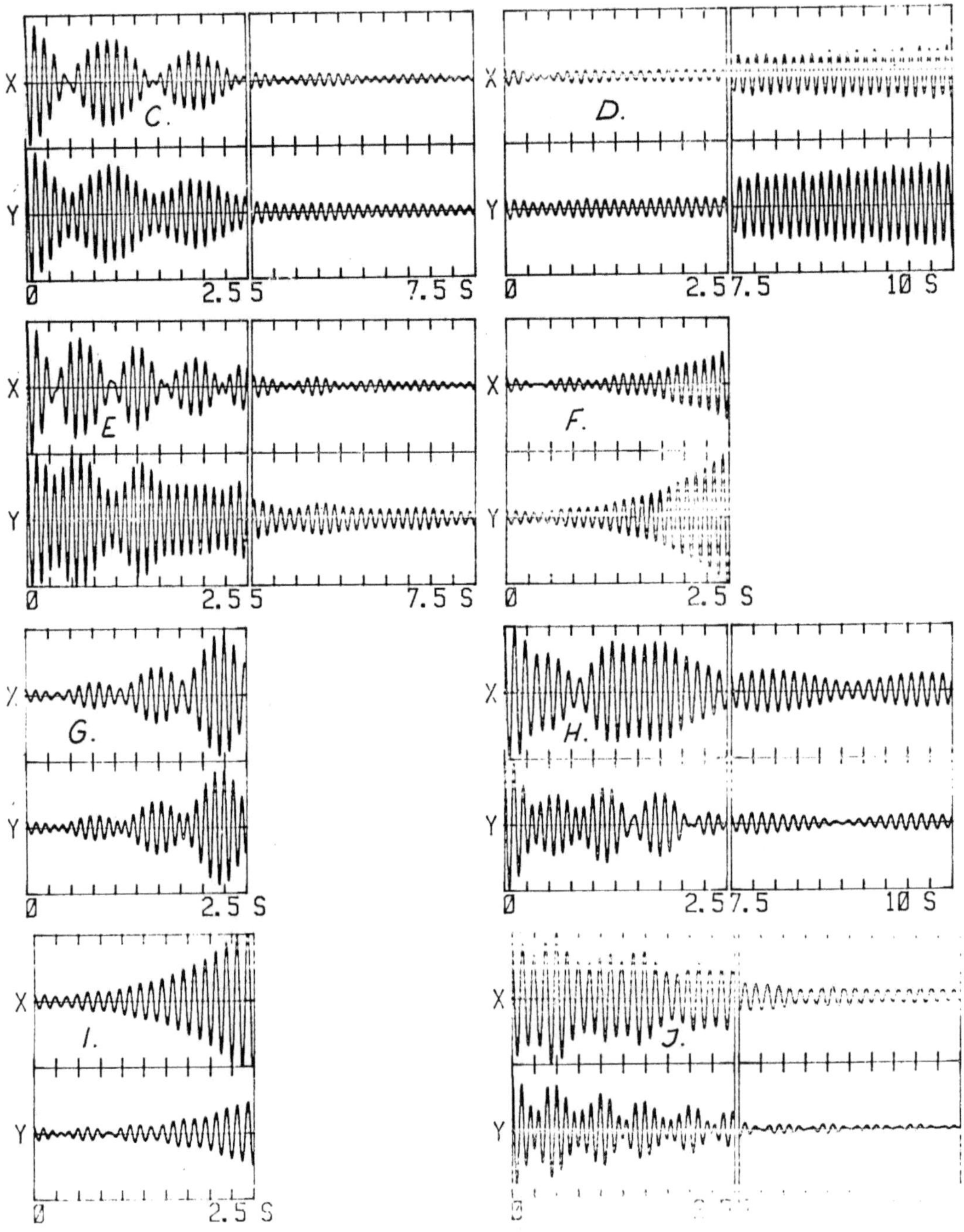

Fig. 12

Disbalance response in some speed range is shown in Fig. 13.

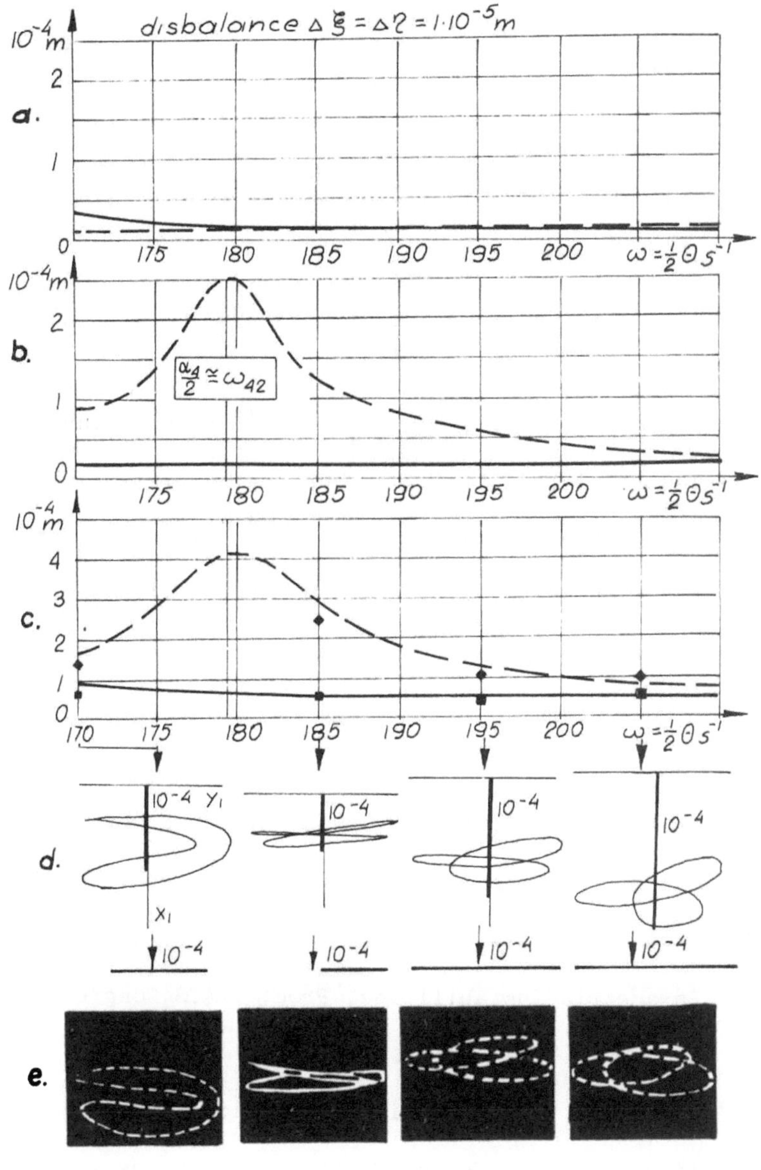

Fig. 13

Fig. 13 gives also comparison of some of the computed (various lines) and experimental (black squares) results. The journal centre loci in Fig. 13d were computer plotted and those in Fig. 13e are experimental.

REFERENCES

(1) Parszewski, Z.A.: Drgania i Dynamika Maszyn, WNT, Warszawa 1982, 420p.

(2) Bishop, R.E.D.: The Vibration of Rotating Shafts, J.Mech.Eng.Sc. 1959, Vol.1 No.1, p.50.

(3) Bishop R.E.D.: Gladwell, G.M.L.: Vibration and Balancing of Unbalanced Flexible Rotors, J.Mech.Eng.Sc. 1959, Vol.1, No.1 p.66.

(4) Hsu, C.S.: On the Parametric Excitation -.. ASME Ser.B, 1963.

(5) Black, H.F.: Parametrically Excited Lateral Vibrations of Asym. Slender Shaft in Asym. Flexible Bearings, J.Mech.Eng.Sc. 11, 1, 1969.

(6) Krodkiewski, J.M., Marynouski,K.O., Parszewski, Z.A.: Instablities of Parametric Rotor-Support Systems. I.Mech.E. Proc. Conf. on Vibrations in Rotating Machines, Cambridge 1980.

(7) Krodkiewski, J.M. Parszewski, Z.A.: Computer Aided Analysis of Rotors Parametric Instabilities, Proc. IMACS Congress Montreal 1982.

(8) Krodkiewski, J.M.: Parszewski, Z.A.: Rotor-Support Parametric Instability Limits in Terms of Rotor Receptances in Rotating Axes, Proc. IFTOMM Rotordynamics Conf. Rome, 1982, p.55-65.

CHAPTER 2.4

FLUID FILM AND INSTABILITY

Z.A. Parszewski

1 Journal Bearing Stability.

Hydrodynamic bearings effect is important in many machines. Besides lowering critical speeds it may lead also to instability of rotor steady motion and hence self excited vibrations.

A hydrodynamic journal bearing is characterised, for steady state (see chapter 2.1) by the nondimensional bearing capacity (Sommerfield's number).

$$S = \frac{P\left(\frac{2r}{D}\right)^2}{DL\mu\omega} \tag{1}$$

as function of journal eccentricity ratio $c = \frac{e}{r}$ and by the locus (Fig. 1) of journal centre equilibrium positions (in steady state). All these give static characteristics of the bearing (Fig. 2). Here P is bearing load. The length of the bearing is L, diamater D and its radial clearance is $R - R_c = r$, η - oil viscosity, ω - rotor angular velocity.

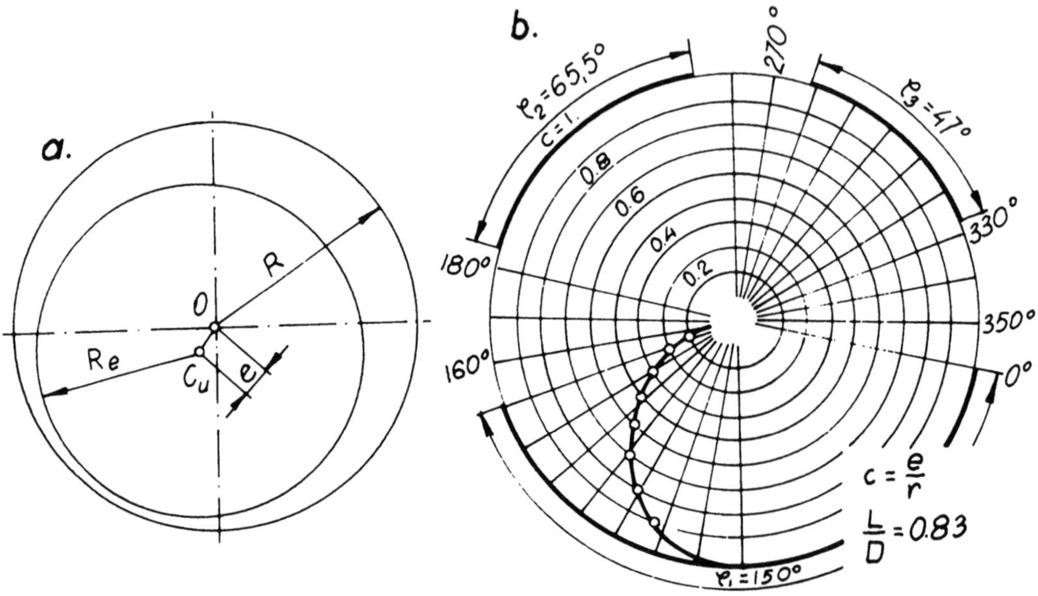

Fig. 1

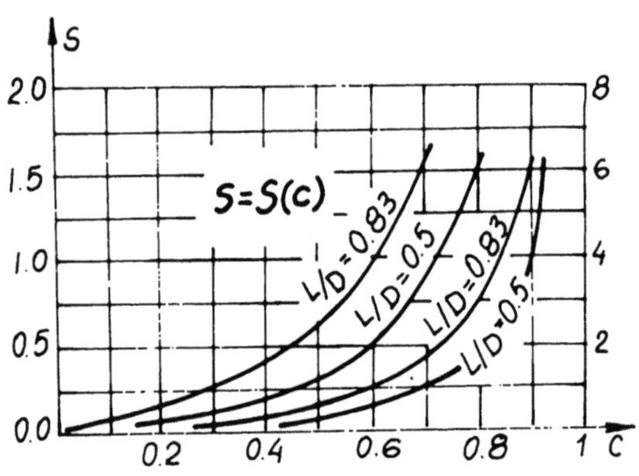

Fig. 2

Dynamic characteristics of the bearing give oil film stiffness K_{xx}, ..., K_{yy} and damping C_{xx}, ..., C_{yy} coefficients, describing (linearized) hydrodynamic forces acting on the journal in case of its small vibrations around its equilibrium position.

$$\Delta F_x = -(\bar{K}_{xx}x + \bar{K}_{xy}y + \bar{C}_{xx}\dot{x} + \bar{C}_{xy}\dot{y})$$

$$\Delta F_y = -(\bar{K}_{yx}x + \bar{K}_{yy}y + \bar{C}_{yx}\dot{x} + \bar{C}_{yy}\dot{y}) \qquad (2)$$

Nondimensional values of those coefficients as functions of eccentricity ratio (at equilibrium position) are given in Fig. 3 and Fig. 4 for a bearing of three partial arcs and $\frac{L}{D} = 0.83$

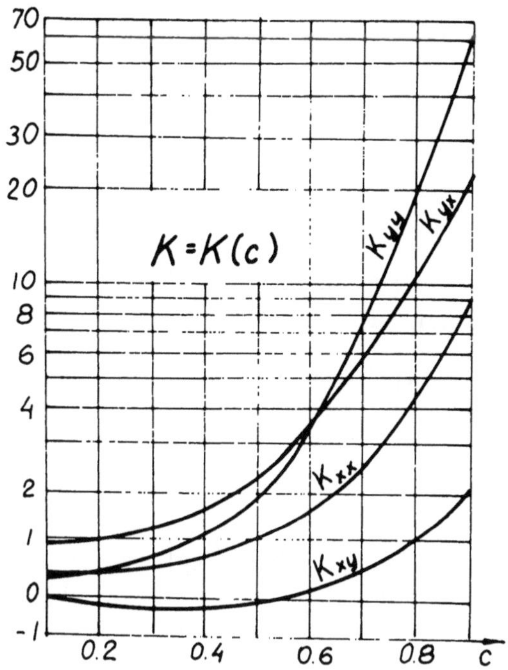

Fig. 3

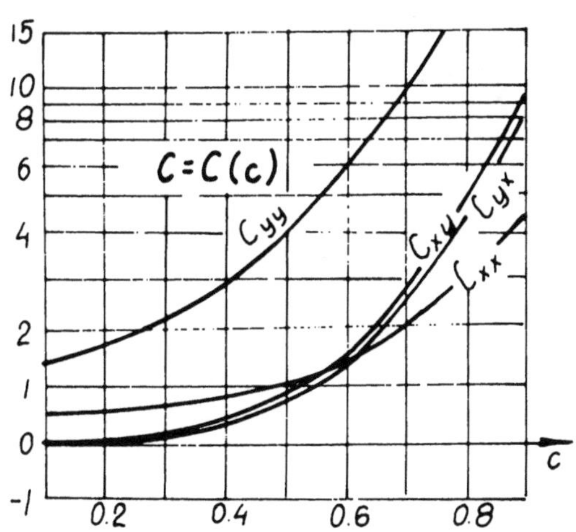

Fig. 4

characterised in Fig. 1 and 2 [1], [2], [4].

The lowest (first) of principal modes of vibrations becomes unstable first (at lowest speed) and it remains unstable with increasing speed when the higher modes become consecutively unstable. The stability threshold or limit of a system is hence equal to that for its first principal mode as was explained in previous point. To characterise the bearing stability it is hence sufficient to consider a symmetric single span rotor supported in two identical bearings. The first principal mode for this system is a symmetric one. If hence shaft deflection is disregarded the problem reduces to the analysis of small vibrations of the journal (with half the rotor mass) in the single bearing.

In nondimensional coordinants and parameters:

$$K = \frac{Sr}{P}\,\overline{K} \qquad C = \frac{Sr\omega}{P}\,\overline{C} \qquad (P = mg)$$

$$\overline{q} = \begin{Bmatrix} x \\ y \end{Bmatrix} \qquad q = \frac{\overline{q}}{r} \qquad \dot{q} = \frac{\dot{\overline{q}}}{r\omega} \tag{4}$$

$$\ddot{q} = \frac{\ddot{\overline{q}}}{r\omega^2} \quad ; \quad \omega_o = \sqrt{\frac{g}{r}}$$

the equation of free vibrations has the form

$$[m]\ \{\ddot{q}\} + [c]\ \{\dot{q}\} + [K]\ \{q\} = 0$$

Here

$$[m] = \begin{bmatrix} S(\frac{\omega}{\omega o})^2 & 0 \\ 0 & S(\frac{\omega}{\omega o})^2 \end{bmatrix} \qquad [K] = \begin{bmatrix} K_{xx} & K_{xy} \\ K_{yx} & K_{yy} \end{bmatrix} \qquad [C] = \begin{bmatrix} C_{xx} & C_{xy} \\ C_{yx} & C_{yy} \end{bmatrix} \tag{5}$$

With partial solutions of equations (4) $\{q\} = \{Q\}\ e^{st}$ the characteristic equation is

$$\begin{vmatrix} S(\frac{\omega}{\omega_o})^2\ s^2 + C_{xx}s + K_{xx} & C_{xy}s + K_{xy} \\ C_{yx}s + K_{yx} & S(\frac{\omega}{\omega_o})^2\ s^2 + C_{yy}s + K_{yy} \end{vmatrix} = 0 \tag{6}$$

of the form

$$s^4 + a_3 s^3 + a_2 s^2 + a_1 s + a_0 = 0$$

Hurwitz's criterion of stability for the 4th order characteristic equation reduces to the unequality

$$a_1 a_2 a_3 - a_0 a_3^2 - a_1^2 a_4 > 0 \tag{7}$$

ensuring negative real parts of all roots of equation (6) .

The stability threshold (the limitng speed ratio) for the described bearing with $\frac{L}{D} = 0.83$ is given in Fig. 5 as function of journal eccentricity ratio c in steady rotation (equilibrium position). This diagram can be used for checking if, for the given bearing type and its work parameters, the journal equilibrium position in steady rotation will be stable.

Example: Consider the generator of 120 MW of Dolmel production. Its bearings are of the type corresponding to Fig. 1 , 2 with the following dimensions.

$L = 314.5$ mm , $D = 380.936 \, {}^{+0.013}_{\;\;0}$ mm , $D_c = 380.43 \, {}^{+0.011}_{-0.014}$ mm

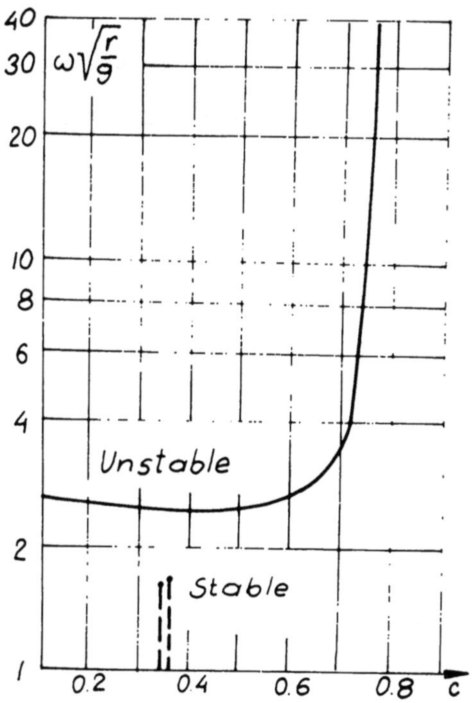

Fig. 5

$\eta = 2.16510^{-2} \frac{Ns}{m^2}$, $n = 3000$ RPM

$m_w = 32500$ kg , $\frac{L}{D} = 0.83$

Load per bearing $P = 0.5\, m_w g = 162500$ N

Sommerfield's numbers for the tolerance limits are

$$S_1 = S_{max} = \frac{P(\frac{2r}{D})^2_{max}}{D_{max} L\eta\omega} = 0.394$$

$$S_2 = S_{min} = \frac{P(\frac{2r}{D})^2_{min}}{D_{min} L\eta\omega} = 0.3396$$

Hence eccentricity ratios, taken from the diagram S = f(c), Fig. 2 are

$c_1 = c_{max} = o.384$ $c_2 = c_{min} = 0.346$

For the working speed is hence

$$\frac{\omega}{\omega_{01}} = \frac{314}{192} = 1.64 \qquad \frac{\omega}{\omega_{02}} = \frac{314}{199} = 1.58$$

This gives sufficient safety margin at the eccentricity ratios in tolerance limits $c_1 = 0.384$ and $c_2 = 0.346$ (Fig. 5).

2 Fluid Film Influence on Critical Speeds.

For assessing oil film influence a symmetric rotor in equal bearings will be considered. Again it is sufficient to consider small vibrations around equilibrium position. The end conditions characterising rotor-bearing interaction reduce to the equality between rotor shear force T and bearing reaction ΔF, [1], [3], [4].

$$\Delta F_x = \pm T_x \quad ; \qquad \Delta F_y = \pm T_y \tag{8}$$

The reaction is given by formula (2) and rotor shear forces at the supports for the symmetric case considered, can be expressed in the form

$$T_x = xN^2 \qquad T_y = yN^2 \tag{9}$$

where

$$N = \begin{cases} \omega_{Fk}\sqrt{1 + 4 \sum\limits_{i=1,3\ldots} \dfrac{\omega_{Fk}^2}{\omega_i^2 - \omega_{Fk}^2} \dfrac{t_{io}^2}{mD_i}} \\ \text{and} \\ \omega_{Fk}\sqrt{\left(\dfrac{2i_c}{\ell}\right)^2 + 4 \sum\limits_{i=2,4\ldots} \dfrac{\omega_{Fk}^2}{\omega_i^2 - \omega_{Fk}^2} \dfrac{t_{io}^2}{mD_i}} \end{cases} \tag{10}$$

Where ω_i are natural circular frequencies of the rotor supported on rigid supports, and

$$t_{io} = \frac{1}{\ell} \sum_{j=1}^{n} m_j y_{ij} \ell_{j\ell} \quad ; \quad D_i = \sum_{j=1}^{n} m_j y_{ij}^2$$

Here m_i are masses of the rotor segments, their deflections y_{ij} in the ith mode and distance $\ell_{j\ell}$ from the support, ℓ is the rotor span. Symmetric rotor can be hence characterised by N given by the formula 10 as function of an argument ω_{Fk}. The characteristics of this type for the 120MW generator rotor is given in Fig. 6. The vertical assimptotes correspond to the ω_i values giving the first three critical speeds of the rotor supported rigidly.

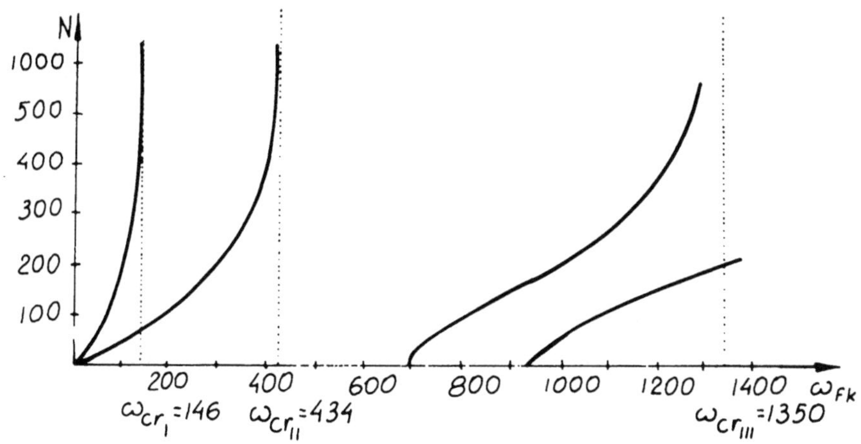

Fig. 6

The actual support conditions (8) with (2) (in non-dimensional form) give the following frequency equation (with damping neglected)

$$\begin{vmatrix} K_{xx} - N^2 & K_{xy} \\ K_{yx} & K_{yy} - N^2 \end{vmatrix} = 0 \tag{11}$$

This equation, for any N values gives the bearing characteristics in the form of an equivalent frequency, being a function of eccentricity ratio $c = \frac{e}{r}$, as the stiffness coefficient of oil film are (Fig. 3).

This type of characteristics is represented in Fig. 7 for the bearings considered in Chapter 2.1. For consecutive eccentricity ratio c values the values of stiffness coefficient were taken from Fig. 3 and corresponding N values computed from equation 11).

The interaction conditions are fulfilled for

$$N_{Rotor} = N_{Bearing} \tag{12}$$

It is sufficient hence to draw in the rotor characteristics (Fig. 6) a horizontal line (Fig. 8) corresponding to the N value obtained from the bearing characteristics (Fig. 7) for the eccentricity ratio c at equilibrium.

This eccentricity ratio c is obtained from Fig. 2 for the Sommerfield's number S calculated for given rotor speed ω.

Abscissas of the intersection points (Fig. 8) give the circular natural frequencies ω_{Fk} of the rotor-bearing system at the assumed speed ω.

Critical speeds can be found by successive approximations ie. taking an obtained ω_{Fk} value for ω and repeating till resulting ω_{Fk} value is sufficiently close to the preceeding one. In case of constant stiffness coefficients of support Fig. 8 gives directly the critical speeds. Fig. 8 shows also directly oil film influence on rotor-bearing natural frequencies and eventually critical speeds.

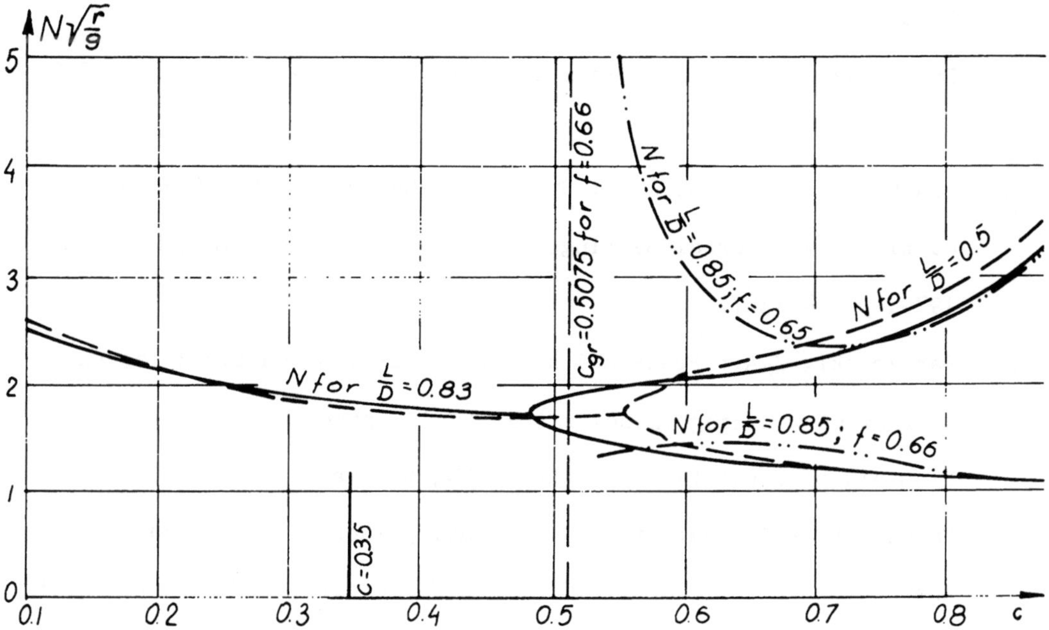

$N\sqrt{\frac{r}{g}}$
5
4
3
2
1
0
0.1
0.2
0.3
0.4
0.5
0.6
0.7
0.8
c
N for L/D = 0.83
c_{gr}=0.5075 for f=0.66
N for L/D = 0.85; f = 0.65
N for L/D = 0.5
N for L/D = 0.85; f = 0.66
c=0.35

Fig. 7

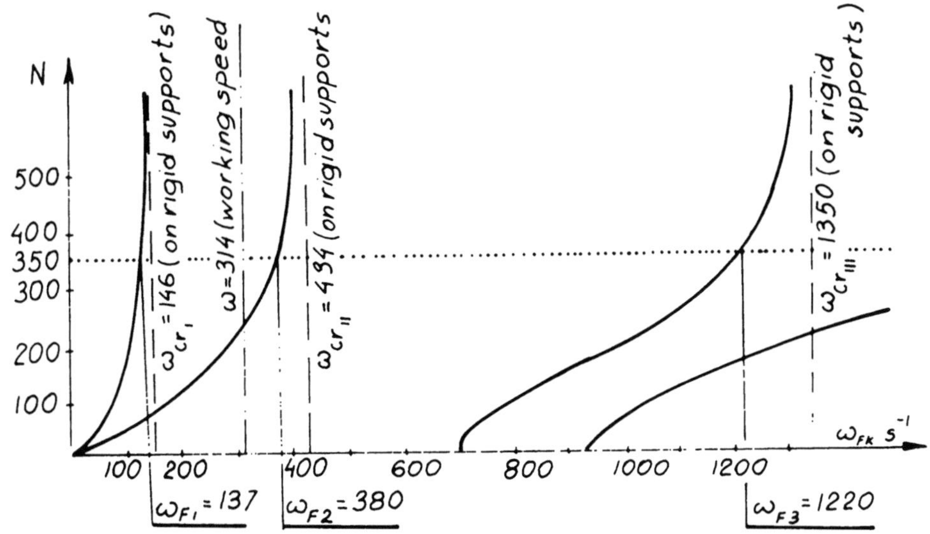

N
500
400
350
300
200
100
100
200
400
600
800
1000
1200
ω_{FK} s^{-1}
ω_{cr_I} = 146 (on rigid supports)
ω = 314 (working speed
$\omega_{cr_{II}}$ = 434 (on rigid supports
$\omega_{cr_{III}}$ = 1350 (on rigid supports)
ω_{F1} = 137
ω_{F2} = 380
ω_{F3} = 1220

Fig. 8

REFERENCES

(1) Parszewski, Z.A. : Drgania i Dynamika Maszyn, WNT, Warszawa 1982, 420p.

(2) Roszkowski, M, Hien, N.H.: Stiffness and Damping Matrices of Multi-Sleeve Journal Bearings, Proc. V World Cong. TMM, Montreal 1979, p1364-1367.

(3) Parszewski, Z.A.: Predkosci krytyczne wirnikow maszyn z lozyskami slizgowymi w szczegoinosci z lozyskami p panewkach czesciowych, Problems of Fluid Flow Machines, PWN, Warszawa 1968 pp.663-683.

(4) Parszewski Z.A.: Vibrations of Rotating Machines, Publication of the School in Power Generation, The University of Melbourne, Mech. Eng. Dept., Melbourne 1983. pp. 5.83-5.109.

CHAPTER 2.5

STABILITY OF ROTORS IN BEARINGS

N.F. Rieger

THE NATURE OF ROTOR INSTABILITY PROBLEMS

Introduction

A rotor is said to be unstable when the shaft orbit increases with time, without apparent limit. Instability begins when the rotor speed exceeds a so-called "threshold" speed. Stable operation usually resumes when the speed is again decreased below this threshold speed. Unstable whirl motions can cause mechanical problems such as rubbing between journal and bearing, seal rubbing, and blade/stator rub contacts, and may result in substantial machine damage. Unstable motions can also themselves introduce additional dynamic forces within the bearing which stabilize the whirling at a limiting whirl radius. Such whirl motions are called "bounded" instabilities.

The most common types of unstable rotor conditions are listed in table 1. Each of these instabilities is listed with the type of rotor-bearing system in which it has most frequently been observed. Typical threshold speed conditions ω_{th} are also indicated, together with the whirl frequency ν at which the rotor tends to precess, once the unstable whirl condition has been established.

Unstable whirling is distinguished from unbalance whirling by the following features:

	Unbalance Whirling	Unstable Whirling
a. Whirl frequency	$\nu = \omega$	$\nu = \omega_c$
b. Threshold speed	none	$\omega_{th} = \omega_{c_1}$,

(where ω_c is the lowest critical speed of the rotor in its bearings).

The above comparison shows that unstable whirl motions are initiated beyond a certain threshold speed, and occur at a frequency ν, which differs from the rotor speed ω. The unbalance whirl frequency is the same as the rotation speed, i.e., $\nu = \omega$. Unstable whirling is initiated at speeds above a certain threshold speed which is never less than the lowest critical frequency of the system. Unbalance whirling has no threshold speed: it can occur at all rotor speeds, and may become resonant with some lateral mode of the system to cause the so-called"bending critical speeds."

The rotor threshold speed is a major operating parameter of the system. Beyond that speed, the rotor whirl orbit radius may grow rapidly with time in an outward spiral until some stable whirl radius is found (bounded instability) or until some constraining surface, such as bearing or seal face, is struck. Unstable motions which are bounded by the bearing or seal dynamic properties are often acceptable operating

TYPE OF INSTABILITY	WHIRL FREQUENCY ν	MACHINE SYSTEMS SUBJECT TO INSTABILITY
Hysteretic instability	$\nu = \omega_c$, (independent of ω above ω_c)	Lightly-damped rotor with shrink or press-fitted parts on a central shaft.
Subharmonic shaft whirl	$\nu < \omega/2$	Rigid rotor in fluid film bearings
Fluid film instability		
. Rigid rotor	$0.45\omega < \nu < 0.50\,\omega$	Rotor in fluid-film journal bearings
. Flexible rotor	$0.38\omega < \nu < 0.45\,\omega$	Machines with gas seals
Parametric instability	$\nu = \omega/2,\ \omega/3,\ \omega/4$	Machines with asymmetric rotor or asymmetric shaft or both.
Transverse-torsional interaction instability	$\nu = \omega$	Machines carrying bladed disks, e.g. steam turbine

ν = whirl frequency ω_c = first bending critical speed ω = running speed

Table 1 Details of Various Instability Conditions

conditions, especially for a rigid rotor, but once the journal contacts a bearing or seal surface, a violent counter-rotating whirl may be set up. This second whirl motion can destroy the bearing surface and damage the journal. Typical whirl orbits of common whirl motions are shown in figure 1.

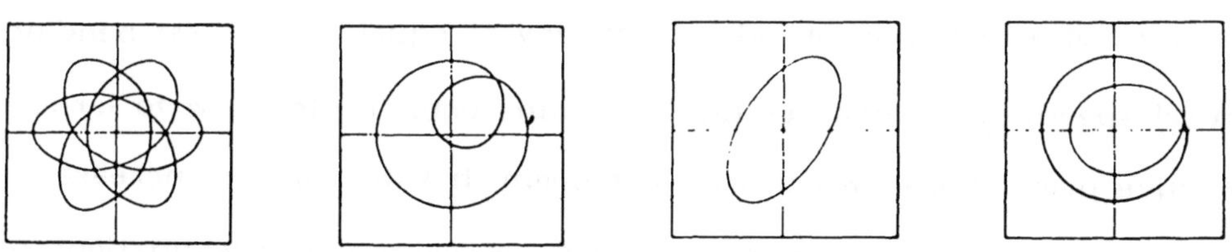

(c) Hysteresis instability, $\nu = \omega_c$. (d) Half frequency bearing whirl, $\nu < \omega/2$. (a) Stable whirl unbalance only, $\nu = \omega$. (b) Dissimilar stiffness whirl instability, $\nu = \omega/2$.

Figure 1 Whirl Orbits Caused by Various Exciting Forces and Unstable Conditions

Types of Rotor Instability

Hysteretic Instability

Hysteretic instability may occur in built-up rotors consisting of disks which have been shrink- or press-fitted upon a central shaft, or which are assemblies of components bolted together. Rubbing between the component parts produces a friction force which tends to suppress rotor whirling at speeds below the first bending critical speed, and to sustain such rotor whirl motions above the bending critical speed. Such whirl motions were first observed by Newkirk [1] and by Kimball [2], and were further investigated by Robertson [3]. In these tests, the frequency of the hysteretic whirl motion was observed to be the same as the first bending critical speed of the rotor. It was first thought that the hysteresis of the material itself was the main cause of this whirling. But measurement of these hysteresis forces by Kimball [4] showed that they were of very small magnitude compared with other forces acting on the system (e.g., rotor weight), and that the instability was more likely caused by rubbing between shrink-fitted parts on the shaft. Above the first bending critical speed, this type of whirling occurs at a frequency ν equal to the first bending critical speed, ω_c. Once started, the whirl orbit could grow to large destructive proportions, which may not diminish with further increase in speed. Since Newkirk's studies in 1925, experimental investigations have been outnumbered by analytical studies. Investigations by Kimball [1], Robertson [2], Kushul [5], and Gunter [6] will be discussed in detail in later chapters of this book. The general conclusion is that hysteretic instability can be successfully suppressed by the addition of a suitable combination of external stiffness and damping to the rotor-bearing system.

Fluid Film Bearing Instability

This well-known form of rotor instability - variously and somewhat loosely referred to as half-frequency whirl, fractional frequency whirl, oil whip, and resonant whipping - is a self-excited whirl of the rotor in its supports induced by cross-coupling effects within the journal bearing fluid film. It was first investigated and reported by Newkirk [1] in 1924, who found that it may begin at speeds above 1.5~2.0 times the first lateral critical speed of the rotor system. Two distinct types of such instability have been observed. The first type, known as "half-frequency whirl", occurs mostly in systems in which the rotor is dynamically rigid and when the whirl frequency is somewhat less than half the running speed. The second type, known as "resonant whip", is more severe. It builds up with a whirl frequency equal to the first bending critical speed of the system. This whirl can appear at rotor speeds 1.5~2.0 times the first bending critical speed of the system. Once started, the whirling involved is strong, and may persist over a wide range of higher speeds. The large rotor whirl amplitudes involved may consequently be very harmful to the system. Resonant whip, of which detailed investigations are presented in later chapters, is more likely to occur in flexible rotors. Research by Newkirk and Lewis [7], Pinkus [8], and Tondl [9] [10] has shown that plain cylindrical bearings are the most likely to become unstable in resonant whipping, and that the most stable bearing types are the tilting pad bearing, the floating-ring bearing, and the three-lobe bearing.

The fundamental conclusion reached by Tondl [11], Lund [12] and others concerning the stability of rotors in fluid film bearings is that rotors can be stabilized by addition of a suitable amount of damping, and by the selection of a bearing type which minimizes the amount of cross-coupled stiffness and damping involved in the system. This conclusion has led to several designs of flexible, damped supports (employing spring-mounted bearings and an oil-film damper) known as squeeze film damper bearings.

Gas Bearing Instability

One limiting factor in the use of high-speed gas journal bearings has been the development of half-frequency whirl motion of the journal within the bearing. The whirl threshold speed represents an upper limit for the speed at which the rotor may be safely operated. Half-frequency whirl, once established, commonly leads to rapid growth of the journal orbit, and possibly to bearing seizure.

Data on dynamic bearing properties is needed to predict the onset of half-frequency whirl. The compressible Reynolds equation with time-dependent terms included for dynamic loading is:

$$\frac{\partial}{\partial x}\left(\frac{h^3\rho}{\mu}\frac{\partial p}{\partial x}\right)+\frac{\partial}{\partial z}\left(\frac{h^3\rho}{\mu}\frac{\partial p}{\partial z}\right)=6U\frac{\partial}{\partial x}(h\rho)+12\frac{\partial}{\partial t}(h\rho) \qquad (1)$$

where x = $a\theta$

θ = central angle from the line of centers

z = axial distance from the middle plane of the bearing

h = $C(1+\varepsilon\cos\theta)$

C = radial clearance

ε = eccentricity ratio

p = absolute pressure

ρ = density

μ = viscosity

U = $a\omega$, peripheral speed of journal due to rotation

Castelli and Elrod [15] developed an analysis in which the equations of motion for the rigid rotor and the compressible Reynolds equation were simultaneously integrated on a digital computer to determine the rotor orbital path. The stability, or instability, of particular cases was established from the growth, stabilization or decay of the orbit. With assumed initial conditions for both the rotor motion and the fluid-film pressure, the influence of incremental displacements on the rotor equations and then on the fluid-film properties was calculated. The calculation was intended to provide data for the next incremental change.

Cheng and Pan [16] applied Galerkin's method to solve the Reynolds' equation with time-dependent effects included, for the case of finite plain cylindrical bearings and other geometries for which a representative (ph) function could be deduced.

Several experimental programs have been conducted to determine the effect of bearing geometry on the threshold of instability. The main results of these investigations are:

- There is a clearance which gives minimum onset speed for a given bearing load, references [16] and [17];
- For a given rotor mass and bearing clearance, the eccentricity ratio at the onset of instability remains virtually constant, even though the bearing load is varied significantly, reference [16];
- As the bearing clearance increases, so does the eccentricity ratio of the onset of instability, reference [17];
- If a gas bearing is designed for high stiffness and large clearance, the threshold speed will be increased, reference [18].

- Increasing the gas pressure in the bearing will increase the onset of instability as well, reference [19].

Kerr's detailed experiments [20], [21], show that by mounting the bearings in O-ring supports, it is possible to proceed through the half-frequency whirl region and operate at much higher speeds without unstable vibration. Some oscillation may occur as, with the increase of speed, the shaft support system passes through its own resonant or unstable regions.

Tondl [22] found that the onset of self-excited vibrations could be efficiently raised by mounting the rotors on an elastically suspended foundation mass with damping. He also determined that the nonlinear components of foundation damping did not cause any additional instabilities.

In general, instability in gas-lubricated bearings closely resembles that found in fluid film bearings.

An example of stabilization of a horizontal symmetrical rotor in a plain cylindrical hydrodynamic gas journal bearing, using a flexible damped foundation has been given by Rieger [23]. The rotor system had the following properties:

Weight	12 lb.	Operating speed	12,000 rpm
Bearing Length	2.0 in.	Dynamic load	14 lb. (on each bearing)
Bearing Diameter	2.0 in.	Ambient pressure	14.7 psi
Radial clearance	0.001 in.	Viscosity	2.8×10^{-9} lb. sec./in.^{2}
Temperature	100°F		

Using instability charts, the threshold speed for this rotor was found to be 7320 rpm. It was therefore expected that the rotor would become unstable before it reached its operating speed. A damped flexible foundation was subsequently added to the analytical model, with foundation stiffness K_f = 600 lb./in. and foundation damping B_f = 0.1 lb.sec./in. Using the same method, it was then found that the threshold speed was still below the operating speed. The rotor would therefore still be unstable at its operating speed.

The next step towards stabilization was to change some characteristics of the bearing and to add more damping while the stiffness of the foundation was decreased, thus:

Bearing length	1.0 in.
Bearing diameter	2.0 in.
Bearing stiffness K_f	350 lb./in.
Bearing damping B_f	0.23 lb./sec./in.
L/D	0.5

These changes considerably improved the stability of the rotor which ran up to 51,000 rpm without encountering any instability problem. That is, the threshold speed was now well above the operating speed.

Gas Seal Induced Instability

Steam turbine stages, compressor rotors, and pump stages may all experience seal-induced instability regions in which the seal dynamic forces may overcome the system damping. Recognition of this has led to the conclusion that a threshold power level may exist for a given operating speed, beyond which speed the rotor may become unstable, even though it operates in bearings which usually stabilize a rotor system. Shaft destabilizing seal forces and blade seal forces are generated, respectively, by shroud and shaft labyrinth seals and by flow action on the blades. Should these forces, acting together in the whirl direction, exceed the damping forces, a high pressure rotor may experience self-excited whirling whose frequency is equal to the lowest natural frequency of the rotor in its bearings.

Several analyses have been made concerning the effect of labyrinth seal forces on a whirling rotor. Different basic assumptions have caused these analyses to differ substantially in their predictions of the rotor instability threshold conditions. Wright [24] conducted an experimental analysis using an apparatus capable of measuring accurately the labyrinth seal forces on a whirling rotor (see figure 2). Wright's model was designed to provide a better understanding of the basic phenomena involved in seal-excited whirling. He showed that the whirl excitation exponent may be negative or positive depending on the type of seal (converging or diverging) and on the type of whirl (forward or backward), and concluded that some control of steam whirl is possible by stiffening the rotor or modifying the bearing. He also suggested that the steam seals should be designed to have negative (stabilizing) forward whirl excitation constants which tend to cancel the positive (destabilizing) forward whirl excitation constants of the blade rows.

Parametric Instabilities

This type of instability arises from the variation in some system parameter with respect to time, such as shaft lateral stiffness. It can occur when an asymmetric rotor is mounted on a symmetric shaft, or when a symmetric rotor is mounted on an asymmetric shaft which has unequal lateral stiffnesses in its two principal stiffness directions. An introduction to this problem has been given by Den Hartog [25] who describes the "flat shaft" problem, and shows that this leads to twice per-rev variation of the

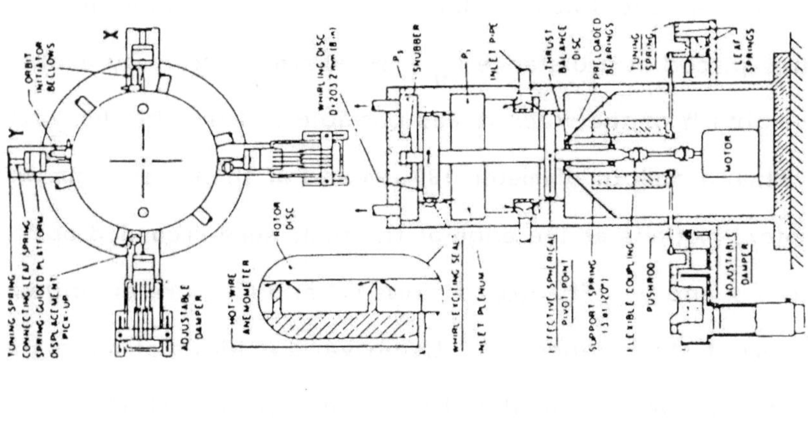

Figure 2 Seal-Excited Rotor Whirl Model

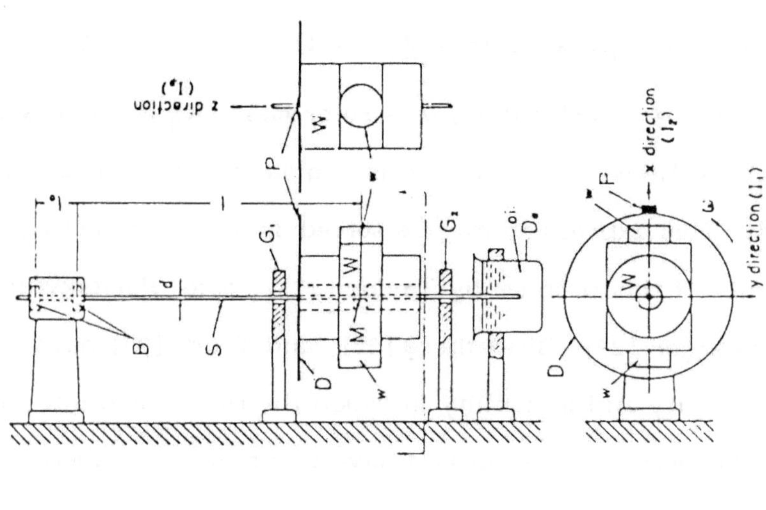

Figure 3 Experimental Apparatus (Yamamoto and Ota)

stiffness term in the rotor equation of motion:

$$M\ddot{R} + (K_0 + K_1 \cos 2\omega t)R = Ma\omega^2(i\omega t) \qquad (2)$$

where R is the rotor c.g. whirl radius, M is rotor mass, K_0 is shaft lateral stiffness, K_1 is the stiffness asymmetry, ω is angular velocity, and a is mass eccentricity. This expression may be solved as a Van der Pol equation [26]. Other studies of this problem by Taylor [27], Foote, Poritsky and Slade [28], Kellenberger [29], Dimentberg [30], and Tondl [31] have indicated regions of stability and instability in which the rotor amplitude will grow with time. Several of these authors have also noted that such shafts also have speed ranges ω in which subharmonics of the dissimilar stiffness effect also induce vibrations at $\omega/4$, $\omega/8$, $\omega/16$, etc.

Yamamoto and Ota [32] have contributed to the investigation of parametric instability of a rotor with dissimilar inertia properties. The experimental apparatus shown in figure 3 was used. A vertical shaft of diameter d = 11.55 mm, and length l = 601.9 mm was driven by a 45 hp DC motor. It was supported at its upper end by two self-aligning double row ball bearings placed at a distance L_o = 36.00 mm apart. By exchanging two attached weights W of the rotor it was possible to vary the dissymmetry parameter Δ. When a disk of diameter 180 mm and of thickness 3 mm was added to the rotor, motions at the edge of the disk were recorded optically in both directions x and y. The guard rings G1 and G2 were arranged to check the increase of shaft deflection. Using various kinds of oil in the damper vessel at the lower end of the shaft to restrain the whirl amplitude, various damping coefficients were obtained.

These authors concluded that for a rotating shaft system carrying an unsymmetrical rotor, instability occurs near the speed range where the sum of two natural frequencies of the system equals twice the rotating speed of the shaft. It was also found that this unstable range may be diminished by adding sufficient viscous damping. In a later experiment, Yamamoto [34] studied the case where the sum of the two system natural frequencies is equal to twice the rotating speed of the shaft. The test rigs used are shown in figures 4 and 5. The unstable vibrations which appeared in these regions were studied for a system consisting of a rotor with unsymmetrical inertia, and a shaft with unequal stiffness. A theoretical investigation, which was verified by this experiment, showed that such unstable vibrations can be removed by selecting a suitable combination of the inequalities in inertia and stiffness, as well as by external viscous damping.

Transverse-Torsional Interaction

Interaction between transverse motions and torsional shaft motions has been shown to induce instability. This problem has been studied by Broniarek [35], Tondl [31], and Smith [36]. Smith [36] investigated a specific problem of synchronous whirl in a group of steam turbines in which the source of the trouble was not excessive unbalance, but an uncommon type of instability. Here, the whirl frequency developed below rotational frequency. A slow build-up of vibrations occurred when the transverse whirl frequency coincided with the running speed of the rotor.

The occurrence of this instability was erratic and unpredictable, but always showed the same general pattern. Large vibrations developed within a specific speed range. Vibration was at first particularly strong in the low pressure turbine, at a frequency of 1-per-rev plus harmonics, especially the three per rev. No low frequency component below 1/rev was present. The harmonics grew much more slowly than the synchronous component, which was strongest in the high pressure turbine. When speed could be varied, the synchronous whirl dropped rapidly in amplitude outside a sensitive speed band between 5 and 15 percent of operating speed, and sometimes even stabilized. Amplitude growth could be started or restarted by a change in the generator load. The different stages of the self-excited vibrations are shown in figures 6 and 7.

Smith concluded that this instability was due to the torsional-transverse interaction. The principal mechanism was resonant vibration of the LP turbine blades.

Causes of Instability

The purpose of this section is to explain some of the most common sources of instability. The approach follows that given by Lund [37] .

Consider the simple symmetric rotor model shown in figure 8, consisting of a shaft with a stiffness of 2 k and a lumped central mass 2 m. The shaft is supported at its ends in bearings with stiffness K, such that the natural frequency of the system is:

$$\omega_n = \sqrt{\frac{k K}{m(k+K)}} \qquad (3)$$

The angular speed of rotation is Ω. External damping is provided at the rotor mass and at the bearings with damping coefficients 2b and B, respectively. (See figure 8.) Although practical rotors are considerably more complex, this simplified model contains the essential features of the problem,and allows useful closed form solutions to be obtained.

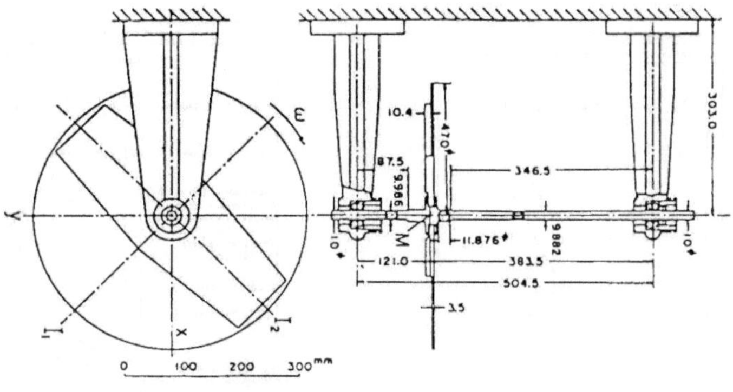

Figure 4 Experimental Apparatus for the Unstable Vibrations near the Major Critical Speed ω_c

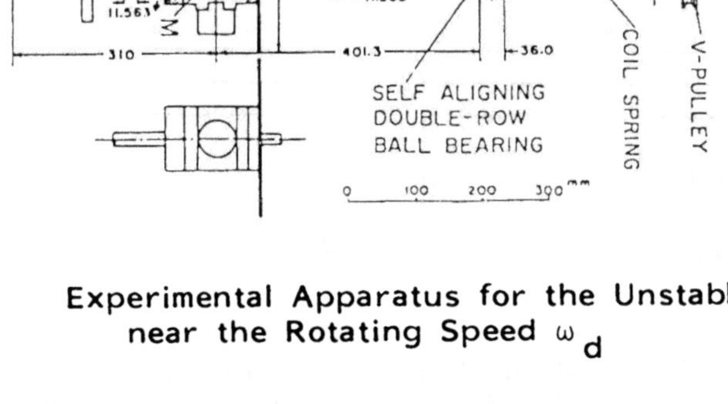

Figure 5 Experimental Apparatus for the Unstable Vibrations near the Rotating Speed ω_d

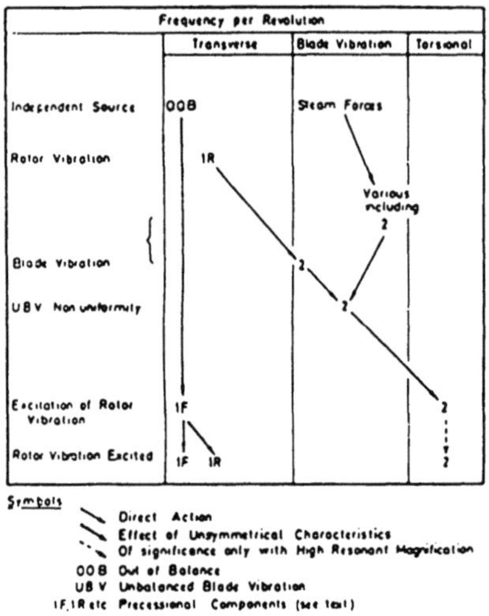

Figure 6

Principal Actions in Initial Build-up of Synchronous Whirl [36]

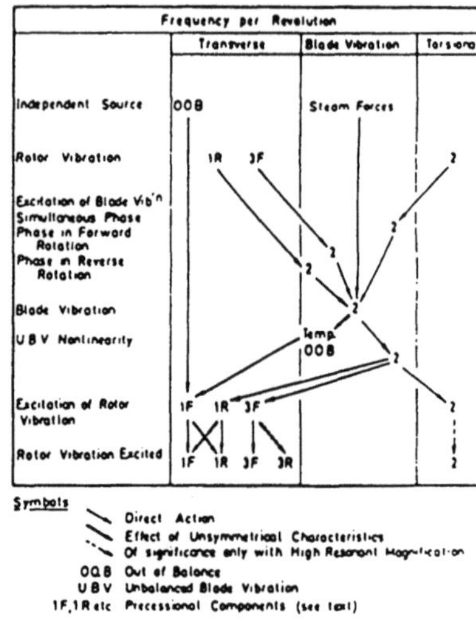

Figure 7

Principal Actions in Later Build-up of Synchronous Whirl [36]

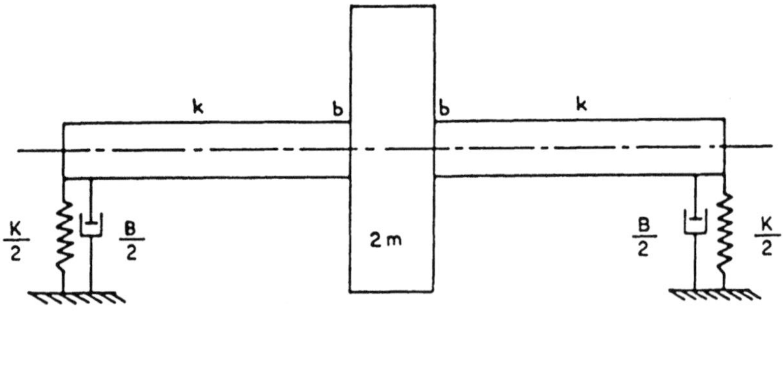

Figure 8 Rotor Model with Flexibility and Damping for Study of Instability Problems, Lund [37]

The system is assumed to be linear, with constant properties. Let (x,y) and (x_1, y_1) be the coordinates of the rotor mass and of the shaft at the supports respectively. Using this simple model, the following cases may be considered:

Hysteretic Instability

Assuming that the energy dissipation is viscous, the shaft may be assigned a damping coefficient c. The equations of motion then become:

$$m\ddot{x}+b\dot{x} = -k(x-x_1) -c(\dot{x}-\dot{x}_1) -\Omega c(y-y_1) = -Kx_1-B\dot{x}_1 \quad (4)$$

$$m\ddot{y}+b\dot{y} = -k(y-y_1) -c(\dot{y}-\dot{y}_1) +\Omega c(x-x_1) = -Ky_1-B\dot{y}_1 \quad (5)$$

where x, y, x_1, and y_1 are the amplitudes at the rotor disk and for the journal in a fixed coordinate system.

Usually, c is small and, assuming the external damping b and B to be equally small, only first order terms in c, b and B need to be retained. With solutions for the amplitude of the form e^{st}, the characteristic equation is obtained as:

$$ms^2 + [b+(\frac{k}{K+k})^2 B + (\frac{K}{K+k})^2 c]s + \frac{kK}{K+k} \pm i\Omega (\frac{K}{K+k})^2 c = 0. \quad (6)$$

Separating this expression into real and imaginary parts, it is found that the whirl frequency ν is equal to ω_n, and the condition which defines the instability threshold speed is:

$$(\frac{K+k}{K})^2 \omega_n b + (\frac{k}{K})^2 \omega_n B > (\Omega-\omega_n)c. \quad (7)$$

In the absence of external damping, the rotor becomes unstable when the first bending critical speed is reached. To operate in a stable manner at higher speeds, external damping, as described in equation (2) is required.

It is assumed in the above analysis that the shaft internal damping is of a viscous nature. The dissipated energy then becomes proportional to frequency. This characteristic, however, is not typical of most shaft materials. Generally, the dissipation is of hysteretic nature, and is independent of frequency. A more realistic model may be introduced by adopting a material property γ as the fixed angle by which the strain lags the stress, such that the hysteresis loop in the stress-strain diagram becomes an ellipse. The dissipated energy is then proportional to sin γ, and hence to the square of the stress amplitude. On this basis, the equations of motion are:

$$m\ddot{x} + b\dot{x} = -k\cos\gamma(x-x_1) - k\sin\gamma(y-y_1) = -Kx_1 - B\dot{x}_1 \quad (8)$$

$$m\ddot{y} + b\dot{y} = -k\cos\gamma(y-y_1) + k\sin\gamma(x-x_1) = -Ky_1 - B\dot{y}_1. \quad (9)$$

As γ is small, cos γ and sin γ may be replaced by 1 and γ respectively. Assuming the external damping to be equally small and retaining only first order terms in γ, b and B, an exponential solution for the amplitudes leads to the following characteristic equation:

$$ms^2 + [b + (\frac{k}{K+k})^2 B]\, s + \frac{kK}{K+k} \pm i\, (\frac{K}{K+k})^2 k\gamma = 0. \quad (10)$$

With $s = i\nu$ at the threshold of instability, it is found that $\nu = \omega_n$ and the rotor is stable when:

$$(\frac{K+k}{K})^2 \omega_n b + (\frac{k}{K})^2 \omega_n B > \gamma k. \quad (11)$$

In contrast to the case of viscous internal damping, where instability can occur only above the first critical speed, hysteretic damping causes the rotor to be potentially unstable at all speeds. Once, however, the required external damping is provided, the rotor is always stable; whereas with viscous damping, there is a finite threshold speed.

Whirl in Hydrodynamic Bearings

For small amplitudes of self-excited whirling induced by hydrodynamically lubricated journal bearings, the equations of motion can be written as:

$$m\ddot{x}+b\dot{x} = -k(x-x_1) = -K_{xx}x_1-B_{xx}x_1 \quad -K_{xy}y_1-B_{xy}\dot{y}_1-B\dot{x}_1 \tag{12}$$

$$m\ddot{y}+b\dot{y} = -k(y-y_1) = -K_{yx}x_1-B_{yx}\dot{x}_1 \quad -K_{yy}y_1-B_{yy}\dot{y}_1-B\dot{y}_1\ . \tag{13}$$

The eight dynamic bearing coefficients, $K_{xx}, \ldots, B_{xx} \ldots$ depend on the particular bearing geometry and, in addition, on the operating conditions, expressed through the Sommerfeld number:

$$S = \frac{1}{2\pi}\,\frac{\mu\Omega DL}{W}\left(\frac{R}{C}\right)^2 \tag{14}$$

where

L	=	bearing length	C	=	radial clearance
D	=	journal diameter	W	=	bearing applied load and
R	=	journal radius	μ	=	lubricant viscosity .

Assuming a short ($L/D \le 1/1$) plain cylindrical journal bearing, operating at sufficiently high speed such that:

$$2\left(\frac{L}{D}\right)^2 S > 1, \tag{15}$$

then the dynamic bearing coefficients are approximately given by:

$$K_{xx} = \frac{4}{\pi}\frac{W}{C} \tag{16}$$

$$K_{xy} = -K_{yx} = \pi^2\left(\frac{L}{D}\right)^2 S\frac{W}{C} = \frac{1}{2}\Omega B_{xx} \tag{17}$$

$$K_{yy} = 2K_{xx}\,\frac{8}{\pi}\,\frac{W}{C} \tag{18}$$

$$B_{xx} = B_{yy} = 2\pi^2 \left(\frac{L}{D}\right)^2 S \frac{W}{\Omega C} = 2\pi\mu L\left(\frac{R}{C}\right)^3 \left(\frac{L}{D}\right) \tag{19}$$

$$B_{xy} = B_{yx} = \frac{8}{\pi} \frac{W}{\Omega C} . \tag{20}$$

At the threshold of instability, the motion is purely harmonic with frequency ω, whereby the equations of motion can be written as:

$$\begin{Bmatrix} (K_{xx}+i\omega B_{xx}-Z) & (K_{xy}+i\omega B_{xy}) \\ (K_{yx}+i\omega B_{yx}) & (K_{yy}+i\omega B_{yy}-Z) \end{Bmatrix} \begin{Bmatrix} x_1 \\ y_1 \end{Bmatrix} = 0 \quad , \tag{21}$$

where $Z = -\dfrac{k(-m\omega^2+i\omega b)}{k-m\omega^2+i\omega b} - i\omega B = \dfrac{k(k-m\omega^2)m\omega^2-(\omega b)^2}{(k-m\omega^2)^2+(\omega b)^2} - i\omega\left[\dfrac{k^2}{(k-m\omega^2)^2+(\omega b)^2} b+B\right].$

(22)

This value must equal the root of the determinant of equation 4 which becomes:

$$Z = \frac{1}{2}(K_{xx}+K_{yy}) + i\left[\omega B_{xx} \pm K_{xy}\sqrt{1+\frac{(\omega B_{xy})^2-\frac{1}{4}(K_{xx}-K_{yy})^2}{K_{xy}^2}}\right] . \tag{23}$$

As ω is equal to $\frac{1}{2}\Omega$ or less, the second term inside the square root is of the order of $(\frac{1}{4})^2$, and can be ignored. With this condition Z is approximately equal to:

$$Z = K + iB_{xx}(\omega \pm \frac{1}{2}\Omega) \quad , \tag{24}$$

where

$$K = \frac{1}{2}(K_{xx}+K_{yy}) = \frac{6}{\pi}\frac{W}{C} . \tag{25}$$

Equating real and imaginary parts of equations 5 and 6, and neglecting the term (ωb^2), the whirl frequency is determined as:

$$\omega = \omega_n = \sqrt{\frac{kK}{m(k+K)}} \, . \qquad (26)$$

The condition for stability becomes:

$$\omega_n B + \left(\frac{K+k}{k}\right)^2 \omega_n b > \left(\frac{1}{2}\Omega - \omega_n\right) B_{xx} = (\Omega - 2\omega_n)\, \pi \mu L \left(\frac{R}{C}\right)^3 \left(\frac{L}{D}\right)^2 . \qquad (27)$$

The whirl motion is a forward precession.

It has been shown by Lund [21] and others that a more effective method than external damping for stabilizing hydrodynamic bearing whirl is to mount the bearing in a flexible damped support. By proper tuning, the instability can be completely eliminated.

Equation 7 shows that in the absence of external damping, this rotor is unstable in cylindrical bearings when the speed exceeds twice the first critical speed. This is true only for sufficiently large values of the Sommerfeld numbers, i.e., for low bearing operating eccentricity values. It should also be noted that the inequality itself is based on several assumptions, and consequently that it will be less valid where these assumptions are violated.

Interaction with Fluid Flow Forces

Any whirl motion of the shaft in rotating machinery affects the flow field of the working fluid and sets up additional forces and moments. Coupling thereby exists between the motion of the shaft and the fluid-dynamic reaction forces which could potentially be destabilizing. One mechanism for such a coupling was suggested by Alford [38]. This applies to axial flow machines where a radial displacement of the wheel center in a stage generates a transverse force proportional to the displacement. Here, the coefficient of proportionality becomes:

$$\kappa = -\frac{d(\frac{\eta}{\eta_o})}{d(\frac{h}{H})}\frac{T_o}{2rH} , \tag{28}$$

where

κ	-	the stage efficiency	H	-	the vane height
T	-	the stage torque	h	-	the tip clearance
r		the pitch radius			

and the subscript o refers to the concentric condition.

Assuming the central rotor mass to be a turbine stage, the equations of motion are:

$$m\ddot{x} + b\dot{x} + \kappa y = -k(x - x_1) = -Kx_1 - B\dot{x}_1 \tag{29}$$

$$m\ddot{y} + b\dot{y} - \kappa x = -k(y - y_1) = -Ky_1 - B\dot{y}_1 . \tag{30}$$

Neglecting second order terms in K, b and B, the characteristic equation becomes:

$$ms^2 + [b + (\frac{k}{K+k})^2 B]s + \frac{kK}{K+k} \pm i\kappa = 0 \tag{31}$$

on substituting an exponential solution for the amplitudes x, y.

Letting s = iν at the threshold of instability, the whirl frequency is found to be equal to ω while the stability condition becomes:

$$\omega_n b + (\frac{k}{K+k})^2 \omega_n B > |\kappa| . \tag{32}$$

At the onset of instability, the whirl motion is a forward precession when κ is positive,and a backward precession when κ is negative.

Rotor Asymmetry

Practical cases of rotor parametric excitation arise from dissimilar lateral stiffness properties of the shaft. Writing the principal lateral stiffnesses of the shaft as K + ΔK and K - ΔK respectively gives the system equations of motion as:

$$m(\ddot{x}-\Omega^2 x-2\Omega\dot{y}) + b(\dot{x}-\Omega y) = -(k-\Delta k)(x-x_1) = -Kx_1 - B(\dot{x}_1-\Omega y_1) \quad (33)$$

$$m(\ddot{y}-\Omega^2 y+2\Omega\dot{x}) + b(\dot{y}+\Omega x) = -(k+\Delta k)(y-y_1) = -Ky_1 - B(\dot{y}_1+\Omega x_1). \quad (34)$$

Introducing the parameters:

$$\alpha = \frac{K}{K+k}\,\frac{\Delta k}{k} \quad (35)$$

$$\beta = \frac{b}{2m} + \left(\frac{k}{K+k}\right)^2 \frac{B}{2m}. \quad (36)$$

An exponential form e^{st} is again assumed for the amplitudes.

As ΔK/K is normally small and assuming that β is also small and of the same order, the above equations can be solved, retaining only first order terms in α and β. The characteristic determinant then has roots as follows:

$$(s+\beta)^2 = -(\Omega^2+\omega_n^2-\beta^2) \pm \sqrt{4\Omega^2(\omega_n^2-\beta^2)+\alpha^2\omega_n^4}. \quad (37)$$

At the threshold of instability, the real part of s is zero. Hence, the imaginary part is also zero according to equation 8 because the right-hand side is real for the assumptions used. This expression can then be solved for the threshold speed, which is given by:

$$\Omega^2 = \omega_n^2 - 2\beta^2 \pm \sqrt{4\beta^4 - 4\omega_n^2\beta^2 + \alpha^2\omega_n^4}\;. \quad (38)$$

The rotor is unstable between the two speeds determined from equation 9. The instability region disappears when:

$$\beta^2 > \frac{1}{2}\omega_o^2\left(1 \pm \sqrt{1-\alpha^2}\right) \simeq \frac{1}{4}\omega_o^2\alpha^2 \quad , \tag{39}$$

or

$$\beta \gtrsim \frac{1}{2}\alpha\omega_o \quad , \tag{40}$$

or

$$\left(\frac{K+k}{K}\right)^2 \omega_o b + \left(\frac{k}{K}\right)^2 \omega_o B \gtrsim \Delta k \quad . \tag{41}$$

Thus the rotor can be stabilized with the required amount of damping either at the rotor mass or at the supports. The instability threshold frequency is equal to the first critical speed of the rotor, and the shaft whirls in the same direction as the rotation.

Sample calculation:

Let
$2m = 600$ kg
$2k = 1.4 \times 10^8$ N·m
$K = 0.9 \times 10^8$ N·m

$$\omega_m = \frac{0.9 \times 0.7 \times 10^8}{(0.9 + 0.7) \times 300} = 362.3 \text{ rad./sec/} \doteq 3460 \text{ rpm} \tag{42}$$

In the first case, the shaft has internal damping equivalent to a logarithmic decrement, δ, of 3.16 percent for the free vibrations of the simply supported rotor. Therefore:

$$c = \frac{\delta}{\pi}\frac{k}{\omega_o} = 1932 \text{ N·sec/m} \ . \tag{43}$$

The required external damping is computed from equation 2 as:

$$b > 611.4 \cdot \left(\frac{\Omega}{\omega_n} - 1\right) \text{ N·sec/m} \tag{44}$$

$$B > 3194 \cdot \left(\frac{\Omega}{\omega_n} - 1\right) \text{ N·sec/m} \ . \tag{45}$$

If, instead, the internal damping is hysteretic, the angle γ by which the strain lags the stress becomes

$$\gamma = \frac{\delta}{\pi} = 0.01 \text{ radians} = 0.57 \text{ deg.} \qquad (46)$$

The stabilizing external damping is obtained by equation 3 as:

$$b \quad 611.4 \ N \cdot sec/m \qquad (47)$$

$$B > 3194 \ N \cdot sec/m. \qquad (48)$$

Next, consider the rotor to be supported in hydrodynamic journal bearings with the following characteristics:

D	0.125m		1.5×10^{2} N.sec./m^2
C	6.25×10^{5}m	W	2942 N
L	0.0625 m	N	6000 rpm

therefore,

$$K = \frac{6}{\pi}\frac{W}{C} = 0.90 \cdot 10^{8} \text{ N/m}, \qquad (49)$$

which yields from equation 7 the external damping required for stabilization as:

$$b > \left(\frac{\Omega}{2\omega_n} - 1\right) \cdot 2.8187 \cdot 10^{5} \ N \cdot sec/m \qquad (50)$$

$$B > \left(\frac{\Omega}{2\omega_n} - 1\right) \cdot 1.4726 \cdot 10^{6} \ N \cdot sec/m\ . \qquad (51)$$

Next, consider the case of a rotor which may experience gas-seal induced whirl. The rotor carries a single turbine stage with a power of 1000 kw at 3460 rpm, and the parameters defined previously as:

$$T_o = 2760 \text{ N.m.}$$

$$r = 0.15 \text{ m}$$

$$H = 0.05 \text{ m}$$

which gives for the coupling coefficient:

$$\kappa = 1.0 \cdot \frac{2760}{2.0.15.0.05} = 1.84 \cdot 10^5 \mathrm{N/m} \quad , \qquad (52)$$

The damping required for stability is:

$$b > 507.9 \ \mathrm{N \cdot sec/m}$$

$$B > 2653 \ \mathrm{N \cdot sec/m} \ .$$

Finally, the variation in lateral stiffness is assumed to be ± 1 percent, such that $\Delta K = 0.7 \times 10^6$ N/M.

In the absence of damping, the rotor is unstable from 3450 to 3469 rpm. The instability zone is eliminated by adding external damping according to equation 9:

$$b > \left(\frac{K}{K+k}\right)^2 \frac{\Delta k}{\omega_n} = 611.4 \ \mathrm{N \cdot sec/m}$$

$$B > \left(\frac{K}{k}\right)^2 \frac{\Delta k}{\omega_n} = 3194 \ \mathrm{N \cdot sec/m} \ .$$

Out of five types of instability considered, hydrodynamic bearing whirl is by far the strongest. Unless amplified or specifically aggravated in a given design, the above calculation shows that the four other forms of instability considered can only occur in very lightly damped systems. Such light damping may result from system design (rolling element bearings), from the environment (cryogenic systems), or in fluid-film bearing systems whose operating conditions approach the hydrodynamic whirl threshold condition. In this latter circumstance, the resulting whirl may be influenced by parametric effects, or the other destabilizing influences referred to in this section.

General Stability Analysis

A general form of the equations for forced damped motion of a multi-mass linear elastic system (unbalanced rotor in fluid-film bearings) is:

$$\begin{aligned}
&M_1\ddot{X}_1 + B_1\dot{X}_1 + K_1X_1 - B_1\dot{X}_2 - K_1X_2 = P_1\cos(\omega t+\phi_1)\\
&-B_1\dot{X}_1 - K_1X_1 + M_2\ddot{X}_2 + (B_1+B_2)\dot{X}_2 + (K_1+K_2)X_2 - B_2\dot{X}_3 - K_2X_3 = P_2\cos(\omega t+\phi_1)\\
&\qquad\vdots\\
&-B_{r-1}\dot{X}_{r-1} - K_{r-1}X_{r-1} + M_r\ddot{X}_r + (B_{r-1}+B_r)\dot{X}_r + (K_{r-1}+K_r)X_r - B_r\dot{X}_{r+1} - K_rX_{r+1} = P_r\cos(\omega t+\phi_r)\\
&\qquad\vdots\\
&-B_{N-1}\dot{X}_{N-1} - K_{N-1}X_{N-1} + M_N\ddot{X}_N + (B_{N-1}+B_N)\dot{X}_N + (K_{N-1}+K_N)X_N = P_N\cos(\omega t+\phi_N)\,,
\end{aligned}$$

where $P_n \cos(\omega t + \phi n)$ represents the rotating unbalance of the system. These equations may be written in matrix form as:

$$\begin{bmatrix} M_1 & 0 & \cdots & & 0\\ 0 & M_2 & \cdots & & 0\\ \vdots & \vdots & M_3 & \cdots & 0\\ \vdots & \vdots & \vdots & \ddots & \vdots\\ 0 & 0 & 0 & \cdots & M_N \end{bmatrix}\begin{Bmatrix}\ddot{X}_1\\ \ddot{X}_2\\ \ddot{X}_3\\ \vdots\\ \ddot{X}_N\end{Bmatrix} + \begin{bmatrix} B_1 & -B_1 & \cdots & & 0\\ -B_1 & B_1+B_2 & -B_2 & \cdots & 0\\ \vdots & & \ddots & & \\ \vdots & & & \ddots & \\ 0 & & & B_{N-1} & B_{N-1}+B_N\end{bmatrix}\begin{Bmatrix}\dot{X}_1\\ \dot{X}_2\\ \dot{X}_3\\ \vdots\\ \dot{X}_N\end{Bmatrix}$$

$$+\begin{bmatrix} K_1 & -K_1 & \cdots & 0 & X_1\\ -K_1 & K_1+K_2 & \cdots & 0 & X_2\\ \vdots & & \ddots & & X_3\\ \vdots & & & \ddots & \vdots\\ 0 & & K_{N-1} & K_{N-1}+K_N & X_N\end{bmatrix} = Re\begin{Bmatrix}P_1e^{\phi_1}\\ P_2e^{\phi_2}\\ P_3e^{\phi_3}\\ \vdots\\ P_Ne^{\phi_N}\end{Bmatrix}e^{i\omega t}$$

or

$$M\ddot{x} + B\dot{x} + Kx = P\cos\omega t = Re\{Pe^{i\omega t}\}$$

Considering now the homogeneous portion of this equation, and introducing the trial solutions:

$$X_1 = x_1 e^{pt}$$

$$X_2 = x_2 e^{pt}$$

$$X_n = x_n e^{pt}$$

gives:

$$\begin{bmatrix} (M_1P^2 + B_1P + K_1) & -(B_1P + K_1) & \cdots & 0 \\ -(B_1P + K_1) & (M_2P^2 + \langle B_1 + B_2\rangle P + \langle K_1 + K_2\rangle) & -B_2P + K_2 \cdots & 0 \\ 0 & \vdots & & \vdots \\ \vdots & & & \\ 0 \cdots & -(B_{N-1}P + K_{N-1}) & & (M_NP^2 + \langle B_{N-1} + B_N\rangle P + \langle K_{N-1} + K_N\rangle) \end{bmatrix} \begin{Bmatrix} X_1 \\ X_2 \\ \vdots \\ X_N \end{Bmatrix} = 0$$

$$\rightarrow [D]\ \{X\} = \{0\}$$

where [D] is the stability determinant. This equals zero for non-trivial solutions.

Expanding this determinant into the characteristic equation gives an expression of the form:

$$A_0 p^{2N} + A_1 p^{2N-1} + A_2 p^{2N-2} + \cdots + A_r p^{2r-1} + \cdots + A_{2N-2} p^2 + A_{2N-1} p + A_{2N} = 0$$

which can be written:

$$\sum_{r=0}^{2N} A_r P^{2N-r} = 0 .$$

The relations between the A_r coefficients in this polynomial and their influence on the stability of the system have been studied by Routh (1877) and Hurwitz (1895). The stability or instability of the dynamical system depends on the nature of the roots of the polynomial. These roots are the roots of the algebraic equation D(p) = 0. In general, the 2N roots will be complex numbers. Since the coefficients A_r are real numbers, the roots of D(p) = 0 occur in complex conjugate pairs of the form:

$$p_r = \alpha_r + i\,\nu_r \;; \quad \bar{p}_r = \alpha_r - i\,\nu_r \;; \quad \text{for } r = 1, 2, \ldots, 2N \quad .$$

If the real parts α_r of all the roots are negative numbers, the solution to this equation will contain decrement factors of the form $e^{\alpha_r t}$, α_r = -ve. If this occurs when the system is disturbed from its equilibrium position (x = 0), it will return to this (stable) position. Alternatively, if one or more of the α_r is positive, any displacement of the system will cause the whirl amplitude to increase indefinitely with time (unstable system).

It is thus clear that for a linear system to be stable, all the roots must have negative real parts. If the system also has on or more of the α_r equal to zero while the other α_r values are negative, then any disturbance will eventually execute sustained oscillations of constant, finite amplitude. Routh has shown that the stability or instability of a system may be determined from a study of the A_r coefficients. Routh's criterion provides a means for determining whether or not an equation has real positive roots without solving it. For example, with an eighth-order system, the procedure is to set up the following array:

A_0	A_2	A_4	A_6	A_8
A_1	A_3	A_5	A_7	
C_1	C_2	C_3	C_4	
D_1	D_2	D_3		
E_1	E_2	E_3		
F_1	F_2			
G_1	G_2			
H_1				

where

$$C_1 = \frac{A_2A_1 - A_0A_3}{A_1} \quad ; \quad C_3 = \frac{A_6A_1 - A_0A_2}{A_1}$$

$$C_2 = \frac{A_4A_1 - A_0A_5}{A_1} \qquad C_4 = A_8$$

$$D_1 = \frac{A_3C_1 - A_1C_2}{C_1} \qquad D_3 = \frac{A_2C_1 - A_1C_4}{C_1}$$

$$D_2 = \frac{A_5C_1 - A_1C_3}{C_1}$$

$$E_1 = \frac{C_2D_1 - D_2C_1}{C_1} \qquad E_3 = D_3$$

$$E_2 = \frac{C_3D_1 - C_1D_3}{D_1}$$

$$F_1 = \frac{D_2E_1 - D_1E_2}{E_1} \qquad F_2 = \frac{D_3E_1 - D_1E_3}{E_1}$$

$$G_1 = \frac{E_2F_1 - E_1F_2}{F_1} \qquad G_2 = E_3$$

$$H_1 = \frac{F_2G_1 - F_1G_2}{G_1}$$

The necessary and sufficient condition of stability is that all of the coefficients of the first column of the array must be positive, i.e.:

$$A_0,\ A_1,\ C_1,\ D_1,\ E_1,\ F_1,\ G_1,\ H_1 > 0 \quad .$$

The following calculation for the steam-excited whirling of a turbine rotor illustrates this procedure:

Consider the steam turbine rotor in fluid-film bearings shown in figures 9 and 10. The rotor is acted on by a steam force G in the direction

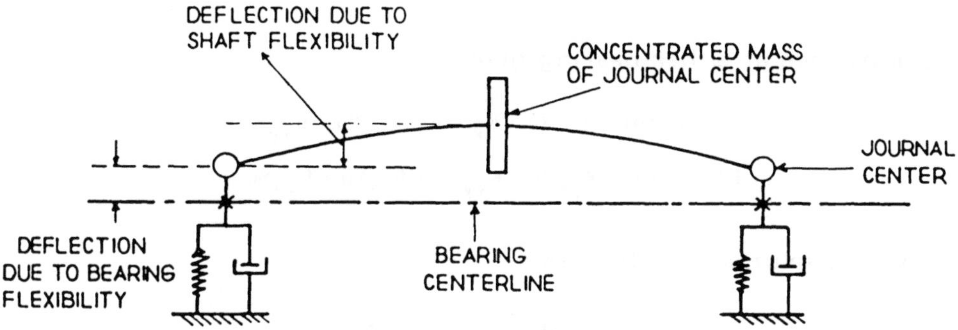

Figure 9 Model of Turbine Rotor System with Steam-Excited Whirl.

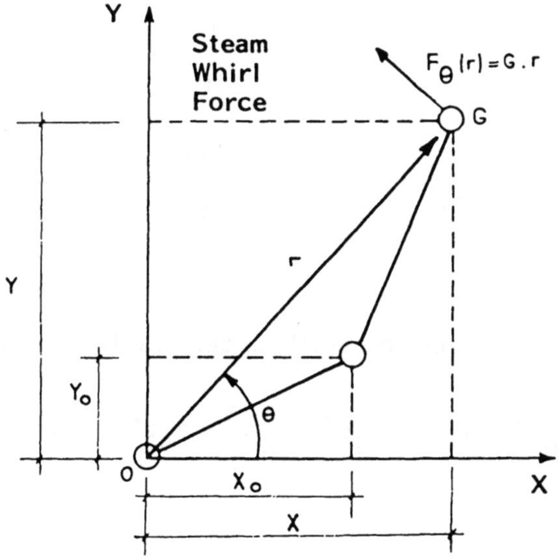

Figure 10 Coordinate System and Steam Forces

of rotation. In the absence of rotor unbalance, the equations of motion of the rotor c.g. are:

$$M\ddot{X} \quad K_s(X \quad Xo) \quad Gr \sin$$
$$M\ddot{Y} \quad K_s(Y \quad Yo) + Gr \cos$$

A force balance at either bearing gives:

$$\tfrac{1}{2} K_s(X \quad Xo) \quad K_{xx}Xo + K_{xy}Yo + B_{xx}\dot{X}o + B_{xy}\dot{Y}o$$
$$\tfrac{1}{2} K_s(Y \quad Yo) \quad K_{yx}Xo + K_{yy}Yo + B_{yx}\dot{X}o + B_{yy}\dot{Y}o$$

introducing solutions of the form:

$$X \quad xe^{st} \qquad Xo \quad xoe^{st}$$
$$Y \quad ye^{st} \qquad Yo \quad yoe^{st}$$

leads to the following determinant of coefficients for system stability:

$$\begin{vmatrix} K_s & G & K_s & 0 \\ G & K_s \quad MS^2 & 0 & \cdot K_s \\ \tfrac{1}{2} K_s & 0 & K_{xx} + \tfrac{1}{2} K_s + B_{xx}S & K_{xy} + B_{xy}S \\ 0 & -\tfrac{1}{2} K_s & K_{yx} + B_{yx}S & K_{yy} + \tfrac{1}{2} K_s + B_{yy}S \end{vmatrix} \cdot 0$$

Expanding this expression gives the sixth-order stability polynomial:

$$a_0 s^6 + a_1 s^5 + a_2 s^4 + a_3 s^3 + a_4 s^2 + a_5 s + a_6 = 0$$

where the constants a_i, i=1,...,6 are given by:

$$a_0 = \bar{M}^2(\bar{B}_{xx}\bar{B}_{yy}-\bar{B}_{xy}\bar{B}_{yx})$$

$$a_1 = \bar{M}^2(\bar{B}_{xx}\bar{K}_{yy} + \tfrac{1}{2}\bar{B}_{xx} + \tfrac{1}{2}\bar{B}_{yy} + \bar{B}_{yy}\bar{K}_{xx} - \bar{B}_{xy}\bar{K}_{yx} - \bar{B}_{yx}\bar{K}_{xy}$$

$$a_2 = \bar{M}^2(\bar{K}_{xx}\bar{K}_{yy} + \tfrac{1}{2}\bar{K}_{xx} + \tfrac{1}{2}\bar{K}_{yy} - \bar{K}_{xy}\bar{K}_{yx} + \tfrac{1}{4})$$
$$+ 2\bar{M}(\bar{B}_{xx}\bar{B}_{yy}-\bar{B}_{xy}\bar{B}_{yx})$$

$$a_3 = 2\bar{M}(\bar{B}_{xx}\bar{K}_{yy} + \bar{B}_{yy}\bar{K}_{xx} + \tfrac{1}{2}\bar{B}_{xx} + \tfrac{1}{2}\bar{B}_{yy} -$$
$$\bar{B}_{xy}\bar{K}_{yx} - \bar{B}_{yx}\bar{K}_{xy}) - \tfrac{1}{2}\bar{M}(\bar{B}_{xx} + \bar{B}_{yy})$$

$$a_4 = 2\bar{M}(\bar{K}_{xx}\bar{K}_{yy} + \tfrac{1}{2}\bar{K}_{xx} + \tfrac{1}{2}\bar{K}_{yy} - \bar{K}_{xy}\bar{K}_{yx} + \tfrac{1}{4}) +$$
$$(\bar{G}^2 + 1)(\bar{B}_{xx}\bar{B}_{yy}-\bar{B}_{xy}\bar{B}_{yx}) - \tfrac{1}{2}\bar{M}(\bar{K}_{xx} + \bar{K}_{yy} + 1)$$

$$a_5 = (\bar{G}^2 + 1)(\bar{B}_{xx}\bar{K}_{yy} + \bar{B}_{yy}\bar{K}_{xx} + \tfrac{1}{2}\bar{B}_{xx} + \tfrac{1}{2}\bar{B}_{yy} - \bar{B}_{xy}\bar{K}_{yx} -$$
$$\bar{B}_{yx}\bar{K}_{xy}) - \tfrac{1}{2}(\bar{B}_{xx} + \bar{B}_{yy}) + \tfrac{1}{2}\bar{G}(\bar{B}_{xy} - \bar{B}_{yx})$$

$$a_6 = (\bar{G}^2 + 1)(\bar{K}_{xx}\bar{K}_{yy} + \tfrac{1}{2}\bar{K}_{xx} + \tfrac{1}{2}\bar{K}_{yy} - \bar{K}_{xy}\bar{K}_{yx} + \tfrac{1}{4})$$
$$- \tfrac{1}{2}(\bar{K}_{xx} + \bar{K}_{yy} + 1) + \tfrac{1}{2}\bar{G}(\bar{K}_{xy} - \bar{K}_{yx}) + \tfrac{1}{4} \quad .$$

The bar over the variables indicates that these quantities have been normalized by the shaft stiffness, i.e.:

$$\bar{K}_{xx} = \frac{K_{xx}}{K_s}, \quad \bar{K}_{xy} = \frac{K_{xy}}{K_s}, \quad \bar{K}_{yx} = \frac{K_{yx}}{K_s}, \quad \bar{K}_{yy} = \frac{K_{yy}}{K_s}$$

$$\bar{B}_{xx} = \frac{B_{xx}}{K_s}, \quad \bar{B}_{xy} = \frac{B_{xy}}{K_s}, \quad \bar{B}_{yx} = \frac{B_{yx}}{K_s}, \quad \bar{B}_{yy} = \frac{B_{yy}}{K_s}$$

Routh's criterion can now be used to investigate the stability of the turbine rotor system. The H.P. steam turbine rotor in figure 9 has the following properties :

Rotor weight	$W_r = 48,046\ lb_f$
First bending critical speed	$N_{ci} = 2349$ rpm
Bearing diameter	$D = 12.0$ in
Bearing length	$L = 8.0$ in
Bearing load	$W = 18058\ lb_f$
Radial clearance	$C = 0.008$ in
Oil viscosity	$\mu = 2.5 \times 10^{-6}$ lb. sec/in^2
Average temperature	$T = 120°$ F
Average speed	$N = 3600$ rpm
Steam force coefficient	$G = 8.16 \times 10^4$ lb/in

The Sommerfeld number is given by:

$$S = \frac{\mu N L D}{W}\left(\frac{R}{C}\right)^2 = \frac{(2.5\times10^{-6})(60)(8\times12)}{18058}\left(\frac{6}{8\times10^{-3}}\right)^2 = 0.47$$

The tilting pad bearing dynamic coefficients are obtained from Lund [39]:

$$\frac{CK_{xx}}{W} = 4.2$$

$$\frac{CK_{yy}}{W} = 0.7$$

$$\frac{CK_{xy}}{W} = \frac{CK_{yx}}{W} = 0$$

$$\frac{C\omega B_{xx}}{W} = 4.3$$

$$\frac{C\omega B_{yy}}{W} = 1.3$$

$$\frac{C\omega B_{xy}}{W} = \frac{C\omega B_{yx}}{W} = 0$$

The procedure now is to seek by trial and error that value of G which will cause the rotor to become unstable at 3600 rpm operating speed. For example, substituting values of G to form the coefficients of the stability polynomial for G (assumed) $= 1.10^6$ lb/in gives:

G=	1000000
A0=	9.66864E-16
A1=	1.16086E-12
A2=	4.58307E-10
A3=	1.03662E-07
A4=	2.44960E-05
A5=	2.09726E-03
A6=	.28669
B1=	3.71968E-10
C1=	3.26653E-08
D1=	9.05557E-06
E1=	1.68392E-04
F1=	.28669

The stability criterion is that the terms A0, A1, B1, C1, D1, E1, F1 in the left column of the array must be of the same algebraic sign for stability. Between $G = 1.5 \times 10^6$ and $G = 1.75 \times 10^6$, we observe:

$G = 1.50 \times 10^6$ lb/in		$G = 1.75 \times 10^6$ lb/in	
G=	1500000	G=	1750000
A0=	9.66864E-16	A0=	9.66864E-16
A1=	1.16086E-12	A1=	1.16086E-12
A2=	4.58307E-10	A2=	4.58307E-10
A3=	1.03662E-07	A3=	1.03662E-07
A4=	2.45742E-05	A4=	2.46250E-05
A5=	2.19112E-03	A5=	2.25212E-03
A6=	.324284	A6=	.332219
B1=	3.71968E-10	B1=	3.71968E-10
C1=	3.26653E=08	C1=	3.26653E-08
D1=	8.96745E-06	D1=	8.91017E-06
E1=	6.54560E-05	E1=	-2.62655E-06
F1=	.314284	F1=	.332219

i.e., the sign of E1 changes between these values. A cross plot of E1 vs G establishes the critical value of the steam coefficient G as 1.72×10^6 lb/in. The stated value of the coefficient for this machine is $G = 8.16 \times 10^4$ lb/in. Thus the machine has ample margin against steam whirl for these conditions.

Computer Calculation of Instability

Several computer programs have been written for the determination of the stability threshold. In these programs, the following techniques have been used to find the instability threshold speed of rotors operating in linear bearings.

a. Modal growth factor analysis, based on sign of real part of complex eigenvalue. Sign change, negative to positive,on any mode signals the onset of instability ;

b. Orbit plots to determine conditions at which the rotor whirl will begin to grow without bound ;

c. Routh-Hurwitz criterion applied to the coefficients of the characteristic polynomial for the system .

Table 2 gives details of some of these programs.

The programs MTI and WPAFB use the complex real part technique and observe sign changes in the modal growth factors. The NASA program and LINK III use the Routh-Hurwitz criterion. An analgous procedure for obtaining whirl orbits is also described in the NASA program report.

The GIBERSON program is used to examine the stability of rotors in nonlinear bearings. This program provides a stability analysis which may include an extremely broad range of system parameters.

All the above calculations consider flexible rotors which operate in damped, flexible bearings. The NASA rotor is a single disk rotor, whereas the ROTDYN rotor may have up to 100 masses, and the CADENSE 25 rotor may have up to 60 mass stations.

	ROTDYN	CADENSE 21,27	MTI-WPAFB	GIBERSON	LINK III
Vendor	Franklin Inst. Research Labs.	MTI-CAD 21,27	MTI-WPAFB	Turboresearch	Com/Code Corp.
Address	Benj. Franklin Pkwy., Phila., PA	968 Albany Shaker Rd. Latham, NY	968 Albany Shaker Rd. Lathan, NY	1440 Phoenix Ave., West Chester, PA	2550 Hunt Ave., Alexandria, VA
Language	Fortran IV	Fortran IV	Fortran IV	Fortran	Fortran IV
Storage, words	65,000 (package)	63,000	2782 Fortran Statements	20K Fortran Statements	
Author	N. Reddi	J. Lund	J. Lund	M. Giberson	P. Y. Chang
Publication	Machine Design 1972	ASME 1967	WPAFB 65-TR-45		
Comment	One Option of Gen. Rotordynamics Program	Specific Program or Part of Package	Specific Programs	Advanced Comprehensive Program Package	One Option of General Package
GENERAL					
Synchronous Response Only?	No	Yes (non-sync. option)	Yes	No	Option
Liquid Mass Rotor?	Yes	No, distributed	Yes	Yes	Yes and distributed
Circular Orbit?	No, computed	No, elliptical	No, elliptical	No, computed	No
Orbital Response Calculation?	Yes	No, optional elliptical display.	No, axes	Yes	No
BEARINGS					
Maximum No.	15	120	25	10	20
Type	Optional	Optional	Optional	Optional	Optional
Linear/non-linear	Linear	Linear	Linear	Non-linear	Linear
Eight-Coeff. Representation	Matrix of coeffs. related to all bearing degrees of freedom	Yes, options for other bearing degrees of freedom	Yes, 8 tilt coeffs. also	No	Yes
Source of Bearing Coeff. Data	Other program or package	Other program or package	Other than program	Routine in program	Other program
ROTOR					
Max. No. Rotor Rotor Stations	100 400 D.O.F.	60		100	Unlimited
Rotor Formulation	Lumped mass finite element	Dist. mass elasticity	Lumped mass	Discrete mass	Lumped and dist. mass elasticity
Shear Effects	Included	Yes	No	No	Yes
Conc. Disks	Yes	Yes	Yes	Yes	Yes
How Input?	Mass, inertia	Mass, inertia	Mass, inertia	Mass, inertia	Mass, inertia
Shaft Taper Included	Yes	No	No		Yes

Table 2 Computer Programs for the Prediction of Instability Threshold

	ROTDYN	CADENSE 21,27	MTI-WPAFB	GIBERSON	LINK III
Gyroscopic Effects:					
Shaft	No	Yes, discrete	Yes, discrete	Yes, discrete	Yes
Disks	Yes	Yes	Yes	Yes	Yes
Translatory Inertia					
Shaft	No	Yes, discrete	Yes, discrete	Yes, discrete	Yes
Disks	Yes	Yes	Yes	Yes	Yes
Axial Thrust	No	No	No		Yes
Internal Damping	No	No	No	Yes	No
PEDESTAL					
Stiffness	Yes	Yes	Yes	Yes	Yes
Damping	Yes	Yes	Yes	Yes	Yes
Mass	Yes	Yes	Yes	Yes	Yes
Discrete/ distributed	Discrete	Both	Discrete	Discrete	Discrete
Input Data Source	Other than program	Other than program	Other than program	Other than program	Other than program
FOUNDATION	Foundation Corresponds to Pedestal	CAD 21, 27	Foundation Corresponds to Pedestal	Special Options: Misalignment, Settlement, etc.	Foundation Corresponds to Pedestal
Stiffness		- Yes			
Damping		- Yes			
Mass		- Yes			
Discrete/Dist		-			
Input Data Source		Other than program			
ENVIRONMENT					
Gas Forces:					
Static	Yes	No	No	Yes	No
Dynamic	Yes	Yes, coeffs.	Yes, coeffs.	Yes	Yes, coeffs.
Gravity	Yes	No	No	Yes	No
Field Force	Yes, linear	No	No	Yes	No
Eccentric Gear Load	Yes	No	No	Yes	No
Arbitrary Rotating Load/Moment	As function of time	Special option	No	Yes, linear or non-linear	No
OUTPUT					
Printout					
-Orbit Details	Yes	Yes	Yes	Yes	Yes
-Response Amp.	Yes	Yes	Yes	Yes	Yes
-Transmitted force	Yes	Yes	Yes	Yes	Yes
-Stress	Yes	Option	No	No	No
Graphics Capability	Yes	Yes	No	Yes	No
-with Prog.	Option	Option	No	Yes	No
-Selected Orbit	Yes	No	No	Yes	No
-Response Amp. vs. Speed	No	Yes	No	No	No
-Transmitted Force	No	Yes	No	No	No

Table 2 Computer Programs for the Prediction of Instability Threshold (continued)

Conclusions

Several different types of rotor instability in bearings have been identified. The experimental data on which these observations are based has been described. Several rotor system models which predict the observed behavior have been demonstrated. A number of computer programs for stability threshold prediction of general multi-mass rotor systems have been detailed, and the general literature of this subject has been specified.

It should be noted that much of this chapter pertains to linear systems, i.e. for systems involving equations with constant coefficients. Good progress can be made using linear stability models, but the prediction of instability of large orbit bounded (or unbounded) whirl orbits is a non-linear problem which must be solved by iterative or analog methods.

References

1. Newkirk, B.L., "Shaft Whipping," General Electric Review (1924).

2. Kimball, A.J., "Internal Friction Theory of Shaft Whirling," General Electric Rep. Vol. 27, p. 266 (1926).

3. Robertson, D., "Hysteretic Influences on the Whirling of Rotors," Proc. Inst. Mech. Eng., Vol. 131, p. 513 (1935).

4. Kimball, A.L., "Measurement of Internal Friction is a Revolving Deflected Shaft," General Electric Review, Vol. 28, p. 554 (1925).

5. Kushul, M.Y., "The Self-Induced Oscillations of Rotors," Consultants Bureau, NY (1964).

6. Gunter, E.G. and Trumpler, P.R., "The Influence of Internal Friction on the Stability of High Speed Rotors with Anisotropic Supports," ASME Paper No. 69-VIBR-2, (1969).

7. Newkirk and Lewis, "Oil Film Whirl, and Investigation of Disturbances due to Oil Films in Journal Bearings," Trans. ASME Vol. 78, p. 21 (1954).

8. Pinkus, O.J., "Experimental Investigation of Resonant Whip," Trans. ASME Vol. 87, p. 975 (1956).

9. Tondl, A., "Experimental Investigation of Self-Excited Vibrations of Rotors due to the Action of Whirling Oil Film in Journal Bearings," Monographs and Memoranda No. 1 (1961).

10. Tondl, A., "Notes on the Problem of Self-Excited Vibrations and Non-Linear Resonances of Rotors Supported in Several Journal Bearings," Wear, 8 (1965).

11. Tondl, A., "Some Problems of Rotor Dynamics," Publishing House of the Czechoslovakian Academy of Sciences, Prague (1965).

12. Lund, J.W., "The Stability of an Elastic Rotor in Journal Bearings with Flexible, Damped Support," Trans. ASME, Journal Appl. Mech., Series E, Vol. 87, p. 911 . (1965).

13. Sternlicht, B., "Gas-Lubricated Cylindrical Journal Bearings of the Finite Length: Part 1-Static Loading," Trans. ASME, Journal Appl. Mech., Paper 61-APM-17 (1961).

14. Sternlicht, B., Poritsky, H., and Arwas, E.B., "Dynamic Stability of Cylindrical Journal Bearings Using Compressible and Incompressible Fluids," First International Symposium on Gas-Lubricated Bearings, Washington, D.C., ONR/ACR-49, p. 119 (October, 1959).

15. Castelli, V. and Elrod, H.G., "Solution of the Stability Problem for 360 Degree Self-Acting, Gas-Lubricated Bearings," Trans. ASME, Jnl. Basic Eng., Series D, Vol. 87, p. 199 (1965).

16. Cheng, H.S. and Pan, C.H.T., "Stability Analysis of Gas-Lubricated, Self-Acting, Plain, Cylindrical Journal Bearings of Finite Length, Using Galerkin's Method," ASEM Paper No. 64-LUBS-5 (1964).

17. Sternlicht, B., "On Stability of Rotors Supported by Gas Bearings," MTI Report prepared for Rotating Machinery for Gas-Cooled Reactor Symposium, April 26, 1962.

18. Taniguchi, O., Tamura, A., and Ano, K., "Experimental Study of Whirl Instability for Externally Pressurized Air Journal Bearings," Bulletin of JSME, Vol. 11, No. 43 (1968).

19. Larson, R.H. and Richardson, H.H., "A Preliminary Study of Whirl Instability for Pressurized Gas Bearings," Journal of Basic Engineering, p. 511 (December, 1962).

20. Kerr, J., "The Onset and Cessation of Half-Speed Whirl in Air-Lubricated Self-Pressurized Journal Bearings," NEL Report No. 237 (1966).

21. Kerr, J., "The Onset and Cessation of Half-Speed Whirl in Air-Lubricated Self-Pressurized Journal Bearings," Paper No. 22, p. 236 (1965).

22. Tondl, A., "The Effect of an Elastically-Suspended Foundation Mass and Its Damping on the Initiation of Self-Excited Vibrations of a Rotor Mounted in Air-Pressurized Bearings," Gas Bearing Symposium, Paper No. 2 (1971).

23. Rieger, N.F., RPI, MTI "Gas-Bearings Design" Course, Section 8 of the Gas Bearing Design Manual, Vol. 1 (1966).

24. Wright, D.V., "Air Model Tests of Labyrinth Seal Forces on a Whirling Rotor," Trans. ASME, Vol. 100, p. 533 (1978).

25. Den Hartog, J.P., "Mechanical Vibrations," Fourth Edition, McGraw-Hill Book Co., Inc, NY (1956).

26. Van der Pol. Phil. Mag. Ser. 7, Vol. 5, p. 18 (1878).

27. Taylor, H.D., "Critical Speed Behavior of Unsymmetrical Shafts," Trans. ASME A71, Jnl. Appl. Mech. A79 (1940).

28. Foote, W.R., Poritsky, H., and Slade, J.J., Jr., "Critical Speeds of a Rotor with Unequal Flexibilities Mounted in Bearings of Unequal Flexibility," Trans. ASME 84 Jnl. Appl. Mech. A77 (1945).

29. Kellenburger, W., "Forced, Double-Frequency, Flexural Vibrations in a Rotating Horizontal Circular Shaft," Brown-Boveri Review, 42, 3, 79 (1955).

30. Dimentberg, F.M., "Flexural Vibrations of Rotating Shafts," Butterworth's Publishing House, London (1961).

31. Tondl, A., "Some Problems of Rotor Dynamics," Publishing House Czechoslovakian Academy of Sciences, Prague (1965).

32. Yamamoto, T., and Ota, H., "On the Unstable Vibrations of a Shaft Carrying an Unsymmetrical Rotor," ASME Paper No. 64-APM-32 (1966).

33. Yamamoto, T., and Ota, H., "The Damping Effect on Unstable Whirlings of a Shaft Carrying an Unsymmetrical Rotor," Memoirs of the Faculty of Engineering, Nagoya University, Vol. 19, No. 2 (November, 1967).

34. Yamamoto, T., and Ota, H., "On the Unstable Vibrations of a Shaft with Unsymmetrical Stiffness Carrying an Unsymmetrical Rotor," ASME Paper No. 68-APM-N (1960).

35. Broniarek, C., Osinski, Z., "Pewien Przypadek Drgan Gietnych Walu Wirujacego Przy Nieliniowej Charakterystyce Materialu," Arch. Bud. Maszyn, t. 10, 4 (1963).

36. Smith, D.M., "Dynamics of Rotors," IUTAM Symposium, Lyngby, Denmark, pp. 524-565 (1974).

37. Lund, J.W., "Some Unstable Whirl Phenomena in Rotating Machinery," Shock and Vibration Digest. (1978).

38. Alford, J.S., "Protecting Turbomachinery from Self-Excited Rotor Whirl," Jnl. Eng. for Power, Trans. ASME 87, pp. 333-344 (1965).

39. Lund, J.W. "Spring and Damping Coefficients for the Tilting Pad Journal Bearing", Trans. ASLE, Vol. 7, (1964).

CHAPTER 2.6

SUPRESSION OF ROTOR INSTABILITY

N.F. Rieger

Procedure for Suppressing Unstable Whirling of Rotors

The following principles are effective for increasing the stable operating speed range of a rotor:

a. raise the lowest critical speed of the system

b. increase the external system damping

The first requirement can be achieved by increasing the bearing radial stiffness or the bending stiffness of the rotor, or both. The second requirement can be met by using a bearing or support type which inherently contains more non-rotating velocity damping. It should be noted that the increase of any damping which rotates with the shaft itself has a stabilizing effect below the bending critical speed, and a destabilizing effect above the bending critical speed.

Stabilization of Rotors in Fluid-Film Bearings

More-Stable Bearing Types

Several investigations have been carried out to assess the relative stability of different bearing types. Pinkus [1] compared the bearing types shown in figure 1. Hydraulically loaded bearings and tilting pad bearings were found to be the most stable, i.e., to give rise to the highest instability threshold speed for the rotor system. Plain cylindrical bearings were the least stable. The self-energizing bearing shown in figure 2 was designed to increase bearing loading by transferring pressure from the lower half of the bearing to the upper half. During testing, the top of the self-energizing bearing was utilized to apply an additional external load to the journal. By closing either valve 1 or valve 2, the bearing could be either self-energized or externally loaded.

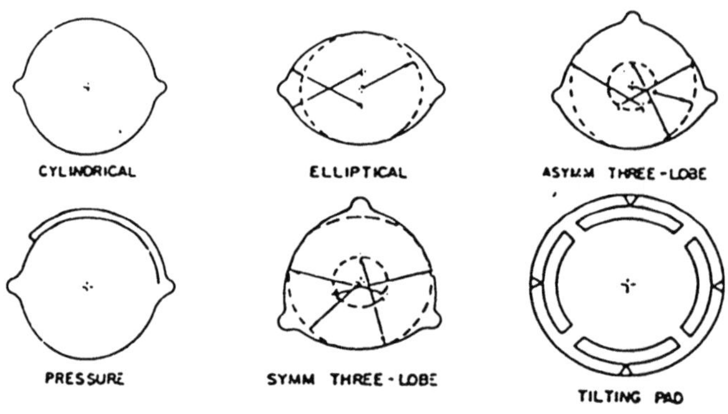

Figure 1 Bearings Tested by Pinkus to Compare Stability [1]

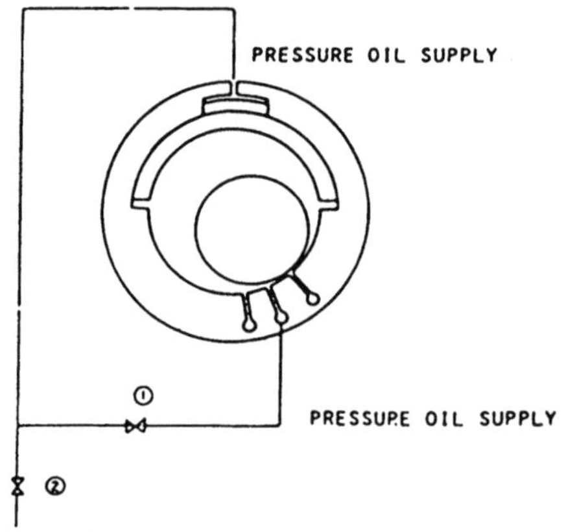

Figure 2 Self-Energizing Bearing Pinkus [1]

Tondl [2] also undertook an experimental investigation to compare several basic types of bearing geometry for their ability to resist initiation of unstable self-excited whirling. Tests were made on cylindrical and elliptical bearings, and also on specific multi-lobe bearing designs, flexible-element bearings, and on two loose-bushing bearing designs. Of these, one design had a cylindrical bushing and the other had a flexible-element loose bushing element.

Tondl's tests showed that self-excited whirling was always very intensive for plain cylindrical bearings. Elliptical bearings exhibited better resistance to initiation of self-excited whirling, i.e., the instability threshold speed was higher than for plain bearings. Tondl found that the multi-lobe bearings shown in figure 3 were also relatively stable. In such bearings

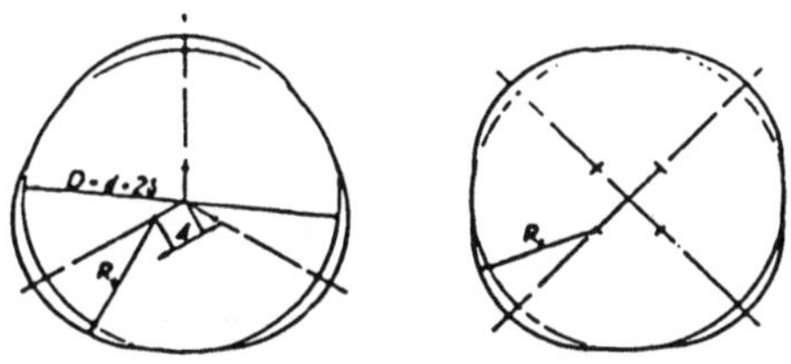

Figure 3 Multi-Lobe Bearings Tested by Tondl [2]

direct contact between the journal and the bushing surface appears to be possible only in small areas. Following the machining of the original bushing, wear was observed in both bushings after a short period of operation.

The three types of flexible element bearings shown in figure 4, were tested by Tondl. These bearings had both three and four identically oriented elements. One additional bearing had three differently oriented elements. With such flexible element bearings, complete suppression of self-excited vibrations was successfully accomplished in certain cases for speeds up to eight times the lowest bending critical speed of the rotor system.

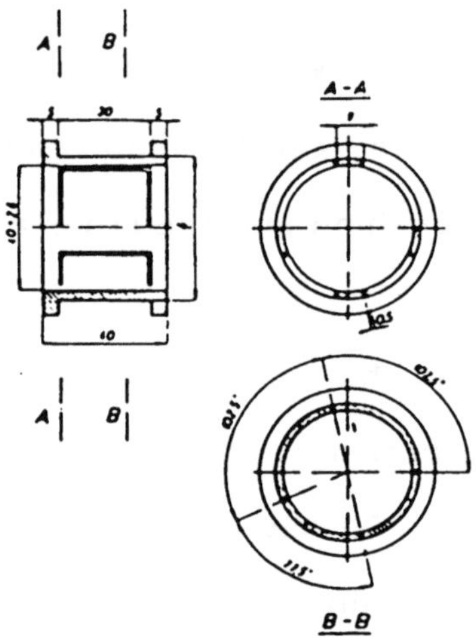

Figure 4 Proportions of Flexible-Element Bearing, Tondl [2]

Tests showed that self-excited whirling of relatively low intensity appeared within a limited range of rotor speeds, the limits of which depended slightly upon the temperature of the outlet oil. Outside this unstable speed range, self-excited whirling could not be induced even by striking the rotor.

Tondl [2] conducted similar tests on bearings with loose bushings. Two differently-designed versions of these bearings were tested. The first had a cylindrical bushing, the second a bushing with a flexible element. Tests were conducted to determine whether this design could be further stabilized by loosening the working bushing and supporting it in a cushion of pressure oil. The objective was both to increase the damping of self-excited vibration and to improve cooling of the working bushing with streaming oil. The tests showed that both goals were attained, particularly in the case of the loose flexible-element bushing where self-excited vibrations occurred only in rare cases, and only when their initiation speed was very high. A picture of these bearings is shown in figure 5. The superior stability of the floating ring (loose bush) bearing is attributed to the large damping effect of the outside cylindrical oil film.

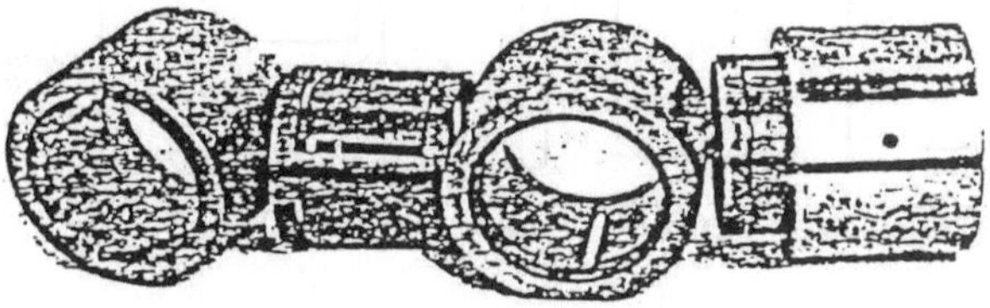

Figure 5 Loose-Bushing Bearing; Loose Bushing is of the Flexible-Element Type, Tondl [2]

Two other bearing types are shown in figure 6. These are the three land bearing and the stabilized bearing, respectively. These bearings also have a practical record for high stability threshold: see Smith [3].

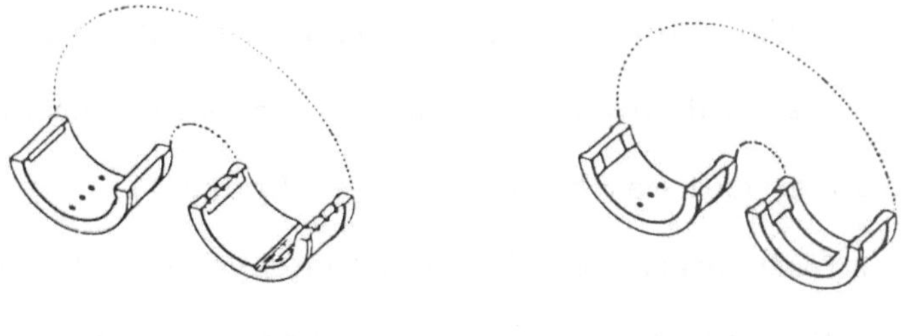

THREE LAND BEARING STABILIZED BEARING

Figure 6 Two Types of Bearings Proposed by Smith [3]

Bearing Groove Modifications

Since the first investigations on instability, it has been found that various modifications to the bearing surfaces favor stable operations at speeds above twice the first critical speed of a rotor system. Newkirk [4] presented the design of a "more stable" grooved bearing. This design is shown in figure 7. The developed bearing surfaces show the system of

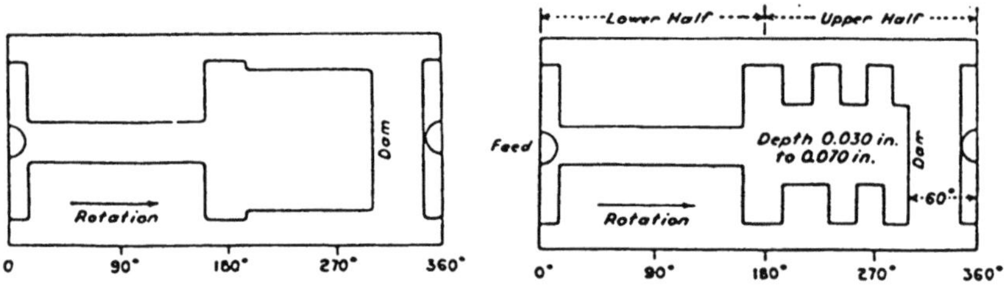

Figure 7 Groove Modifications by Newkirk [4]

grooves which proved most satisfactory during testing. Oil entering at the horizontal joint on the downgoing side of the bearing surface is pumped by action of the shaft through the central peripheral groove to the dam at the end of the groove, where a considerable hydrodynamic pressure builds up, espeçially if the peripheral speed of the journal is high. The upper half of the bearing distributes this pressure, figure 7. If the load on the bearing is sufficient to insure downward pressure under all circumstances, the bands in the upper half may be omitted (see figure 7).

Figure 8 from reference [4] shows the development of hydrodynamic pressure along the peripheral groove as a function of the angular distance from the point of entrance of oil. Figure 9 shows the variation in pressure developed at the dam as a function of journal speed. The oil is supplied to the bearing with a pressure of five or ten pounds per square inch to cause it to enter in sufficient quantity. There is no passage provided for oil exit, consequently end leakage carries away all the oil that passes through the bearing. It is essential that the bearing run full of oil. The amount of oil which must be supplied therefore depends on end leakage, which in turn depends on bearing diameter, clearance, pressure developed by the pumping action of the journal, and viscosity of oil. Figure 10 from reference [4] shows the amount of oil supplied, together with the pressure developed at the dam in tests of a 2.0 in. journal at 30,000 rpm with a 5.5 mil diametral clearance. The grooves must be able to carry enough oil to supply the end leakage, and to cause the bearings to run full.

Sherwood [5] examined the cylindrical bearing design shown in figure 11, and proposed some bearing modifications to increase the stability of the rotor-bearing system. Increasing the bearing oil film pressure is one way to raise the oil film whirl threshold speed. One method is to load the top of the bearing, to increase downward pressure on the load-bearing surface journal, as mentioned previously. The other method involves reducing the load-carrying area of the bearing, i.e., by shortening the bearing length, by decreasing the diameter, or by increasing the size of the damper at the bearing split. Sherwood showed that reducing the length of a bearing has more effect on increasing the bearing oil film pressure than might initially be expected; see reference [5]. In bearings where the L/D ratio is about 1.0, further reduction of length appreciably increases bearing oil film pressure because the effective load carrying area is strongly influenced by end leakage. Figure 11 shows a dammed groove bearing which has also been shown to increase stability. Figure 12 from reference [5] indicates a preference for using larger values of eccentricity ratio to increase the instability threshold speed. Effects of a change in bearing diameter can be explored in several ways. If other factors are constant, decreasing diameter increases bearing pressure, which may be effective in

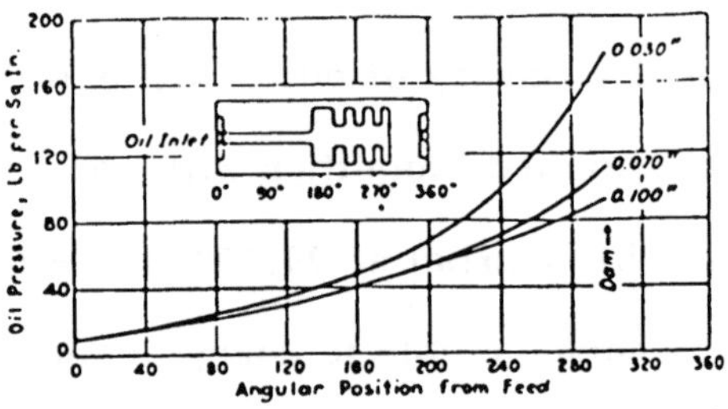

(Bearing, 2 in. X 2½ in.: clearance, 2.75 parts per thousand; peripheral velocity of journal, 15,700 ft (4785 m) per min. Groove system in bearing surface developed is indicated in insert.)

Figure 8 Oil Pressure in Circumferential Groove for Various Depths of Groove, Newkirk [4]

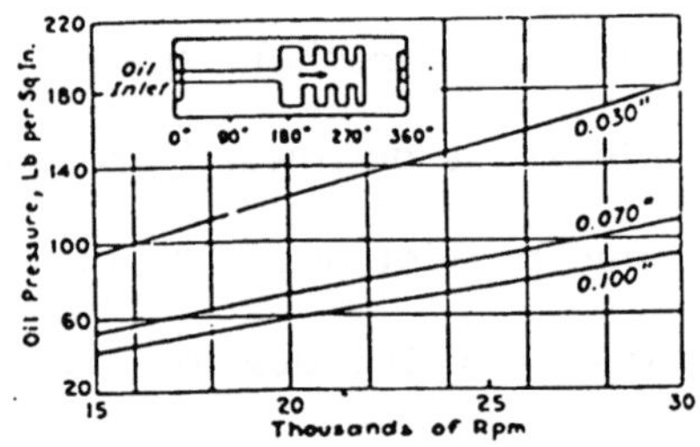

(Bearing, 2 in. X 2½ in.: clearance 2.75 parts per thousand. Groove system in bearing surface developed is indicated in insert.)

Figure 9 Pressure at Dam Versus RPM for Various Depths of Groove, Newkirk [4]

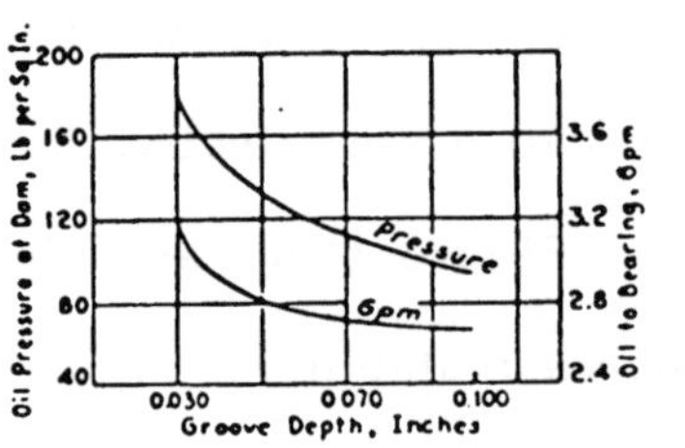

(Shaft running at 30,000 rpm. Bearing 2 in. X 2½ in. Clearance 2.75 parts per thousand.)

Figure 10 Oil Pressure and GPM of Oil to Bearing versus Groove Depth, Newkirk [4]

conferring stability. In considering the effect of changes in bearing clearance, the stabilizing effect of an increase in C/D has already been noted. Therefore, it appears that a decrease in diameter would have a stabilizing

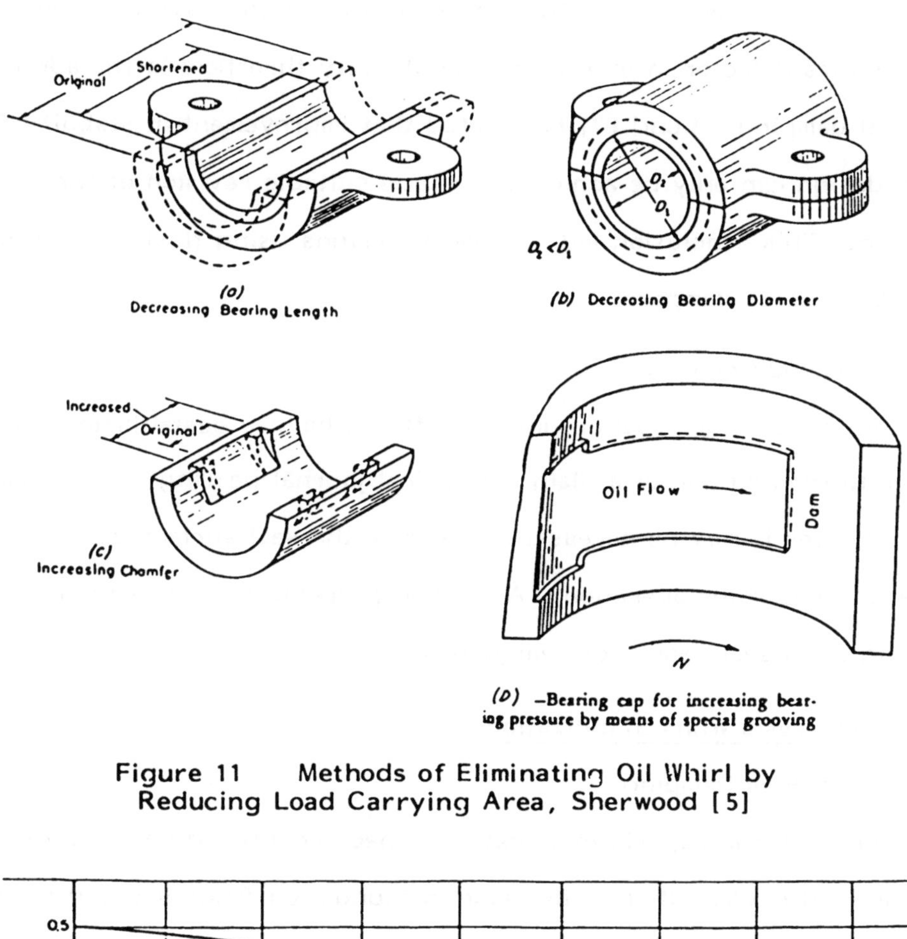

Figure 11 Methods of Eliminating Oil Whirl by Reducing Load Carrying Area, Sherwood [5]

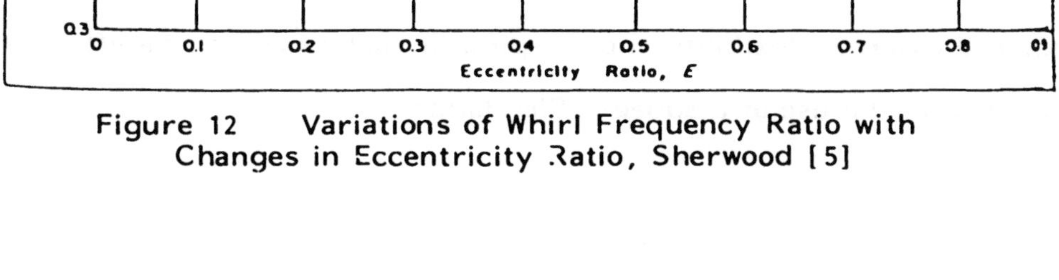

Figure 12 Variations of Whirl Frequency Ratio with Changes in Eccentricity Ratio, Sherwood [5]

effect on the rotor. Most of these conclusions were also confirmed in subsequent tests conducted by Hori [6].

Foundation Modifications

Later experiments and theoretical investigations have determined that the significant factor in increasing the stability threshold is the addition of external damping. Lund [7] has stated that improvement in stability for a given design can only be achieved by some suitable revision of the support damping. This conclusion led to several designs using flexible mountings and film squeeze dampers.

Seal Modifications

An annular gas seal around a shaft can have a destabilizing effect similar to the addition of a plain cylindrical journal bearing at that location. In such cases it may be necessary to modify the seal surface by grooving to achieve stable rotor operation. A common procedure consists in changing an existing plain seal design to a labyrinth seal.

Hysteresis Whirl Stabilization

Fluid-Film Damping

From all the experiments that have been conducted on rotor hysteresis instability it is clear that an increase in foundation flexibility and foundation damping can significantly increase the stability threshold speeds of lightly damped hysteretic rotors. This fact was identified by Newkirk [8] as early as 1925, and has been amply verified by subsequent experience. The required external flexibility and damping in such cases can be efficiently provided by the use of a squeeze film damper.

Raising the System Critical Speed

As all rotor instabilities are associated with the lowest critical speed of the rotor system, raising the first critical speed tends to increase the instability threshold speed. This can be done by "stiffening" the system, and also by decreasing the rotor mass. Increasing the bearing radial stiffness and increasing the transverse stiffness of the rotor are also effective in raising the system critical speed. Imposing a small amount of angular misalignment in the bearings exerts a similar stiffening effect on the system.

Rotor Structural Changes

Certain structural changes to the rotor have been shown to promote increased stability in the case of hysteretic problems. It has been shown by Newkirk [8], Kimball [9], Robertson [10] and others, that shrink fits may cause whirl instability. However, these investigations raised the threshold speed by increasing the shrink pressure and decreasing the shrink length (see figure 13).

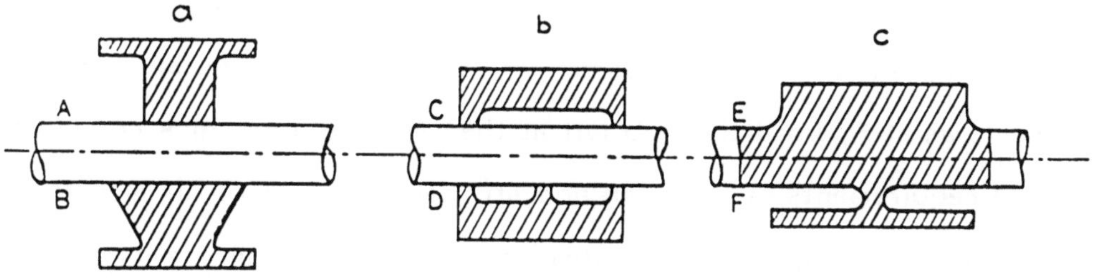

Figure 13 Types of Hubs and Bosses Tested by Robertson [10]

For long shrink fits such as compressor wheels and impeller spacers, it is important that these components should be undercut along the central region of the inner bore so that the contact area is restricted to the ends of the shrink fit.

Robertson [10] also suggested several designs of hubs and bores which have been effective in reducing internal friction effects. These are shown in figure 13. The length of the hub on the shaft should be no longer than necessary for secure fixing. If, for some reason, a greater length is necessary, the hub should then be undercut as shown at position A in figure 13. The construction shown at position C in figure 13, with bearing strips at the ends only, should be avoided in rotors which are run above their bending critical speed. With this pattern the central portion of the shaft may bend more than the hub, thereby increasing the amount of rubbing, or of "working" at the ends. If a long hub base is necessary on the shaft, and if it is not convenient to fit the hub along its whole length, it should have a bearing strip at the middle as well as at the ends, in order to stiffen the shaft. This is shown at position D. The hub should be carried on a base which is solid with the shaft, and has ample fillet radii in the corners. The base stiffens the shaft locally and should have as large a diameter as possible (see E in figure 13). Where possible the base should be undercut, as at F, until the web left at the center is just thick enough to meet the requirements of strength and stiffness. In one of his models, Kimball [9] successfully employed this arrangement to eliminate hysteretic effects from the fit of a flywheel on a shaft.

Squeeze Film Dampers

Many investigators have concluded that stability of rotors operating above their critical speed can only be ensured by an adequate selection of support stiffness and support damping. Lund [11] and others have investigated certain designs of flexible bearing mountings with squeeze film dampers. Figure 14 shows an example of a flexibly-mounted squeeze film damper. This arrangement is called a damped flexible support. Slots have been cut into

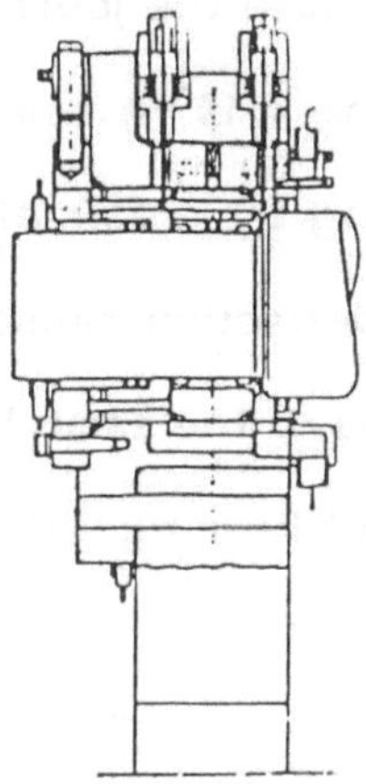

Figure 14 Cross Section of Bearing Arrangement with Damped Flexible Support

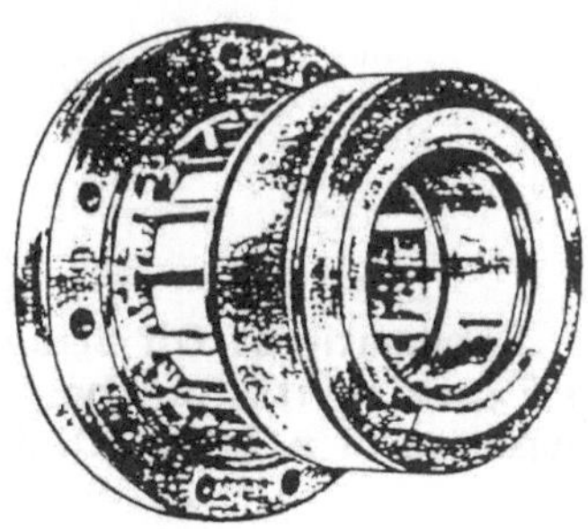

Figure 15 View of Flexible Support Structure for Journal Bearing with Squeeze Film Damper

the bearing shell, leaving 16 axial spokes to support the bearing sleeve, figure 15. The composite radial stiffness of the spoke in the design tested was 4.10^7 n/m. The clearance space between the bearing sleeve and the force gage ring shown is used as a squeeze film damper. This damper is fed with oil under pressure from eight equally spaced holes in the midplane, and is sealed off at the ends by piston ring type seals. The damper oil is the same as that used to lubricate the journal bearing. The damping coefficient was calculated to be 5.10^4 n.s/m. The damper bearing is centered in the force gage ring by a pair of differential adjustment screws to compensate for the static deflection caused by the load on the bearing.

Cunningham [12] has used the same kind of device with the same principle, but applied to a ball bearing, as shown in figures 16 and 17.

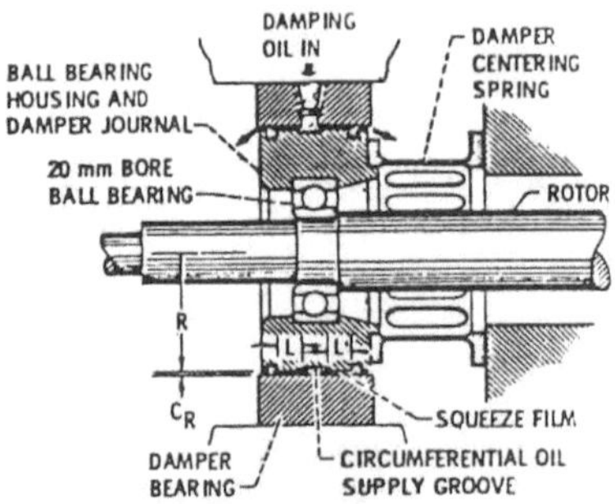

Figure 16 Schematic of Oil Squeeze Film Damper used in Experiments for the Steady-State Response of a Three Disk Rotor

This design has been used to minimize the steady-state unbalance response of a three-disk flexible rotor. During this investigation, the oil squeeze-film dampers of this design have been proven successful in reducing the amplitudes of motion for nonsynchronous whirl.

DeChoudhury and Gunter [13] reported on another squeeze film damper bearing design. When operated on a rigid support system, the rotor was unstable at an operating speed of 11,300 rpm. The whirling component had an angular velocity of 4300 rpm. Placing the rotor on a squeeze film damper support centered by rubber "O" rings effectively stabilized the rotor with a

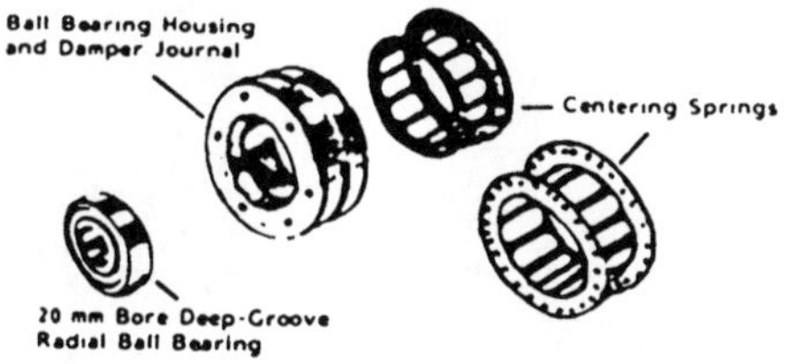

Figure 17 Components of a Flexible Damped Support

small whirling component. Figures 18 and 19 show rotor orbits for both cases. After a comprehensive investigation, these investigators stated that for a system to be stable there is an optimum support damping for any given support stiffness.

Some other experiments have been conducted by Darlow and Smalley [16] on the use of stabilizing dampers on super-critical power transmission shafts (figure 20). The hollow test shaft was 3.66 m long and 7.62 cm in diameter with a 3.175 mm wall thickness and was supported at both ends by disk-type flexible couplings. To minimize support damping effects, the couplings were attached to hand-mounted spindles. When first tested without such dampers, at speeds above the first critical, the shaft became unstable at 1700 rpm with a whirl frequency equal to the first critical speed. These results are shown in figure 21.

Based on rotor dynamic analysis, a stiffness of 7.10^3 n/m was adopted, using an "O" ring as the spring member. The squeeze-film damper was

Rotor Orbits of Turbo Compressor Before and After Stabilization N=11,300 RPM, First Critical N_1 =4,300 RPM

Figure 18

Unstable Rotor Orbit, Discharge Pressure P_D=175 PSIG, Large N_1 Component, DeChoudhury [13]

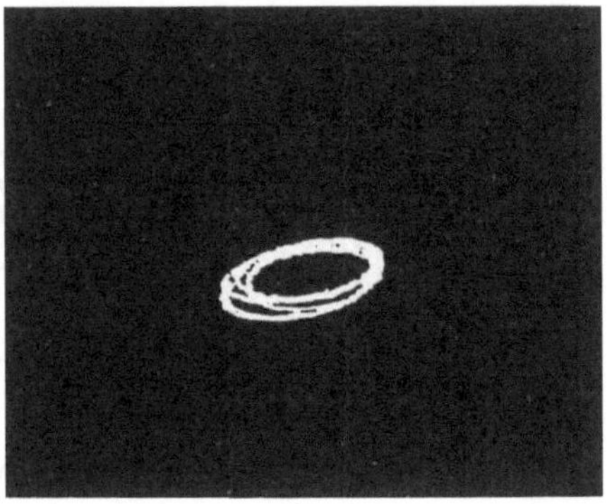

Figure 19

Stabilized Rotor Orbit, P_D=650 PSIG, Small N_1 Component, DeChoudhury [13]

designed to be a sealed damper with no circulation of the oil. In this way, the damper would require no support hardware (such as oil supply pumps) which would prohibit its use in helicopter or other space- and weight-limited applications. A picture of the test rig for this modern damper is shown in figure 20. The damper is evacuated, filled and sealed, and pressurized by the use of a bladder. The damper is predicted to generate less heat than the bearings, so the dissipation of heat is not expected to be a serious problem. The "O" ring retainers are radially adjustable, so that the damper may be centered manually, in order to compensate for static deflection of the "O" rings. The damper was designed to achieve a level of damping in the range of 8750 n-s/m for silicone oil with a viscosity of about 80 centistokes, using the short bearing theory and assuming no cavitation. The radial clearance is 0.635 mm and the length and diameter are 7.67 and 10.16 mm respectively. Although the damper was designed to allow for two "O" rings, it was used initially with just one, in order to obtain a low parallel

Figure 20 Squeeze Film Damper Installed in Supercritical Power Transmission Shaft Test Rig

support stiffness of 7×10^3 n/m for a continuously supported "O" ring. Testing of this device showed that instability was successfully eliminated throughout the range of operating speeds to 7000 rpm.

The conclusions reached by Darlow and Smalley during this investigation were that this damper was extremely effective in eliminating an incipient

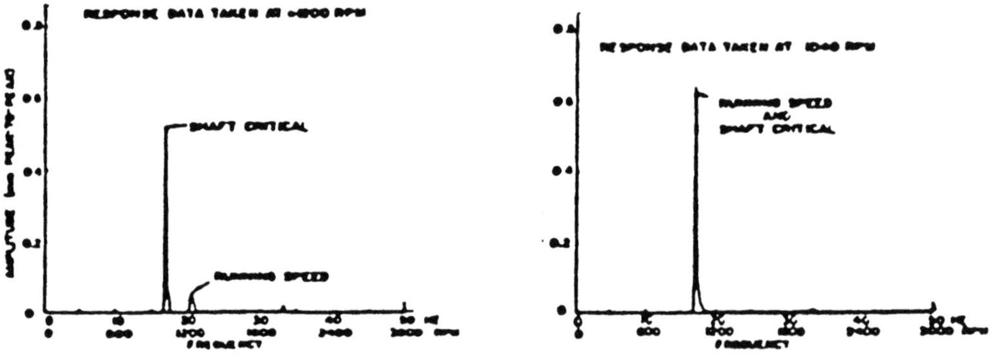

Figure 21 Frequency Spectrum Plot of Test Shaft Vibration with No Damper Running above First Critical Speed

instability in the test shaft. With the damper installed, the test shaft, which had been marginally balanced through one critical speed without the damper, was easily balanced through three critical speeds.

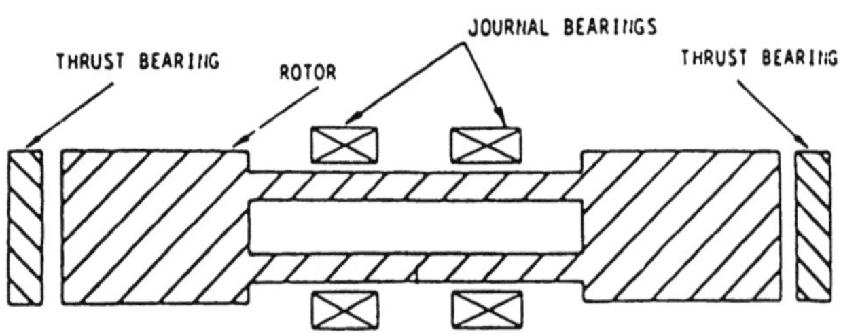

Figure 22 Effect of Thrust Bearing on Instability

Earlier tests on similar systems consisting of long, thin shafts in end bearings were conducted by Voorhees and Meechan [23]. These investigators found it impossible to adequately balance such systems for smooth operation. They concluded that it was essential to provide a suitable damper in order to run their shafts through several critical speeds.

Other Stabilization Procedures

Misalignment of Bearings

Misalignment of bearings has been demonstrated to be an effective method for rotor stabilization by Newkirk [14], Pinkus [1], Hori [6], and Tondl [15]. Misalignment first increases the critical speed by stiffening the bearing then raises the fluid pressure inside the bearing, giving better squeeze-film effects. Misalignment will also cause the bearing to operate at a greater eccentricity ratio, which is always consistent with greater stability, and with smaller effective diametral clearance.

Gashed Shaft

Rotors cross-sections with different principal stiffness values are another cause of rotor instability. An expedient frequently used to improve the stability of such systems is to "gash" the shaft, i.e., to machine a series of slots in the cross section. This reduces the higher principal stiffness and serves to obtain the same shaft stiffness in all transverse directions. For a shaft carrying a keyway, it is suitable to have another keyway 90° apart from the first one in order to avoid the dissimilar shaft stiffness instability

Thrust Bearing

Newkirk [14] recognized that thrust bearings are a very good way of suppressing the conical mode of instability. This was later verified by Cundiff [17]. Since a thrust bearing exhibits no radial stiffness or radial damping properties, the other shaft dynamical properties of critical speed in the later or radial modes will not be affected. The threshold speed is raised by increasing the system damping. The stabilizing effect of a thrust bearing is shown in figure 22.

Computer Programs for Instability Threshold Prediction

General purpose computer programs for instability threshold prediction have been written by Lund [18], Giberson [10], Gunter [20], Pan [21], and others. A specific study of the effects of damping, seal effects and shaft hysteresis on the stability of a sample two-mass rotor in end bearings has been made by Rieger and Thomas [22].

Consider the two-mass flexible rotor in fluid-film bearings shown in figure 23. The disks are rigid and have no rotary or polar inertia. The shaft is massless and may be of any profile provided the transverse stiffness, K, is known or may be found. Both bearing forces are identical and

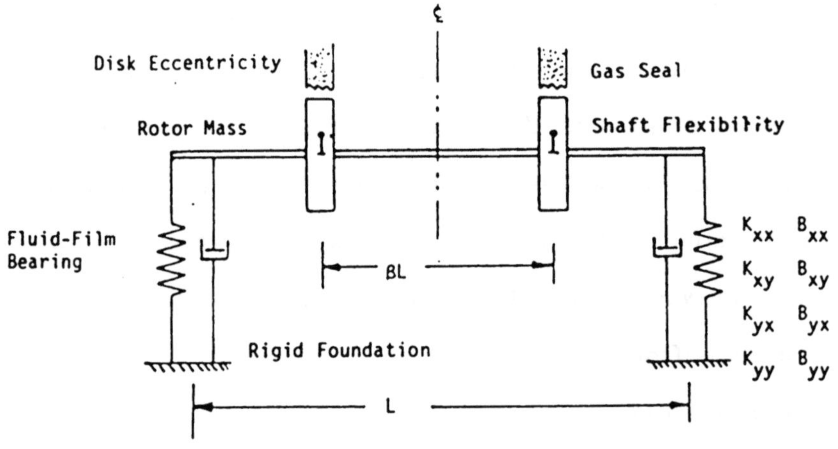

Figure 23 Two Mass Flexible Rotor in Fluid-Film Bearings

linear with rotor displacement and velocity from the equilibrium position of the journal in each bearing. The eight-coefficient bearing force representation includes both direct and cross coupled effects, as follows:

$$F_x = K_{xx}X + K_{xy}Y + B_{xx}\dot{X} + B_{xy}\dot{Y}$$

$$F_y = K_{yx}X + K_{yy}Y + B_{yx}\dot{X} + B_{yy}\dot{Y}$$

Gas seals are located at both disks. Forces arising from disk eccentricity with stator clearance are given by:

$$F_s = iGr \qquad i = \sqrt{-1}$$

where G is the gas seal coefficient and r is the whirl radius. Structural hysteresis effects occur due to the shrink fit of the disk on the shaft, and are given by:

$$F_h = H\xi$$

where H is the hysteresis coefficient and ξ is the whirl velocity in rotating coordinates. The rotor is axisymmetric and system effects are symmetric about mid-span. The equations of motion for either disk, including the shaft, seal, and hysteresis effects, are:

$$M\ddot{X}_1 + H\dot{X}_1 + K(X_1 - \beta X) + (G + \omega H)Y_1 = Ma\omega^2 \text{ Cos } \omega t$$

$$M\ddot{Y}_1 + H\dot{Y}_1 + K(Y_1 - \beta Y) - (G + \omega H)X_1 = Ma\omega^2 \text{ Sin } \omega t + Mg$$

A force balance at either bearing gives:

$$K(X_1 - \beta X) = K_{xx}X + K_{xy}Y + B_{xx}\dot{X} + B_{xy}\dot{Y}$$

$$K(Y_1 - \beta Y) = K_{yx}X + K_{yy}Y + B_{yx}\dot{X} + B_{yy}\dot{Y}$$

To obtain a solution assume that:

$$X = x_0\, e^{st}$$

$$Y = y_0\, e^{st}$$

where $$s = \alpha + i\nu$$

Substituting in the above expressions and simplifying gives:

$$\begin{vmatrix} (Ms^2+Hs+K)\Lambda^*_{xx}+(Gs+H\omega)\,\Lambda_{yx}-K^2\beta & Ms^2+H\omega+K)\,\Lambda_{xy}+(Gs+H\omega)\Lambda^*_{yy} \\ (Ms+Hs+K)\Lambda_{yx}-(Gs+H\omega)\Lambda^*_{xx} & Ms^2+Hs+K)\Lambda^*_{yy}-(GS+H\omega)\Lambda_{xy}-K^2\beta \end{vmatrix} = 0$$

where
$$\Lambda^*_{xx} = K_{xx} + \beta K + sB_{xx}$$

$$\Lambda_{xy} = K_{xy} + sB_{xy}$$

$$\Lambda_{yx} = K_{yx} + sB_{yx}$$

$$\Lambda^*_{yy} = K_{yy} + \beta K + sB_{yy}$$

To obtain the roots of this determinant a convenient procedure is to expand and form the corresponding polynomial in S, viz:

$$A_1S^6 + A_2S^5 + A_3S^4 + A_4S^3 + A_5S^2 + A_6S + A_7 = 0$$

The coefficients A_i of this polynomial are complex algebraic expressions.

A computer program was written to solve the stability polynomial for the lowest complex eigenvalue. Values of M, K, G, and H were held constant for given cases. The program increments as explained earlier. Thus corresponding values of the speed-dependent bearing coefficients are selected from an input table. Instability occurs when the real part of the eigenvalue changes sign from negative to positive.

Sample Calculation

The compressor rotor shown in figure 24 can be used to demonstrate the use of this procedure. The rotor weighs 1000 lb (M=1.39 lb sec^2/in per bearing) and operates in two 4.0 diameter by 2.0 in. long plain cylindrical bearings of diametral clearance 0.004 in. Oil of viscosity 22 cP at 110 °F and 4.84 cP at 200°F is supplied to each bearing at a rate of 1.25 gal/min at operating speed under 20 lb/in^2 gage pressure at 120°F.

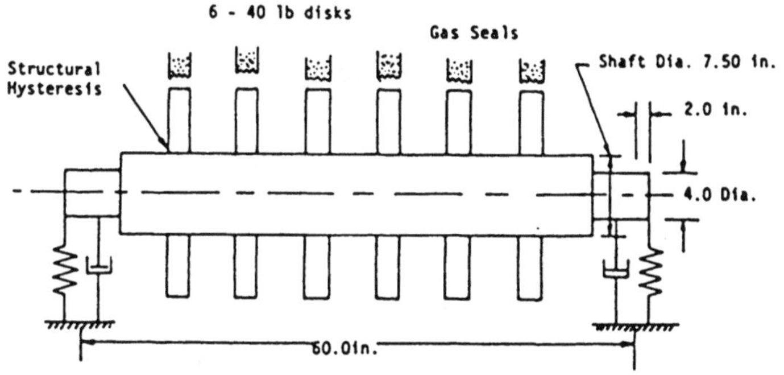

Figure 24 Compressor Rotor in Bearings with Seat and Structural Hysteresis Effects

Parametric studies of the relative significance of bearing, seal, and structural hysteresis effects on the compressor rotor system stability were made. The system threshold speed without seals or hysteresis (i.e., bearing instability) was first determined to be 9200 rpm with the rotor whirling in its translatory mode. The gas seal coefficient was then introduced and increased in value. A decrease in threshold speed was observed. Lastly, the effect of increasing the structural hysteresis was studied, with gas seal forces operating. For known threshold values, the value of H was added and new threshold speeds were found. The results of these studies are shown in figure 25 in which the system threshold speed N_T is plotted

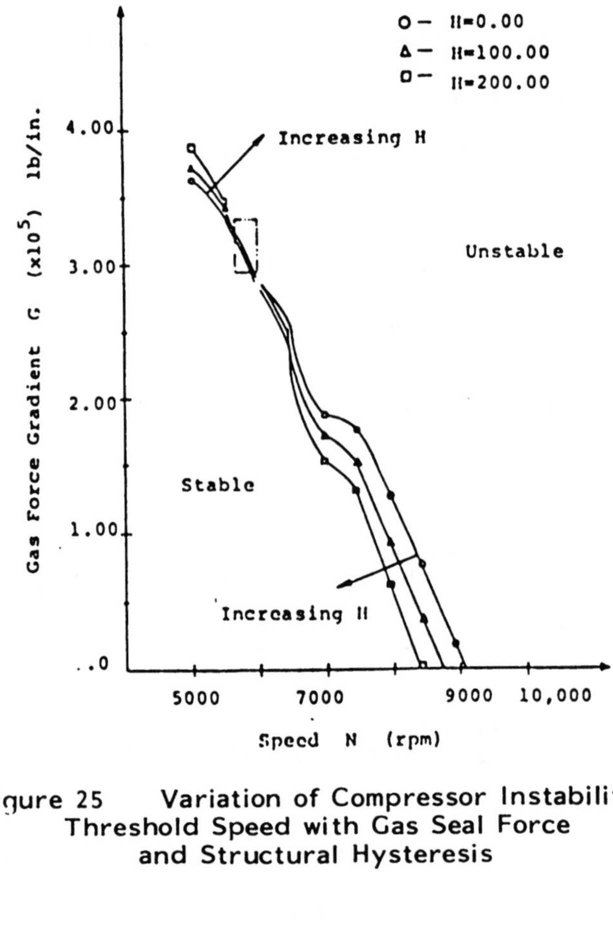

Figure 25 Variation of Compressor Instability Threshold Speed with Gas Seal Force and Structural Hysteresis

against gas seal coefficient G, with structural hysteresis coefficient H as parameter. It is evident that the threshold of instability is depressed with increase of gas seal coefficient. The effect of hysteresis is mixed: below the system fundamental critical speed (5850 rpm) it raises the threshold speed, and above 5850 rpm it depresses the threshold speed, as anticipated from previous experience. Structural hysteresis has a minor effect on system instability threshold compared with gas seal effects where the rotor system incorporates fluid film bearings.

An analog stability study of this compressor rotor system was also undertaken. The analog circuit is shown in figure 26. Bearing coefficients corresponding to each speed were introduced by hand adjustment of circuit

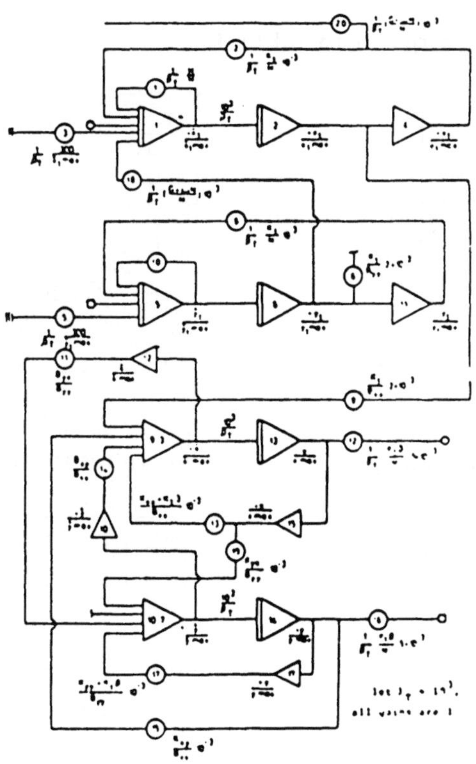

Figure 26 Linear Analog Circuit [22]

resistors. The gas seal coefficient G and the hysteresis coefficient H were held constant for each condition and the corresponding threshold speed was found by observing orbit growth or decay at selected speeds.

The influence of gas seals $G=1.50 \times 10^5$ lb in and hysteresis H=200 lb in are shown in figures 27 and 28 below and above the threshold speed. The analog orbits show that the tendency towards instability is a steadily developing trend. Below the threshold speed the rotor takes increasingly longer periods of time to establish a stable orbit. Above the threshold, orbit growth becomes increasingly more rapid as speed is further increased.

A comparison of threshold speed results obtained by both the digital and analog methods are given in table 1.

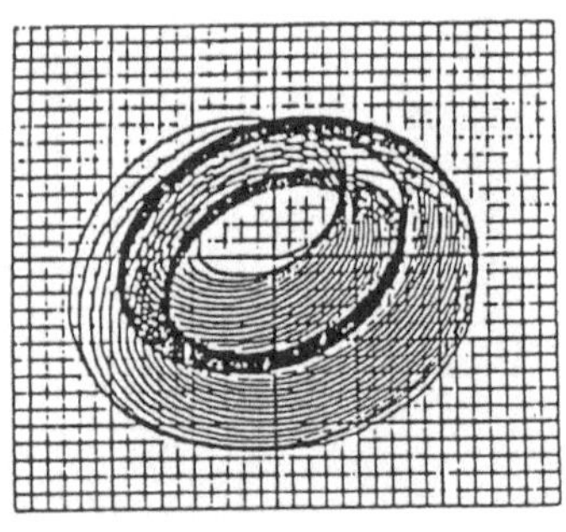

(a) N = 9100 rpm. Below Threshold Speed.

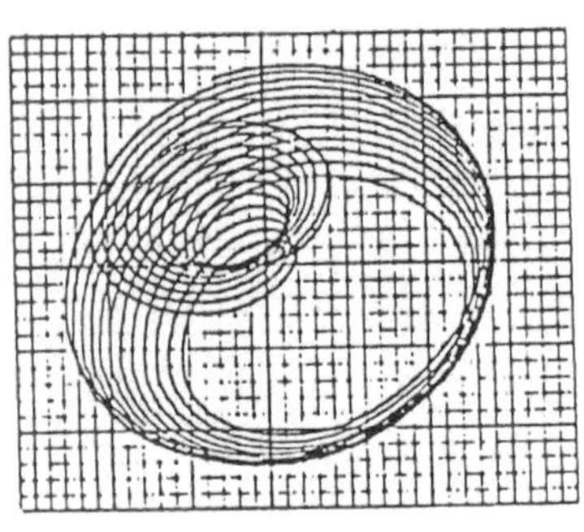

(b) N = 9200 rpm. Slightly Above Threshold Speed.

Figure 27
Journal Whirl Orbits for G+0 and H=0, Linear Short Bearings Analysis

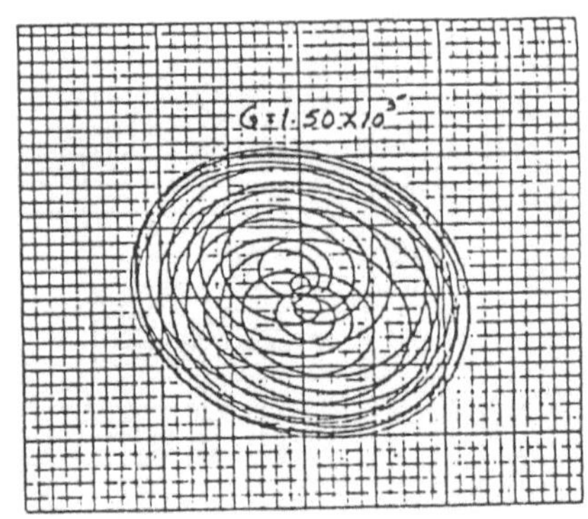

(a) $G = 1.50\ 10^5$ lb/in. Below Threshold Speed.

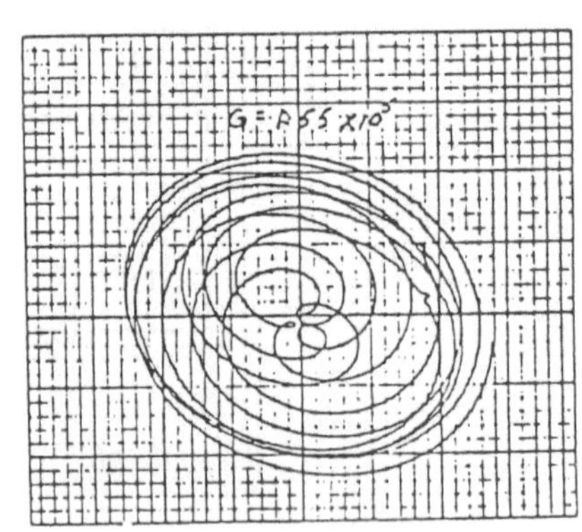

(b) $G = 1.55\ 10^5$ lb/in. Above Threshold Speed.

Figure 28
Journal Whirl Orbits where H=200 and N=7000 RPM

Comparison of Analog and Digital Threshold Speed Values					
Analog			Digital		
G	H	Threshold Speed	G	H	Threshold Speed
lb/in x 10^5	lb sec/in	rpm.	lb/in. x 10^5	lb sec/in	rpm.
0	0	9100-9200	0	0	9100-9200
0	100	8700-8800	0	100	8800-8900
0	200	8500-8600	0	200	8500-8600
3.60-3.65	0	5000	3.60-3.65	0	5000
2.85-2.90	0	6000	2.85-2.90	0	6000
1.85-1.90	0	7000	1.85-1.90	0	7000
1.20-1.25	0	8000	1.25-1.30	0	8000
.15- .20	0	9000	.15- .20	0	9000
3.70-3.75	100	5000	3.70-3.75	100	5000
2.80-2.85	100	6000	2.85-2.90	100	6000
1.75-1.80	100	7000	1.70-1.75	100	7000
.90- .95	100	8000	.90- .95	100	8000
3.85-3.90	200	5000	3.85-3.90	200	5000
2.75-2.80	200	6000	2.80-2.85	200	6000
1.50-1.55	200	7000	1.50-1.55	200	7000
.60- .65	200	8000	.60- .65	200	8000

Table 1 Variation of Threshold Speed with Compressor System Effects

Conclusions

Of the two procedures which are known to be effective in the suppression of instability, i.e., raising the critical speed and increasing the negative real part of the whirl exponent, the second procedure is the easiest to achieve in practice. This procedure can be instituted most efficiently by simultaneously adding external flexibility and damping in such a way that the damping properties of the system support are optimized.

The squeeze film damper is an effective solution for many rotor instability problems. When this device cannot be used, various other solutions may be possible, such as using a more stable bearing type, by bearing modifications, such as grooving, slight misalignment, small structural changes, etc.

As a general rule in the stability problem, every change which makes the rotor support system undergo additional forcing motion (bearing misalignment, larger bearing eccentricity, etc.) is a step towards stabilizing the rotor by raising the instability threshold speed.

Table 2 provides a summary listing of various kinds of rotor instability and methods which have been used effectively to overcome them.

INSTABILITY	DETAILS	SUPPRESSION PROCEDURES
Fluid Film	Bearing whirl or whip	- Add support damping and flexibility - Change bearing type. Stable bearings are: • Tilting pad (most stable) • Floating ring • Pressure dam bearing • Axial groove • Three lobe • Elliptical (less stable) - Pressurize the bearing upper half - Increase the journal operating eccentricity - Increase the critical bending speed of the rotor - Decrease the bearing length
	Gas seal whirl and steam whirl	- Use labyrinth-type seal - Increase bearing operating eccentricity - Improve seal operating concentricity
Hysteretic instability	Instability caused by shrink fit hysteresis	- Use small length,tight shrink fits - Use a rigid rotor construction - Increase external damping and support flexibility - Use squeeze film damper support - Use previous design that has been proven efficient - Undercut long disk contact surfaces
	Instability caused by material hysteresis	- Add support damping and flexibility - Raise bending critical speed - Use squeeze-film support
Parametric Instability	Dissimilar shaft stiffness	- Increase bearing damping - Relieve greater stiffness with circumferntial gashes - Stiffen shaft. Raise bending critical speed.
	Shaft carrying an unsymmetrical rotor	- Increase bearing damping - Re-establish circumferential symmetry (e.g. two rotors 90° apart , 3-bladed propeller, etc.)

Table 2 Causes of Instability and Methods of Suppression

References

1. Pinkus, O.J., "Experimental Investigation of Resonant Whip," Trans. ASME, Vol. 87, p. 975 (1956).

2. Tondl, A., "Notes on the Problem of Self-Excited Vibrations and Non-Linear Resonances of Rotors Supported in Several Journal Bearings," Wear, 8 (1965).

3. Smith, D.M., "Journal Bearings in Turbomachinery," Chapman and Hall, Ltd. (1970).

4. Newkirk, B.L. and Grobel, L.P., "Oil Film Whirl - A Non-Whirling Bearing," Trans. ASME, Vol. 56, p. 607 (1934).

5. Sherwood, R.S., "How to Prevent Oil Film Whirl in Journal Bearings," Machine Design, p. 163 (December, 1953).

6. Hori, Y., "A Theory of Oil Whip," Journal of Applied Mech., p. 189 (June, 1959).

7. Lund, J.W., "Stability and Damped Critical Speeds of a Flexible Rotor in Fluid-Film Bearings," Journal of Engineering for Industry, Trans. ASME, p. 509 (May, 1976).

8. Newkirk, B.L., "Shaft Whipping," General Electric Rev. (1924).

9. Kimball, A.J., "Internal Friction Theory of Shaft Whirling," General Electric Report, Vol. 17, p. 266 (1926).

10. Robertson, D., "Hysteretic Influences on the Whirling of Rotors," Proc. Inst. of Mech. Eng., Vol. 131, p. 513 (1935).

11. Lund, J.W. and Tonnessen, J., "Some Experiments on Instability of Rotors Supported in Fluid-Film Bearings," ASME Paper No. 77-DET-23 (1977).

12. Cunningham, R.E., "Steady-State Unbalance Response of a Three-Disk Flexible Damped Support," ASME Paper No. 77 DET-102.

13. DeChoudhury, P. and Gunter, E.J., "Dynamic Stability of Flexible Rotor-Bearing Systems," Report No. ME-6060-106-700, University of Virginia (1970).

14. Newkirk and Lewis, "Oil Film Whirl, an Investigation of Disturbances due to Oil Films in Journal Bearings," Trans. ASME, Vol. 78, p. 21 (1951).

15. Tondl, A., "Some Problems of Rotor Dynamics," Publishing House of the Czechoslovak Academy of Sciences, Prague (1965).

16. Smalley, A.J., Darlow, M.S., Paper included in "Topics in Fluid Film Bearings and Rotor Bearing System Design and Optimization," ASME (1978).

17. Cundiff, R., "Thrust Bearing Effects on Rotor Stability," MTI Report No. AFAPL-TR-65-45, Part IX.

18. Lund, J.W., "The Threshold of Instability of a Flexible Rotor in Fluid-Film Bearings," MTI Unpublished Technical Memorandum (1971).

19. Giberson, N.F., "Non-Linear Dynamics and Stability of Rotor-Bearing Systems," Rotordynamic Symposium and Workshop, Rochester Institute of Technology (1975).

20. Gunter, E.J., "The Influence of Internal Friction on the Stability of High Speed Rotors with Anisotropic Supports," ASME Paper No. 69.

21. Pan, C.H.T., "On the Time Dependent Effects of Self-Acting Gas Bearings," Mech. Techn. Inc. Report 62TR1, ONR Contract No. 3730(00), (February, 1962).

22. Rieger, N.F., Thomas, C.B., Jr., "Some Recent Computer Studies on the Stability of Rotors in Fluid-Film Bearings," International Union of Theoretical and Applied Mechanics, Proc. Dynamics of Rotors Symposium, Lyngby, Denmark, p. 436 (August 12-16, 1974).

23. Voorhees, J.E., Mellor, CC., and Dubensky, R.G., "The Dynamic Behavior of Hypercritical-speed Shafts," Army Research Office, Proc. of the Army Conf. on Dynamic Behavior of Material and Structures, held at Springfield Armory, Springfield, Mass., Sept. 26-28, 1962.

CHAPTER 2.7

HYSTERETIC INSTABILITY

N.F. Rieger

Nature of Hysteretic Whirling

Hysteretic whirling is a self-excited instability usually associated with built-up rotors in lightly damped bearings, in which the rotor whirl amplitude increases with time. It is independent of the state of balance of the rotor and it commences at speeds above a certain threshold speed. It most frequently arises from rubbing between assemblies of shrink- or press-fitted rotor components, e.g., compressor wheels on a central shaft. It can also arise from internal hysteresis of the shaft material. As many modern rotors are complex assemblies of components, the conditions under which assembly or material hysteresis may give rise to unstable whirling are of great interest (see figure 1).

Experimental Observations of Hysteretic Whirling

The first recorded observation of hysteretic instability was made by Newkirk [1] [1] in 1924 while investigating the cause of a series of bearing failures of blast furnace compressors. Newkirk constructed an experimental

1. Numbers in brackets designate references at the end of this section.

rotor test rig to simulate a real compressor unit and extensive testing led him to the following conclusions:

- Neither the threshold speed nor the whirl amplitude is affected by the state of balance of the rotor.
- The whirling always occurs above the first bending critical speed of the rotor, and never below this speed.
- The frequency of whirling is constant, regardless of the rotor speed, and equal to the first bending critical speed of the system.
- Whirling is encountered only with built-up rotors.
- An increase of the foundation flexibility increases the threshold speed at which the whirling commences.
- Distortions or misalignment of the bearings increases the whirl threshold speed.
- Introduction of damping into the foundation increases the whirl threshold speed.
- A small disturbance is sometimes required to initiate the whirl motion in a well-balanced rotor.

Newkirk found that increased foundation flexibility would improve the rotor stability, and that no bearing damping was needed to suppress whirling below the bending critical speed. Further, friction damping in the spring mounted bearing was capable of suppressing unstable rotor whirl motion above the bending critical speed.

Kimball suggested that forces normal to the plane of the deflected rotor could be produced by the hysteresis of the material undergoing cycles of stress reversal [3]. He postulated that, above the rotor critical speed, internal rotor friction would sustain the whirl. Newkirk concluded that some type of friction forces could be developed by the rubbing of a disk shrunk onto a shaft.

The presence of material hysteresis in a rotating deflected shaft as it undergoes alternating stress cycles of compression and tension is shown in figure 2. This action causes the shaft to deflect sideways. By measurement of the shaft vertical inclination angle, Kimball determined that the ratio of the internal friction forces to the elastic shaft forces for most ferrous and non-ferrous materials was in the order of 2×10^{-3}. The small order of magnitude of the friction forces observed by Kimball led Newkirk to conclude that the chief cause of the instability in his model was the friction created by the shrink fits of the impellers and spaces on the shaft.

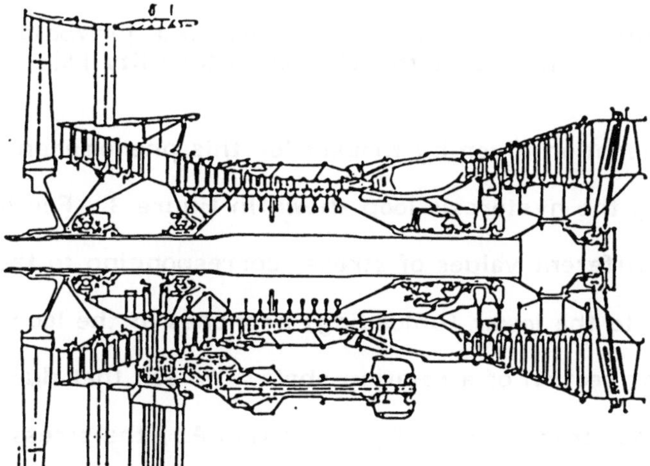

Figure 1 Schematic of an Aircraft Engine Showing Complex Rotor Assembly

This conclusion was confirmed by an experiment conducted by Kimball [4] on a special test rotor with rings on hubs on the shaft. When all shrink fits were removed from the rotor no whirl instability would develop, whereas, with the rings on the shaft, it became unstable above the first bending critical speed. Kimball first demonstrated that with rotors operating above their bending critical speed, long clamping fits commonly lead to instability problems.

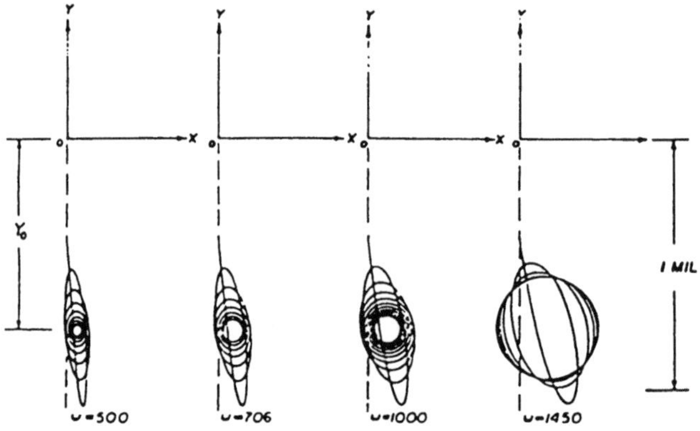

Figure 2 Whirl Orbits of a Balanced Horizontal Rotor Below the Threshold Stability [5]

The explanation given by Kimball for this phenomenon can be demonstrated using the hysteresis loop shown in figure 3. For the same strain there are two different values of stress, corresponding to the upper (loading), and to the lower (unloading) branches of the loop. Figure 3(c) shows the cross section of a rotating shaft. We see that during the motion of a surface fiber from position A_1 to position A_2, the stress varies from tension to compression. Consequently, the upper branch of the loop must

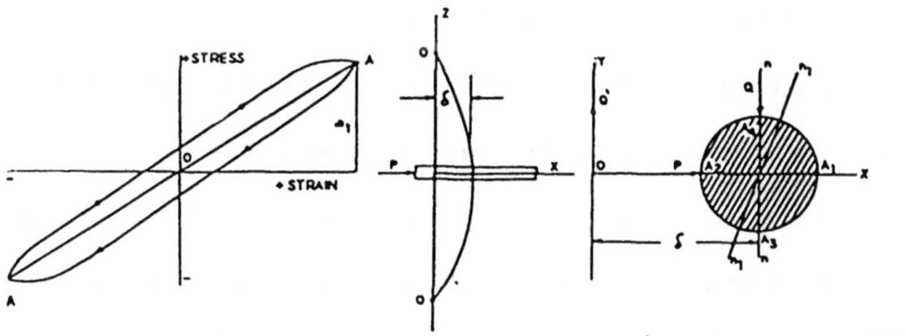

A) HYSTERESIS LOOP B) DEFLECTED SHAFT C) SIDE VIEW OF ROTATING DEFLECTED SHAFT

Figure 3 Whirling of a Rotating Shaft Caused by Hysteresis, Kimball [4]

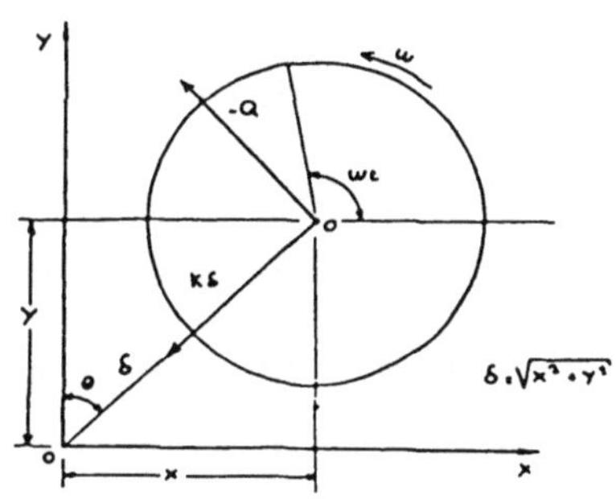

Figure 4 Model Used by Kimball [4]

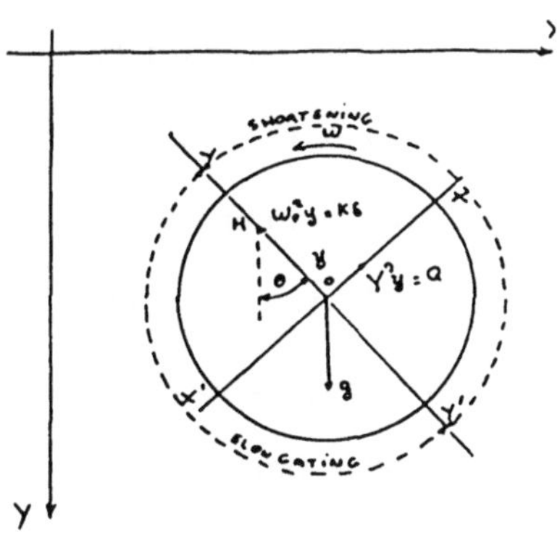

Figure 5

Model Used by Robertson [5] for Hysteretic Deflection of a Rotating Shaft

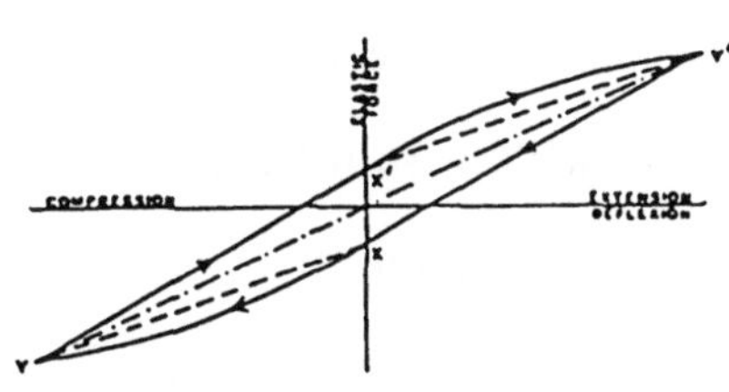

Figure 6

Exaggerated Hysteresis Loop [5]

be used. In the same way, the lower branch of the loop must be used during the motion from A_2 to A_1. From this, it follows that the hysteresis effect may be taken into account by superposing these strains on the steady bending strains. This causes additional fiber stresses, which are positive below the horizontal diameter A_1A_2, and negative above A_1A_2. This system of stresses is the same as would be generated by the bending of the shaft in the yz plane, in addition to the bending about A_3A_4 due to rotation. The combined bending stresses thus produce a bending moment in a plane which is inclined to the xz plane, i.e., with its neutral axis n_1n_1, instead of nn in figure 3(c). This additional bending effect is then equivalent to a force Q as shown in figure 3(c), the effect of which is to cause the rotor to whirl, under the forces shown in figure 4.

Robertson [5] later gave an explanation for shaft hysteretic whirling which differs somewhat from that of Kimball and Newkirk. In figure 5, the center line of the bearings passes through H, and the center of the shaft is deflected to O by the weight of the rotor. The neutral axis of the section is XOX, at right angles to HO; the lower half of the section is tension and the upper half in compression. The stresses now consist of those which would occur with perfect elasticity, plus the discrepancy stresses arising from hysteresis. The former gives an elastic force along OH, and the latter a hysteretic force along OZ at right angles to OH. Figure 6, which shows an exaggerated hysteresis loop for a particular fiber on the shaft, X, Y, X', Y', shows the point in the cycle corresponding to the passage of the fiber across the axes similarly lettered in figure 5.

The discrepancy between the loop and the mean elastic line is a tensile stress striving to lengthen the fiber in the part Y X Y'. Thus the discrepancy stresses are striving to elongate the right hand side of the shaft and to shorten the left hand side. They try to bend the shaft to the right, but as the weight prevents further movement in that direction, they give a hysteretic force acting forwards along OZ, i.e., directly opposite to the rotation of the shaft. With a deflection Y, the magnitude of the hysteretic force per unit mass of rotor may be written:

$$\delta_h = \gamma^2 Y$$

where δ_h is the hysteretic force per unit mass of rotor and γ is the hysteretic coefficient, which is not necessarily a constant. The resultant of this force and the direct elastic force must be vertical, as it has to balance the weight, therefore,

$$(\omega_o^4 + \gamma^4)^{\frac{1}{2}} y = g$$

$$\mathrm{Tan}\ \Theta = \gamma^2/\omega_o^2$$

where ω_o is the critical speed, g is gravitational acceleration, and Θ is the inclination of the deflection to the vertical. The horizontal deflection of the center of the shaft is:

$$Y \sin \Theta = \gamma g/(\omega_o^4 + \gamma^4) = (\gamma^2/\omega_o^4)\ g$$

Since the bearing forces act through H, whereas the weight acts through O, there is a torque opposing the rotation of the shaft whose amount is:

$$\tau = MgY \sin \Theta = Mg^2\gamma^2/\omega_o^4$$

where τ is the torque opposing the rotation of the shaft, due to hysteresis, and M is the mass of the rotor. The work done against this torque is converted into heat by the hysteresis. The amount of hysteresis work per cycle is:

$$W = 2\pi g^2 M\gamma^2/\omega_o^{\,4}$$

A measurement of the horizontal and vertical components of the deflection, or of its inclination to the vertical determines $\gamma^2/\omega_o^{\,2}$.

Even short, highly stressed shrink fits are not entirely devoid of problems [5]. Provided the rotor is given a sufficiently large initial disturbance or displacement to initiate relative internal slippage in the fit, even small, tight shrink fits may develop whirl instability.

Robertson, following Kimball [3], made a study of the hysteretic transient shaft whirling. In the absence of hysteretic effects, the transient whirl decays under the action of friction. When hysteretic effects exist, there is an additional force normal to the radius whose amount per unit mass has already been written as $\gamma^2 Y$. Adding this to the friction force $-2\alpha\omega_t jY$, we now have:

$$-j\,(2\alpha\,\omega_t \pm \gamma^2)Y = -j(2\alpha/\omega_t \pm \gamma^2/\omega_t^{\,2})\omega_t^2 Y, \qquad (j = \sqrt{-1})$$

where α is the friction factor, ω_t is the threshold speed for hysteretic driving of the transient whirl, and the factor j denotes "turn 90 degrees forward" (positive complex operator). The positive sign with this operator applies for shaft speeds below ω_t, which is "practically identical" with the critical speed ω_o, and the negative sign applies above the critical speed.

If the shaft runs faster than the transient, i.e., where γ^2/ω_t^2 exceeds $2\alpha/\omega_t$, the net force drives the transient rather than retarding it. A net driving force speeds up the transient slightly and thereby causes the centrifugal force to exceed the elastic force, causing an outward acceleration which soon produces an outward velocity whose Coriolis force, combined with the whirl friction, balances the driving force. If this continues, the outward velocity may lead to a large shaft whirl radius. The ultimate result will depend upon the way in which α and γ^2 change with the radius of the transient. When both these terms change at the same rate as the whirl radius, their difference retains the same sign. If the friction is the greater, the transient will decay and the rotor motion will be stable. But if the hysteretic force is the greater, the transient will grow and the shaft will become unstable. If they are equal at one radius, they will be equal at any radius and the transient whirl radius magnitude is then indefinite and intrinsically unstable.

If α is small but increases more rapidly than γ^2, the whirl will grow, up to that radius at which they become equal, and will continue at that magnitude. But if γ^2 grows more rapidly than α, the conditions are reversed. Below the point at which friction and hysteretic forces are equal, friction will be the greater and the whirl will decay. But beyond that point, the forward hysteretic force will exceed the friction and the excess will increase as the whirl grows. Thus the shaft is stable for small disturbances, and unstable for large ones.

Robertson [5] reported that he frequently observed the sustained transient, but was never able to control it with certainty. In reference [5] he describes a simple method for experimentally producing it. An observation mark is placed on the rotating disk to follow the whirl orbit. When no transient exists, the observation mark traces out a circle, whose radius is the vector sum of its own eccentricity and the whirling radius of the disk.

In the case of the sustained transient, the circle grows and decays with the change of these vectors between the transient and the steady radius of the spot. If the mark is then viewed under constant illumination while the speed is a little above the critical value, the path of the spot is seen as a waxing and waning circle. But when the speed is considerably greater, two spirals are seen, a left-handed one for whirl growth and a right-handed one for whirl decay. This double spiral has the form of a trochoid.

Robertson [5] finally concluded that a similar effect can be produced by any friction which opposes a change of the deflection of the shaft, such as the friction which exists at the connections of flexible couplings, and even in rigid couplings. The friction of this shaft through the bearing as it deflects can also lead to hysteretic whirling.

Practical instability caused by friction is rarely discussed in the literature. In particular, there is very little information available on rotor behavior in the unstable region. However, in 1964, a translation (from the Russian) of Kushul's work [6] on self-induced oscillations of rotors became available. Kushul's experiments concerned the motion of some high-speed textile spindles which presented instability problems. The spindles were composed of a built-up structure of a long wooden spindle inserted over a thin steel shaft. With such a continuous shrink fit, troubles were encountered above the rotor's first critical speed. Some typical rotor orbits obtained by Kushul above the stability threshold are shown in figure 7. Unable to monitor rotor motion electronically, he used an optical system: a fine needle was attached to the spindle end, and shaft orbit times were obtained by photographing the resulting end motion under a microscope. Figure 8 shows a sketch of his device. The orbits shown in figure 7 are of importance as they illustrate the conclusions that the rotor precession rate is approximately equal to the rotor bending critical speed, and also that the precession rate is constant over a large speed range.

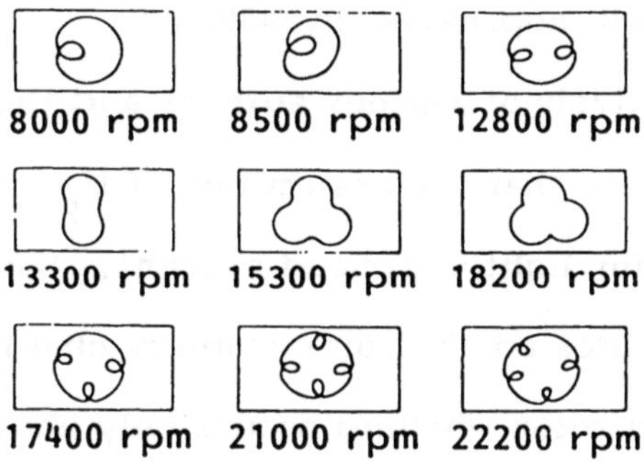

Figure 7 Typical Hysteretic Whirl Orbits, Kushul [6]

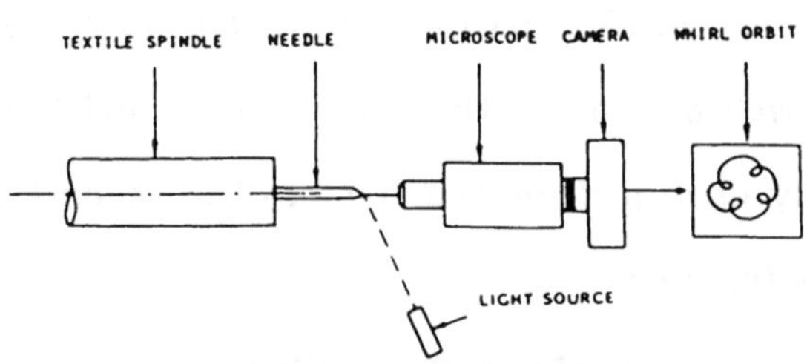

Figure 8 Experimental Device for Visualization of Textile Spindle Motion [6]

The critical speed of the spindle was determined to be about 4300 rpm. The fact that the rotor nonsynchronous precession rate remains constant can be easily verified by inspection of the various whirl patterns of figure 7. For example, at the speed range of 8000 to 8500 rpm, the rotor orbit forms one stationary internal loop. This indicates that the whirl ratio is one-half, that is, the precession speed ω_p is approximately 4700 rpm.

At 17,800 rpm, a stationary orbit with two internal loops is formed, which indicates a whirl ratio of one-third ω_p = 4300 rpm. Also at 21,000 to 22,000 rpm, a stationary pattern is formed with four internal loops to indicate a one-fifth ratio. Likewise, the rotor nonsynchronous precession rate at 27,000 rpm is still approximately 4300 rpm.

Based on the analysis of various authors and the experimental observation of Kushul, it is clear that the assumption of $\omega_p = \omega_c$ = constant has considerable justification for lightly-damped systems. Some of the major conclusions that Kushul states on the rotor stability characteristics are:

- Self-excited rotor whirling occurs only above the first bending critical speed.
- The whirl frequency remains almost constant at all speeds and is close to the first natural frequency of the spindle. In certain cases, well above the threshold, the whirl frequency can abruptly change from the first to the second order spindle natural frequency.
- The use of an elastic support by itself, without any increase in damping force, does not reduce the self-excitation.
- External damping improves the rotor stability.
- The most effective means to control the instability consisted of a spring-loaded bushing and damping sleeves. No dangerous self induced vibrations were observed with any spindle with this type of bushing.

Later experiments conducted in 1969 by Gunter [7] and in 1975 by Lund [8] using analog computer simulation led to the conclusion that, in the absence of external damping, a symmetric flexible foundation will reduce the rotor first bending critical speed and also increase the instability threshold speed. Both investigators fully agreed with their predecessors on the fact that the stability threshold can be greatly improved by adding external non-rotating damping.

Theory of Hysteretic Whirling

Consider the single-disk horizontal rotor shown in figure 9 which runs at constant speed under the influence of viscous damping, gravity, and disk unbalance. The equation of motion for the disk c.g. is:

$$M\ddot{r} + B\dot{r} + Kr = Mg + Ma\omega^2 e^{i\omega t}$$

where M is the mass of the disk, B is the velocity damping coefficient of the bearings and surroundings, assumed linear and viscous for convenience, K is the shaft flexural stiffness, a is the disk eccentricity, and r is the whirl radius of the shaft center given by:

$$r = x + iy \qquad i = \sqrt{-1}$$

where x and y are the Cartesian coordinates of the shaft center at any time t.

Now transform this expression into rotating coordinates using the following expression:

$$r = \zeta e^{i\omega t}$$

where:

$$\zeta = \xi + i\eta$$

This yields:

$$M(\ddot{\zeta} + 2i\omega\dot{\zeta} - \omega^2\zeta) + B(\dot{\zeta} + i\omega\zeta) + K\zeta = Mge^{-i\omega t} + Ma\omega^2$$

Internal friction Coulomb damping is also assumed to act at the shaft-disk interface. This occurs as an inwardly directed force (-P), along the

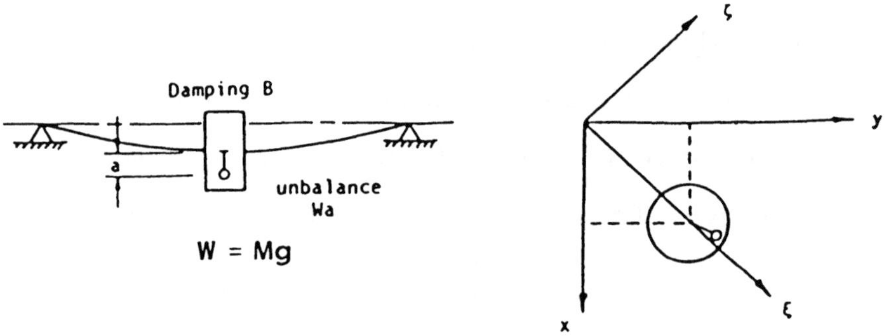

Figure 9 Single Mass Unbalanced Rotor

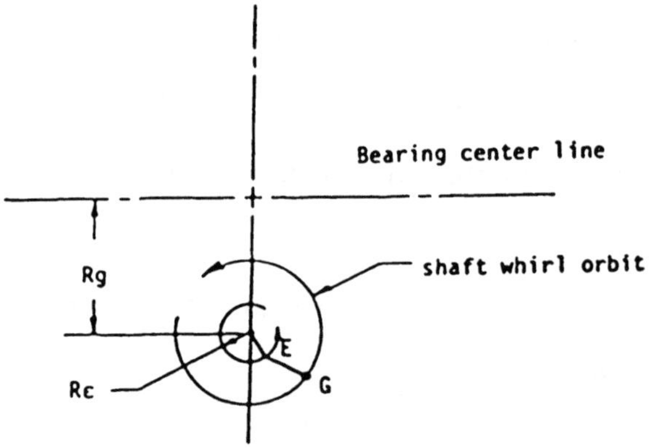

Figure 10 Graphical Representation of Shaft Whirl Orbit about Deflected Position

rotating radius vector ζ. This force may now be included in the mechanics of the rotor motion by adding it into the previous equation:

$$M(\ddot{\zeta} + 2i\omega\dot{\zeta} - \omega^2\zeta) + B(\zeta + i\omega\zeta) + K\zeta - P = Mge^{-i\omega t} + Ma\omega^2$$

The true nature of this force is difficult to determine or specify for any particular case. For the present analysis, it is convenient to assume that P is a linear function of the rotating radial velocity $\dot{\zeta}$. Though the results of this assumption describe certain observed rotor phenomena, the mechanism is not strictly correct, as Robertson [5] and Tondl [9] have observed. On writing $P = h\dot{\zeta}$, the equation of motion with viscous and coulomb damping becomes:

$$M(\ddot{\zeta} + 2i\omega\dot{\zeta} - \omega^2\zeta) + B(\dot{\zeta} + i\omega\zeta) + K\zeta - h\dot{\zeta} = Mge^{-i\omega t} + Ma^2$$

introducing the quantitites: $\gamma = B/M$, $\sigma = h/M$, $\omega_c^2 = K/M$, gives:

$$\ddot{\zeta} + 2i\omega\dot{\zeta} - \omega^2\zeta + \gamma(\dot{\zeta} + i\omega\zeta) + \sigma\dot{\zeta} + \omega_c^2\zeta = ge^{-i\omega t} + a\omega^2$$

Transforming back into non-rotating coordinates (X, Y) gives:

$$\ddot{r} + \gamma\dot{r} + \sigma(\dot{r} - i\omega r) + \omega_c^2 r = g + a\omega^2 e^{i\omega t}$$

The complete solution to the above equation consists of a complementary function (damped transient) and a particular integral of the form:

$$r = r_e^{i\omega t} + r_g$$

This corresponds to the unbalance term and the gravity deflection term of the shaft. Solving for the coefficients r_e and r_g gives:

$$r_\varepsilon = \frac{a\omega^2}{(\omega_c^2 - \omega^2) + i\gamma\omega} = \frac{a\omega^2(\omega_c^2 - \omega^2 - i\gamma\omega)}{(\omega_c^2 - \omega^2)^2 + \gamma^2\omega^2} = R_\varepsilon e^{-i\phi}$$

$$r_g = \frac{g}{\omega_c^2 - i\sigma\omega} = \frac{g(\omega_c^2 + i\sigma\omega)}{\omega_c^4 + \sigma^2\omega^2} = R_g e^{i\Psi}$$

where

$$R_\varepsilon = \frac{a\omega^2}{(\omega_c^2 - \omega^2)^2 + \gamma^2\omega^2} \; ; \quad \operatorname{Tan}\phi = \frac{\gamma\omega}{\omega_c^2 - \omega^2}$$

$$R_g = \frac{g}{\omega_c^4 + \sigma^2\omega^2} \; ; \quad \operatorname{Tan}\Psi = \frac{\sigma\omega}{\omega_c^2} \; .$$

These results are illustrated in figure 10.

Stability of Whirl Motion

Stability of the rotor is determined by whether the transient term grows or decays with time. The general form of the transient is:

$$r = r_1 e^{i\lambda_1 t} + r_2 e^{i\lambda_2 t} \quad ,$$

where r_1 and r_2 are constants of integration, and λ_1, λ_2 are the roots of the characteristic equation, found by substitution of r in the equation of motion:

$$\lambda^2 - i\lambda(\gamma + \sigma) - \omega_c^2 + i\sigma\omega = 0 \quad ,$$

hence:

$$\lambda_{1,2} = \frac{1}{2} i(\gamma + \sigma) \pm \sqrt{\omega_c^{\,2} - \frac{1}{4}(\gamma + \sigma)^2 - i\sigma\omega)} \quad .$$

Writing: $\omega_o^{\,2} = \omega_c^{\,2} - \frac{1}{4}(\gamma + \sigma)^2$ where ω_o is the natural frequency of the damped vibration of the rotor, and observing that usually $\omega_o^{\,2} \gg \sigma\omega$ we may write the radical as:

$$\sqrt{\omega_o^2 - i\sigma\omega} \approx \omega_o - \frac{1}{2} i\sigma\omega \quad .$$

This gives the following expressions for λ_1 and λ_2

$$\lambda_1 = \omega_o + \frac{1}{2} i(\gamma + \sigma - \sigma \frac{\omega}{\omega_o})$$

$$\lambda_2 = -\omega_o + \frac{1}{2} i(\gamma + \sigma + \sigma \frac{\omega}{\omega}) \quad .$$

The complementary function thus has the form:

$$r = r_1 \exp\{i\omega_o - \frac{1}{2}\langle\gamma + \sigma(1 - \frac{\omega}{\omega_o})\rangle\}t + r_2 \exp\{-i\omega_o - \frac{1}{2}\langle\gamma + \sigma(1 + \frac{\omega}{\omega_o})\rangle\}t.$$

A study of the indices of this expression shows the conditions under which the rotor will be stable. First notice that as long as:

$$\gamma + \sigma - \sigma\frac{\omega}{\omega_o} > 0$$

i.e.

$$\omega < \omega_o(1 + \frac{\gamma}{\sigma})$$

both terms in the transient amplitude expression will go to zero with increasing time, or

$$\lim_{t \to \infty} r = 0.$$

However, if the inequality does not hold, but

$$\omega > \omega_o(1 + \frac{\gamma}{\sigma})$$

then

$$\lim_{t \to \infty} r_1 = \lim r_1 \exp\{-\frac{1}{2}(\gamma + \sigma - \sigma\frac{\omega}{\omega_o}) + i\omega_o\}t = \infty$$

$$\lim_{t \to \infty} r_2 = \lim r_2 \exp\{-\frac{1}{2}(\gamma + \sigma + \sigma\frac{\omega}{\omega_o}) - i\omega_o\}t = 0$$

i.e., the first term tends to infinity while the second decays to zero; thus

$$\lim_{t \to \infty} r = \infty$$

It follows that, if the inequality

$$\omega < \omega_o(1 + \frac{\gamma}{\sigma}) = \omega_o(1 + \frac{B}{h})$$

where γ = B/M (Viscous damping factor), and

σ = h/M (hysteretic damping factor)

is not satisfied, the hysteretic whirl will be unstable.

This leads to the following conclusions:

1. The criterion for stability of a rotor with viscous and Coulomb damping is the inequality given above, which determines the hysteretic whirl threshold speed.

2. Above the whirl threshold speed, the rotor will whirl at its damped natural frequency ω_o with radius increasing with time.

3. The threshold of hysteretic whirl is raised in proportion to the amount of viscous damping present in the system.

Hysteretic Whirling of a Rotor in Flexible Supports

A more complete study for hysteretic whirl of a simple rotor in a flexible foundation has been conducted by Gunter [7]. This study considers the bearing mass, stiffness and damping of the foundation. A single mass rotor on an elastic foundation, shown in figures 11, 12, and 13, was investigated.

The general equations of rotor motion are obtained by Lagrange's method. The only quantity which presented some difficulties is the internal damping due to friction forces. These forces cannot be derived from a potential function (otherwise the system would be inherently stable) but can be obtained from a dissipation function of the proper form.

The notations used by Gunter for this analysis are the following:

A = rotor amplification factor = ω_{cro}/D_2, (DIM)
C_1 = foundation damping coefficient, lb-sec/in.
C_2 = rotor internal damping coefficient, lb-sec/in.
D_1 = foundation damping = C_1/m_2; rad/sec
D_2 = rotor internal damping = C_2/m_2; rad/sec
D = damping ratio = D_2/D_1
D = dissipation function
e_μ = displacement of rotor mass center from shaft elastic centerline, in.
ϕ = polar moment of inertia
K_1 = isotropic foundation stiffness, lb/in.
K_2 = rotor stiffness, lb/in.
K_x = foundation stiffness in horizontal direction, lb/in.
K_y = foundation stiffness in vertical direction, lb/in.
L = Lagrangian = $T - V$
m_1 = foundation mass, lb-sec^2/in.
m_2 = rotor mass, lb-sec^2/in.
q_r = generalized coordinate
R = flexibility ratio = K_2/K_y or K_2/K_1 for symmetric support

t = time
T = kinetic energy
T = rotor torque
V = potential energy
X_1 = support horizontal displacement, in.
X_2 = rotor horizontal displacement relative to the support, in.
X = rotor absolute horizontal displacement $x_1 + x_2$
Y = rotor absolute vertical displacement = $y_1 + y_2$, in.
α = foundation flexibility ratio = K_x/K_y (DIM)
δm = mass ratio = m_1/m_2, (DIM)
ω = rotor angular velocity, rad/sec
ω_s = rotor stability threshold, rad/sec
ω_{cro} = rotor critical speed on rigid supports $\sqrt{K_2/m_2}$
ω_{cr} = system critical speed on flexible supports = $\omega_{cro}\sqrt{1/1 + R}$
ω_{cx} = rotor critical speed in the horizontal direction, rad/sec
ω_{cy} = rotor critical speed in the vertical direction, rad/sec
θ = angular rotor coordinate
k = K/m_2, equation (21)
ϕ = rotor attitude angle = $\tan^{-1} X_2/Y_2$
$\dot{\phi}$ = rotor precession rate
ρ_2 = $\sqrt{X^2{}_2 + Y^2{}_2}$
F_{qr} = generalized force

The following assumptions apply to this analysis:

- The gyroscopic forces are ignored;
- The characteristics and displacements of each bearing housing are identical;
- The relative shaft bearing displacements are negligible in comparison to the absolute rotor and support displacements;
- There is no acceleration of the rotor (constant velocity);
- There is no gravitational force included.

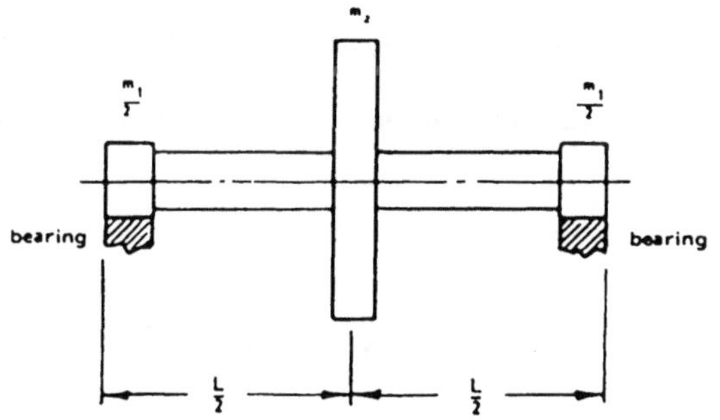

Figure 11 Single Mass Rotor with Massive Bearings [7]

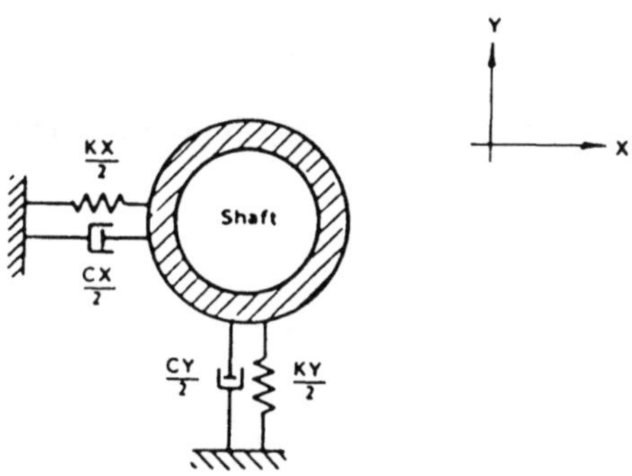

Figure 12 Front View of Bearing Showing the Vertical and Horizontal Coefficients [7]

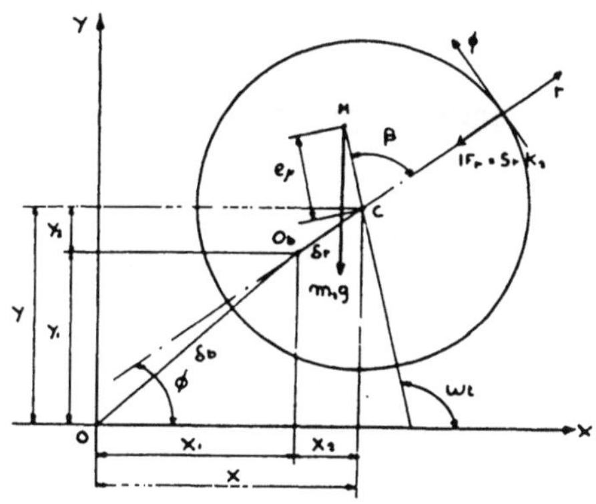

Figure 13 Geometry of Rotor Whirl Motion [7]

With these assumptions, the system will be fully described by four coupled second order equations of motion. We have for the position vectors:

bearing mass:

$$\text{Vector } OM_1 = \vec{P}_1 = X_1 \vec{n}_x + Y_1 \vec{n}_y$$

rotor mass:

$$\text{Vector } OM_2 = \vec{P}_2 = (X_1 + X_2 + e_\mu \cos \omega t)\, \vec{n}_x + (Y_1 + Y_2 + e_\mu \sin \omega t)\, \vec{n}_y$$

Replacing ωt by Θ gives five degrees of freedom to the system and hence five equations of motion are required to completely describe the system. This gives for the velocities:

bearing mass:

$$\vec{V}_1 = X_1 \vec{n}_x + Y_1 \vec{n}_y$$

rotor mass:

$$\vec{V}_2 = (\dot{X}_1 + \dot{X}_2 - e_\mu \dot{\Theta} \sin \Theta)\, \vec{n}_x + (\dot{Y}_1 + \dot{Y}_2 + e_\mu \Theta \cos \dot{\Theta})\vec{n}_y$$

Therefore, the kinetic energy is given by:

$$T = \frac{1}{2} \left\{ M_2 \left[(\dot{X}_1 + \dot{X}_2 - e_\mu \dot{\Theta} \sin \Theta)^2 + (\dot{Y}_1 + \dot{Y}_2 + e_\mu \dot{\Theta} \cos \Theta)^2 \right] + m_1 \left[\dot{X}_1^{\,2} + \dot{Y}_1^{\,2} \right] + \Phi \dot{\Theta}^2 \right\}$$

and the potential energy is:

$$V = \frac{1}{2} \left[K_x X_1^2 + K_y Y_1^2 \right] + \frac{1}{2} K_2 \left[X_2^2 + Y_2^2 \right]$$

The external damping is given by the dissipation function:

$$D_1 = \frac{1}{2} C_1 \left[\dot{X}_1^2 + \dot{Y}_1^2\right] \quad .$$

The dissipation function for internal damping is:

$$D_2 = C_2 \left[\frac{\dot{X}_2^2 + \dot{Y}_2^2}{2} + \omega\,(Y_2 \dot{X}_2 - X_2 \dot{Y}_2)\right] \quad .$$

Applying Lagrange's equation:

$$\frac{d}{dt}\left(\frac{\partial L}{\partial \dot{q}_r}\right) - \frac{\partial L}{\partial q_r} + \frac{\partial D}{\partial \dot{q}_r} = F_{qr} \quad .$$

where L = T-V, yields the following five generalized equations of motion:

$$X_1: \quad m_1 \ddot{X}_1 + m_2\left[\ddot{X}_1 + \ddot{X}_2 - e_\mu \ddot{\theta} \sin\theta + e_\mu (\dot{\theta})^2 \cos\theta\right] + C_1 \dot{X}_1 + K_x X_1 = 0$$

$$Y_1: \quad m_1 \ddot{Y}_1 + m_2\left[\ddot{Y}_1 + \ddot{Y}_2 + e_\mu \ddot{\theta} \cos\theta - e_\mu (\dot{\theta})^2 \sin\theta\right] + C_1 \dot{Y}_1 + K_y Y_1 = 0$$

$$X_2: \quad m_2\left[\ddot{X}_1 + \ddot{X}_2 - e_\mu \ddot{\theta} \sin\theta - e_\mu (\dot{\theta})^2 \cos\theta\right] + C_2\left[\dot{X}_2 + \omega Y_2\right] + K_2 X_2 = 0$$

$$Y_2: \quad m_2\left[\ddot{Y}_1 + \ddot{Y}_2 + e_\mu \ddot{\theta} \cos\theta - e_\mu (\dot{\theta})^2 \sin\theta\right] + C_2\,(\dot{Y}_2 - \omega X_2) + K_2 Y_2 = 0$$

$$\theta: \quad I \ddot{\theta} + m_2\left[-(\ddot{X}_1 + \ddot{X}_2)\, e_\mu \sin\theta + (\ddot{Y}_1 + \ddot{Y}_2)\, e_\mu \cos\theta + e_\mu^2 \ddot{\theta}\right] = T \quad .$$

By neglecting the acceleration of the rotor, these equations can be reduced to four equations:

$$(1 + \delta m)\,\ddot{X}_1 + \ddot{X}_2 + D_1 \dot{X}_1 + \omega_x^2 X_1 = e_\mu\, \omega^2 \cos \omega t$$

$$(1 + \delta m)\,\ddot{Y}_1 + \ddot{Y}_2 + D_1 \dot{Y}_1 + \omega_y^2 Y_1 = e_\mu\, \omega^2 \sin \omega t$$

where $\delta m = \frac{m_1}{m_2}$

$\omega_x^2 = \frac{kx}{m_2}$

$$\ddot{X}_1 + \ddot{X}_2 + D_2\dot{X}_2 + D_2\,\omega Y_2 + \omega_2^2 X_2 = e_\mu\,\omega^2 \cos\omega t \qquad \omega_y^2 = \frac{k_y}{m_2}$$

$$\ddot{Y}_1 + \ddot{Y}_2 + D_2\,\dot{Y}_2 - D_2\,\omega X_2 + \omega_2^2 Y_2 = e_\mu\,\omega^2 \sin\omega t \qquad \omega_2^2 = \frac{k_2}{m_2} = \omega_{cro}^2$$

If the rotor total damping forces are considered small in comparison to the shaft elastic restoring forces, and if the bearing housing mass is neglected, then one of the displacement variables may be eliminated to obtain a single fourth order equation in either X or Y to represent the system.

Eliminating γ yields:

$$\ddddot{X} + \left[c_x + c_y\right]\dddot{X} + \left[\omega_{cx}^2 + \omega_{cy}^2 + c_x c_y\right]\ddot{X} +$$

$$\left[\omega_{cx}^2\, c_y + \omega_{cy}^2\, c_x\right]\dot{X} + \left[\omega_{cx}^2\,\omega_{cy}^2 + \omega^2\mu_x\mu_y\right]X = 0$$

where:

$$\begin{cases} c_x = \mu_x + \nu_x \\ c_y = \mu_y + \nu_y \end{cases}$$

and: ω_{cx} = natural system resonance frequency for the X direction

$$= \sqrt{\frac{K_2\,K_x}{M_2\,K_2 + K_x}} = \omega_{cro}\sqrt{\frac{a}{R + a}}$$

ω_{cy} = natural system resonance frequency for the Y direction

$$= \sqrt{\frac{K_2\,K_y}{M_2\,(K_2 + K_y)}} = \omega_{cro}\sqrt{\frac{1}{1 + R}}$$

ω_{cro} = rotor natural resonance frequency on rigid supports.

$$\mu_x = \frac{C_2}{M_2}\left(\frac{K_y}{K_x + K_2}\right)^2 = D_2\left(\frac{1}{a + R}\right)^2$$

$$\mu_y = \frac{C_2}{M_2}\left(\frac{K_y}{K_y + K_2}\right)^2 = D_2\left(\frac{1}{1 + R}\right)^2$$

$$\nu_x = \frac{C_1}{M_2}\left(\frac{K_2}{K_x + K_2}\right)^2 = D_1\left(\frac{R}{a + R}\right)^2$$

$$\nu_y = \frac{C_1}{M_2}\left(\frac{K_2}{K_y + K_2}\right)^2 = D_1\left(\frac{R}{1 + R}\right)^2$$

$$D_1 = \frac{C_1}{M_2},\quad D_2 = \frac{C_2}{M_2},\quad R = \frac{K_2}{K_y},\quad a = \frac{K_x}{K_y}$$

Applying the Routh-Hurwitz stability criterion for 2N = 4, the rotor threshold stability is obtained after some algebraic manipulations as follows:

$$\omega_s = \omega_{cro}\sqrt{F_1 + F_2}$$

where:

$$F_1 = \left(\frac{\omega_{cro}}{D_2}\right)^2 \frac{\left[R^2 + D\right]\left[R^2 + D\,a^2\right] R^2\,(a + R)^2\,(1 + R)^2\,(1 - a)^2}{\left\{a\left[(1 + R)^2\,(R^2 + D\,a^2) + (a + R)^2\,(R^2 + D)\right]\right\}^2}$$

$$F_2 = \frac{(R^2 + D)\,(R^2 + D\,a^2)}{D^2\,a^2}\left[\frac{a(R + a)\,(R^2 + D) + (R + 1)\,(R^2 + D\,a^2)}{(R + 1)^2\,(R^2 + D\,a^2) + (R + a)^2\,(R^2 + D)}\right]$$

Following this investigation, a digital computer program was developed to calculate the threshold of stability by the general Routh procedure outlined previously. Since the coefficients of the characteristic equation are speed dependent, an iterative approach was employed to obtain the

Gunter used the results of his investigation to draw some charts corresponding to various cases of rotor support properties. Figures 16 and 17 represent the stability characteristics of a rotor on a symmetric foundation. The use of this chart is illustrated by an example. The rotor characteristics are the following:

m_2 = 0.25 $lb/sec^2/in.$ (96.6 lb. rotor)

k_2 = shaft stiffness = 250,000 lb/in

K_1 = support stiffness = 250,000 lb/in

R = 1

D1=D2 = 200 rad/sec; $D = \frac{D1}{D2} = 1$

ω_{cro} = k_2/m_2 = 1000 rad/sec - rotor natural frequency (considered on rigid supports)

ω_{cr} = 706 rad/sec = system natural frequency

A = critical speed amplification factor = $\omega_{cro}/D_2 = 5$

The rotor stability threshold speed on rigid supports is 1000 rad/sec. Figure 14 shows that for D = 1 and R = 1; the stability threshold is raised to 1414 rad/sec. If the support damping D1 were zero, the stability threshold would be thereby reduced.

For unsymmetric bearing support, figure 15 from [7] shows the stability threshold for stiffness values of R = 0.1, 1.0, and 10.0 with zero foundation damping. Examination of the stability curve for R = 10 shows the influence of small changes in D on stability. At low values of R, the vertical foundation stiffness is much stiffer than that of the rotor. Very little increase in the stability threshold speed is obtained by varying the horizontal foundation stiffness.

threshold speed. In the computer program, F_1 is determined by the difference between the Routh coefficients D22 and E2. Figure 14 gives plots of these two functions for various values of external damping and over a range of rotor speeds.

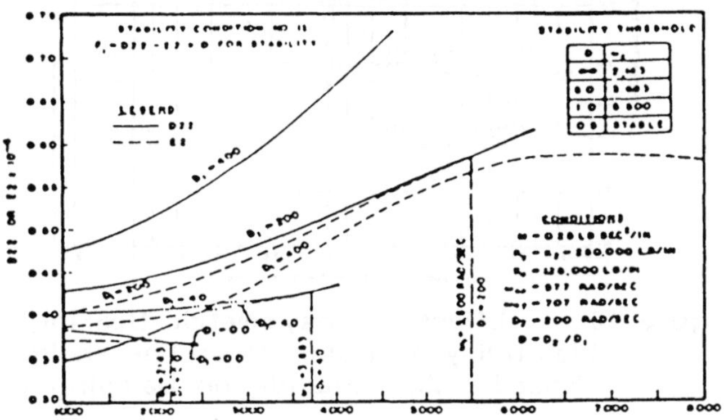

Figure 14   Routh Stability Coefficients at Various Rotor Speeds [7]

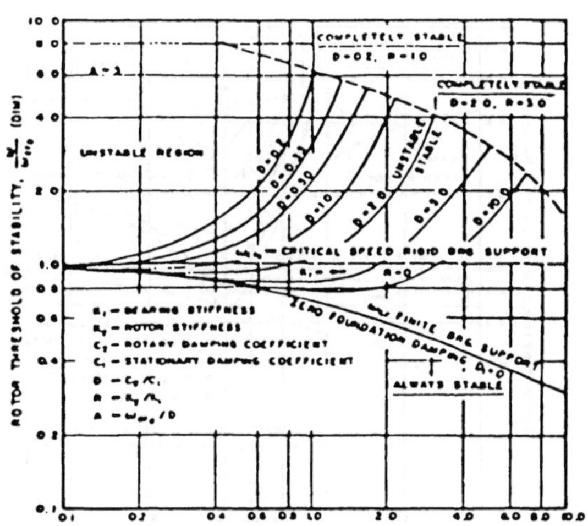

Figure 15   Effect of External Damping and Bearing Flexibility on the Rotor Whirl Threshold Speed Symmetric Support [7]

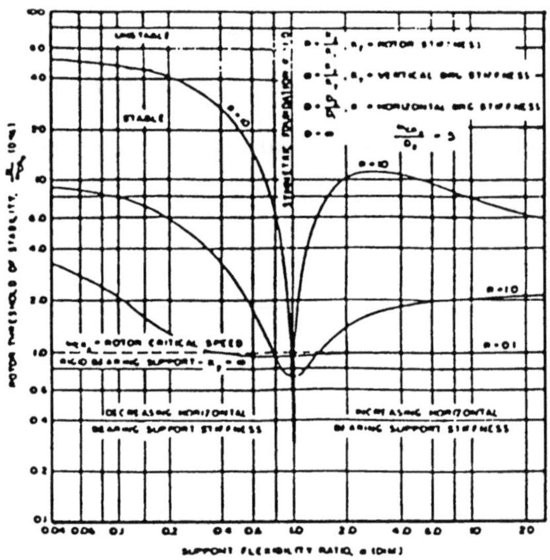

Figure 16 Effect of Unsymmetric Bearing Support Flexibility on Rotor Whirl Threshold Speed - Zero Foundation Damping [7]

Figure 17 represents the rotor stability characteristics for R = 1, and A = 5 for various values of support damping.

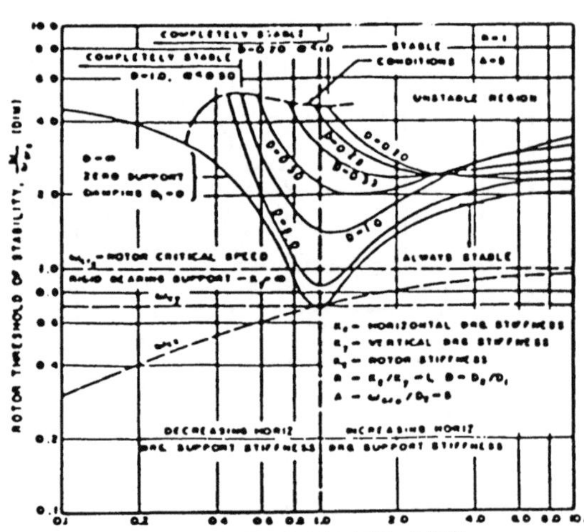

Figure 17 Effect of Unsymmetric Bearing Support Flexibility on the Rotor Whirl Threshold Speed for R=1.0 and A=5.0 General System [7]

To more clearly illustrate the rotor stability characteristics in the absence of foundation damping, a three-dimensional stability model was developed by Gunter [7]. This is shown in figure 18.

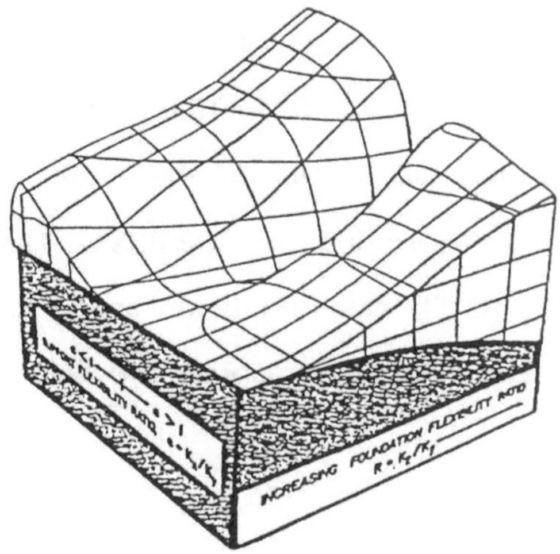

Figure 18 Topographical Model of Rotor Stability Characteristics with Zero Foundation Damping [7]

It clearly shows that reduction of horizontal bearing flexibility and increase in foundation damping produce a rapid rise in the stability threshold. This is illustrated by an example:

The characteristics of the system are the following -

m_2 = 0.25 lb-sec^2/in (96.6 lb rotor)

m_1 = 0 $\qquad$ $D_1 = D_2$ = 200 rad/sec

$K_2 = K_y$ = 250,000 lb/in $\qquad$ ω_{cx} = 577 rad/sec

K_x = 125,000 $\qquad$ ω_{cy} = 707 rad/sec

This example is identical to the previous one, except the support stiffness in the horizontal direction has been reduced to one-half of the previous case. This increases the rotor stability threshold from 1414 to over 3200 rad/sec.

Figure 19 shows the variation of rotor stability for a range of values of internal damping, in a system with no external damping. It shows that large increases in rotor stability are possible by the introduction of bearing asymmetry. It further shows the important conclusion that the larger the internal friction of the rotor, the less effective the bearing asymmetry on improving the stability. These results were later verified by Rieger and Thomas [10] in an analog computer study of the influence of hysteresis, viscous damping and seal effects on rotor stability threshold.

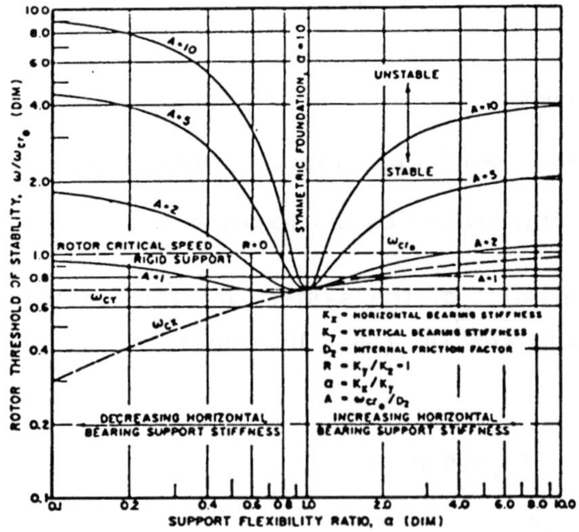

Figure 19 Effect of Unsymmetric Bearing Support on the Rotor Whirl Threshold Speed for Various Values of Internal Friction - Zero Foundation Damping [7]

Conclusions

Internal friction between the press- or shrink-fitted parts of a built-up rotor, and internal hysteresis of the material, are the two main causes of hysteretic instability. Internal rotor damping tends to cause unstable, nonsynchronous precession above the rotor bending critical speed where no external damping acts on the system. The hysteresis whirl frequency is approximately equal to the rotor bending critical speed. This frequency remains constant over a wide speed range.

One important aspect of this phenomenon is the influence of foundation flexibility and damping on the instability threshold speed. A symmetric flexible foundation will reduce the rotor critical speed, and will also reduce the whirl threshold in the absence of external damping. If external damping is added, the stability threshold can be greatly increased. To introduce foundation damping, it is necessary to permit some foundation motion. This lowers the system critical speed, and hence the whirl threshold speed. However, the threshold speed is increased significantly through the introduction of foundation damping. Of these two opposed effects, the influence of foundation damping is more significant than the foundation stiffness decrease.

References

1. Newkirk, B.L., "Shaft Whipping," General Electric Review (1924) .

2. Newkirk, B.L., and Taylor, H.D., "Shaft Whipping due to Oil Action in Journal Bearings," General Electric Review (1925).

3. Kimball, A.J., "Internal Friction Theory of Shaft Whirling," General Electric Report, Vol. 27, p. 266 (1926).

4. Kimball, A.K., "Measurement of Internal Friction in a Revolving Deflected Shaft," General Electric Review, Vol. 28, p. 554 (1926).

5. Robertson, D., "Hysteretic Influences on the Whirling of Rotors," Proc. Inst. Mech. Engrs., Vol. 131, p. 513 (1935).

6. Kushul, M.Y., "The Self-Induced Oscillations of Rotors," Consultants Bureau, New York (1964).

7. Gunter, E.G. and Trumpler, P.R., "The Influence of Internal Friction on the Stability of High Speed Rotors with Anisotropic Supports," ASME Paper No. 69-VIBR-2 (1969).

8. Lund, J.W., "Stability and Damped Critical Speeds of a Flexible Rotor in Fluid-Film Bearings," Journal of Engineering for Industry (May, 1976), p. 509.

9. Tondl, A., "Some Problems of Rotor Dynamics," Publishing House of the Czechoslovakian Academy of Sciences, Prague, Czechoslovakia (1965).

10. Rieger, N.F. and Thomas, C.B., "Some Recent Computer Studies on the Stability of Rotors in Fluid Film Bearings," International Union of Theoretical and Applied Mechanics, Proc. Dynamics of Rotors Symposium, Lyngby, Denmark (August 12-16, 1976), p. 436.

CHAPTER 2.8

FLUID INSTABILITY

N.F. Rieger

Nature of Fluid Film Instability

Fluid-film instability may occur in high speed rotating machinery whose rotor-bearing system includes fluid-film journal bearings. This instability is independent of rotor unbalance. It occurs as a self-excited whirl motion above a certain threshold speed, and is induced by the action of hydrodynamic forces within the journal bearings. The rotor journals then whirl in a circular or elliptical orbit within the bearing clearance, about the steady-state equilibrium position. The speed beyond which the shaft becomes unstable and the whirl orbit begins to increase in diameter with time is defined as the threshold speed. Fluid-film instability is strongly influenced by the type of bearings which support the rotor, and by the relative flexibility of the rotor itself.

Types of Fluid Film Instability

Although Newkirk and Lewis [1], Pinkus [2], Hori [3], Tondl [4], Sternlicht [5], and Smith [6] have all classified unstable whirling of rotor-

---

1. Numbers in brackets designate references at end of the section.

bearing systems in various ways, it can be concluded that there are two basic types of unstable rotor whirl:

- Subharmonic whirling (half-frequency, fractional-frequency, low-speed whirl, light-load instability)
- Resonant whipping.

It should be noted that there is general agreement that light-load instability and fractional-frequency whirl are of the same general nature as half-frequency whirl. In all these cases the whirl is usually bounded and its frequency may vary from slightly less than half running speed to one-third of running speed.

In resonant whipping the threshold speed occurs at around twice the bending critical speed of the rotor system. The whirl frequency is typically the lowest bending critical speed of the system.

The first class of whirling has been observed mostly with rigid rotors in plain fluid-film bearings, while resonant whipping--a violent, dynamic whirling of the rotor in its bearings--is more likely to occur with flexible rotors in plain, fluid-film bearings.

Gas seals have also been known to contribute to instability in rotor systems. They tend to further destabilize the rotor in its bearings. The same classification of whirling stated above still applies, and the same threshold speed and whirl frequency properties apply.

Experimental Observations of Fluid-Film Whirl

Oil whip was first identified in 1924 by Newkirk [7] who first made an experimental study of this problem. A parametric study of both rigid rotor instability and flexible rotor instability was later conducted by Newkirk and Lewis [1]. The following properties of unstable motion were observed:

- Whirling occurred at a frequency close to half the running frequency (usually below). It was sharply resonant in a narrow speed range and built up only when the natural frequency of the system (critical speed) was close to half the running speed. This instability could occur for any running speed over twice the first critical speed of the rotor.

- The second type of disturbance (resonant whipping) was also observed at speeds above twice the first critical speed of the rotor system. This disturbance was more severe, and it built up with a frequency equal to the first critical frequency of the rotor. It appeared only for speeds above twice the first critical speed. The main characteristic of this disturbance was that, once started, it persisted at all higher speeds. (These unstable whirl motions could be of a large amplitude and consequently could be very harmful if allowed to persist.) Whirling was independent of the state of balance of the rotor, and the threshold speed was not influenced by external disturbances such as shocks.

These tests (on three rotors and five types of bearings) were run with oils of various viscosities to study the conditions associated with the range for stable operation, using cylindrical bearings at speeds above twice the lowest critical speed of the system. It was concluded that short bearings, rather than large clearance ratios and moderate unit bearing loads, tend to favor a range of stable operation. In certain instances, this stable range may extend up to more than five times the lowest critical speed. Slight bearing misalignment can also cause a significant increase in the threshold speed. In a subsequent paper, Newkirk [8] reviewed results obtained earlier with a flexible rotor whose lowest critical speed was 1210 rpm (see figure 1). Within the speed range 2300-5000 rpm, the rotor whirled with a frequency around 1250 rpm (see figure 2). The severity of the whirl increased with increasing speed. This result was

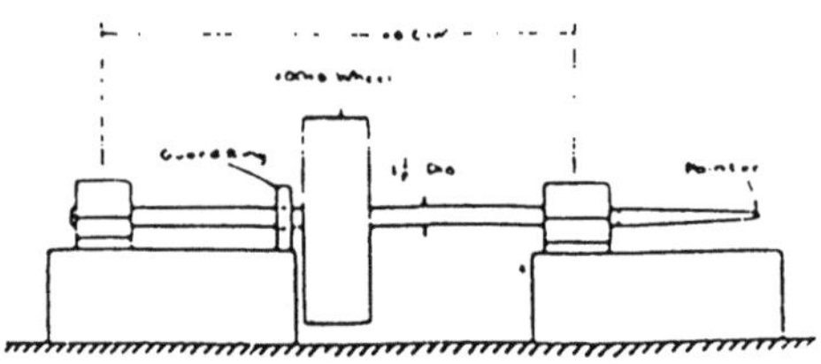

Figure 1

Rotor Model Used by Newkirk [8]

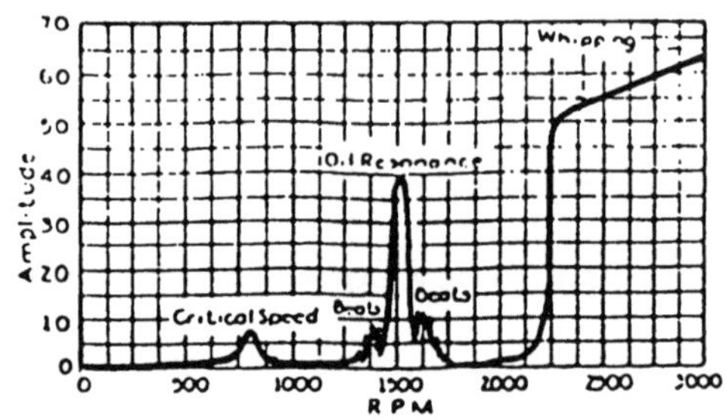

Figure 2

Frequency Plot of Rotor Model [8]

compared with those obtained using a very stiff rotor for which there was no discernible (bending) critical speed up to 30,000 rpm. This shaft whirled at low speeds with a frequency slightly less than one-half the running speed. But the stiff rotor whirl died out at higher speeds, which varied from 7000 to 18,000 rev/min. (Low viscosity oil gave the higher limit.) Newkirk concluded that rotor bending flexibility was the key factor in explaining why the performance of these two machines was so different.

A more extensive investigation was carried out in 1956 by Pinkus [2] on two rotors having relatively light bearing loads (23.4 and 8 lb/in^2, respectively) and reasonably high critical speeds (4000 and 6100 rev/min, respectively). These two rotors are shown in figure 3. The objective of this investigation was to compare the relative stability of several bearing types. Plain cylindrical, axial groove, elliptical, "pressure-dam," three lobe, tilting pad, and hydraulically loaded bearings were studied. Generally, the testing procedure was to increase shaft speed while recording data on:

- Inlet and outlet oil temperature
- Frequency of shaft vibration
- Frequency and amplitude of vibration
- Inlet oil pressure

Pinkus observed that cylindrical bearings were the least stable and hydraulically loaded bearings the most stable types. He also noted that, with sufficiently high applied hydraulic pressure, all whipping could be

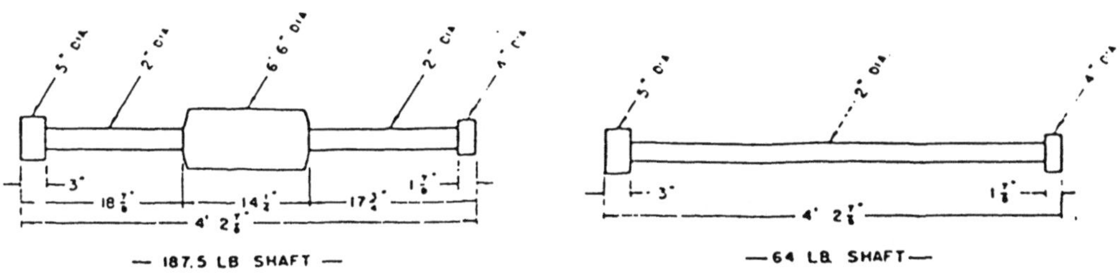

Figure 3 Dimensions of Test Shafts [2]

suppressed. Amplitude vs. speed results obtained in this investigation are shown in figure 4. With the more flexible shaft, the initial amplitude peak corresponds to the rotor bending critical speed. Resonant whipping sets in at approximately 1.6 times the system bending critical speed. It persists with a whirl frequency equal to this critical speed, and may tend to disappear around 3.5 times this critical speed. The large amplitude build-up in the same zone is the second system bending critical speed. The stiffer shaft shows an unbalance whirl peak followed by a steady build-up to full whipping amplitude at around three times the system bending critical speed with no tendency for the whirl to diminish, in this case, up to four times the bending critical speed. In both instances the shaft resonant whipping frequency was the system bending critical speed, as shown in figures 4 and 5.

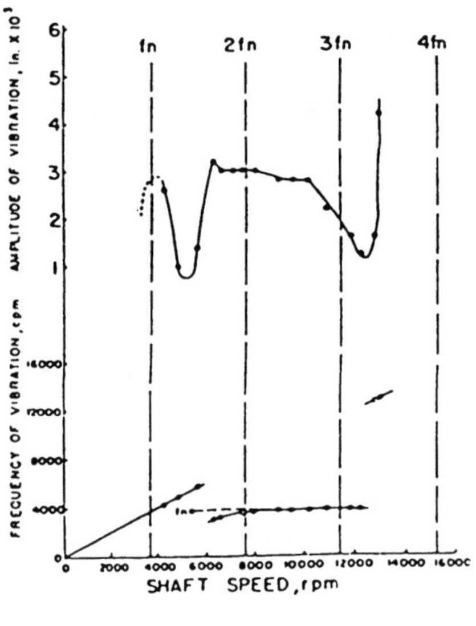

Figure 4

Typical Whip Curves
for 187 lb. Shaft [2]

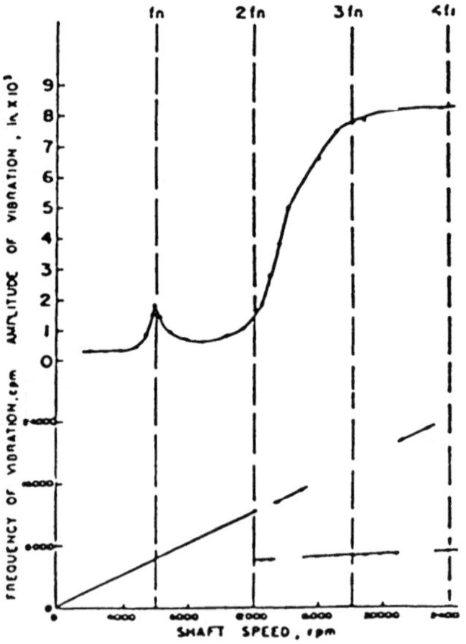

Figure 5

Typical Whip Curves
for 64 lb. Shaft [2]

In 1959, Hori [3] made an independent investigation of rotor system instability, including a theoretical analysis and some experimental studies. He examined the influence of different variables on the resonant whipping, operating speed, bearing load, oil viscosity, and of design variables such as rotor dimensions, journal eccentricity, flexibility of the rotor, bearing misalignment and shape. He, too, reached the conclusion that it was possible to stabilize a rotor by increasing the eccentricity of the journal within the bearing clearance; by using a shorter bearing; and by using a less viscous lubricant. He also reported that it was possible to decrease the amplitude of the whipping motion by making the oil force larger, by using a longer bearing, and by using a more viscous lubricant. This is the opposite of the preceding statement, but the difference can be explained by the increased squeeze-film action of the bearing, once the violent whipping motion is established. These conclusions were drawn for rotors operating at speeds considerably higher than twice the first bending critical speed.

An excellent discussion of most important test findings was given by Tondl in several publications between 1961 and 1968, references [4],[9], and [10], and in his book [11]. Tondl conducted a comprehensive test program in which he examined the influence of many bearing types on the stability of single-disk rotors. Tondl's experimental rig is shown in figure 6, and the types of bearings tested by him are presented in table 1.

Tondl summarized by concluding that: to avoid unstable whirling, cylindrical bearings cannot be used at rotor speeds higher than twice the first bending critical speed of the system; elliptical bearings exhibited greater resistance against the initiation of self-excited vibrations than cylindrical bearings; the threshold speed and the amplitude of whirling became higher where the ratio of horizontal and vertical clearances was greater. He recommended the use of elliptical bearings for machines whose operational speed does not exceed three times the first critical speed of the system. Higher whirl threshold speeds were obtained with flexible element bearings. Tondl described several cases of complete suppression of self-excited vibrations within the test speed range of up to eight times the first bending critical speed of the rotor. His results (amplitude versus frequency) for the various bearing types are shown in figures 7, 8, 9, and 10. Tondl also found that in addition to the lower stability limit (threshold speed), there is also an upper limit of the speed interval in which the self-excited vibrations were initiated. He observed both

No	Type of bearing	Schematic representation of bearing	Dimensions	mm	in	Designation
1	Cylindrical	 diameter of journal d = 40 mm 1.575 in	D	40.3	1.587	2δ – 30
2			D	40.4	1.591	2δ – 40
3	Elliptical	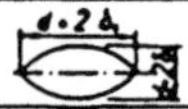	$2\delta_1$ $2\delta_2$	0.2 0.4	0.0079 0.0158	
4		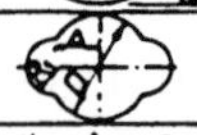	D Δ R_1	40.2 0.628 19.572	1.583 0.0247 0.771	
5	With flexible elements	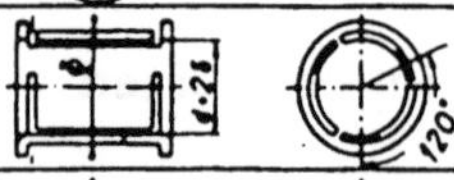	2δ ϕ	0.2 43.8	0.0079 1.724	2δ – 20
6		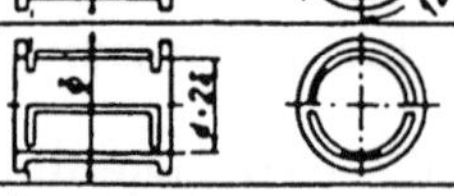	2δ ϕ	0.2 44	0.0079 1.732	2δ – 20
7	Loose bushing with flexible elements	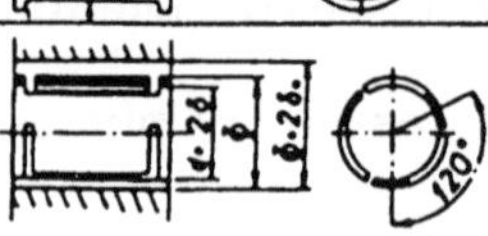	2δ ϕ $2\delta_0$	0.2 44 0.3	0.0079 1.732 0.0118	2δ – 20 2δ – 30
8			2δ ϕ $2\delta_0$	0.3 44 0.35	0.0118 1.732 0.0138	2δ – 20 2δ – 35

Table 1 Types and Principal Dimensions of Bearings Tested [11]

linear and non-linear unstable whirling. Under non-linear whirling conditions the onset speed appeared to be affected to some degree by the lack of refinement in the balancing of the rotor. This leads to bounded instability amplitudes. Observations were made that, after having reached a peak shortly above the threshold speed, the whirl amplitude actually diminished with further increase in rotor speed, e.g., references [2] and [9]. Due to the non-linear motion associated with the oil film at large amplitudes, such rotors have been operated up to several times their threshold speed. Concerning the effect of oil viscosity on threshold speed, Tondl concluded that both low and high viscosity oils can be advantageous, depending on the type of bearing used and the operating conditions. The best result, i.e., the most stable operation, was obtained with the loose flexible element bearing (figure 11), for which instability was observed only occasionally.

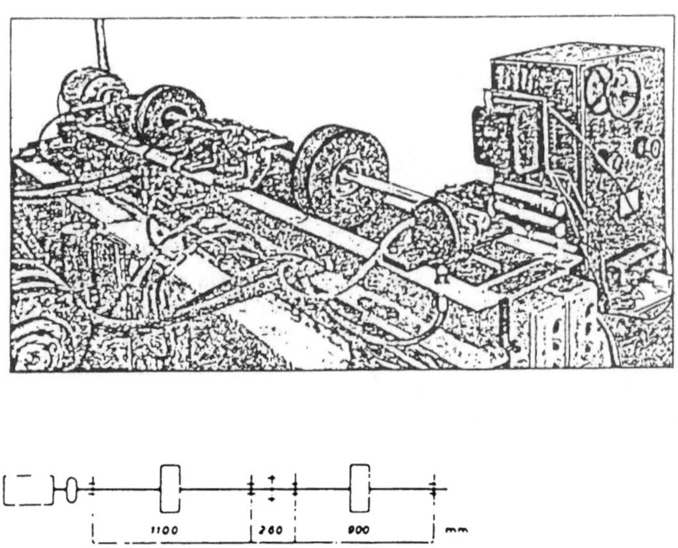

Figure 6 Experimental Rig Used by Tondl [9]

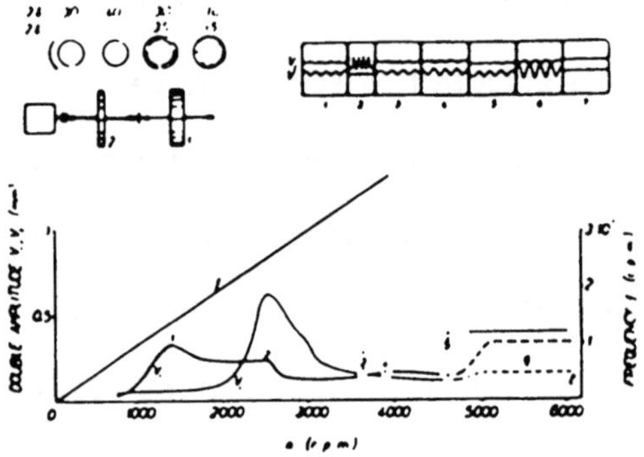

Figure 7

Two Cylindrical Bearings and Two Bearings with Flexible Elements, Tondl [10]

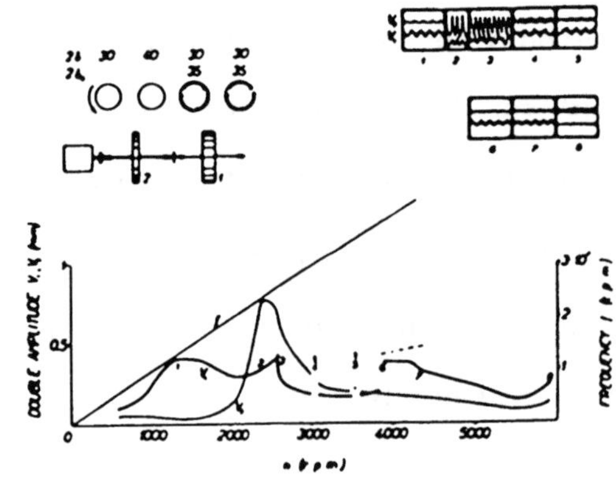

Figure 8

Same Case as Figure 7 but Greater Rotor Unbalance [10]

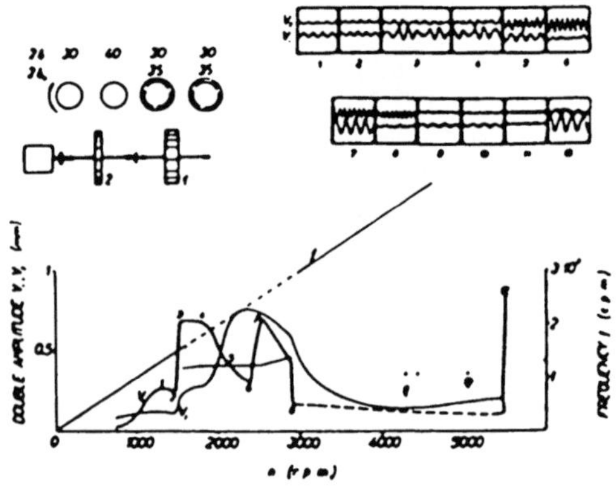

Figure 9

Two Cylindrical Bearings and Two Bearings with Flexible Elements with Opposite Direction of Rotation [10]

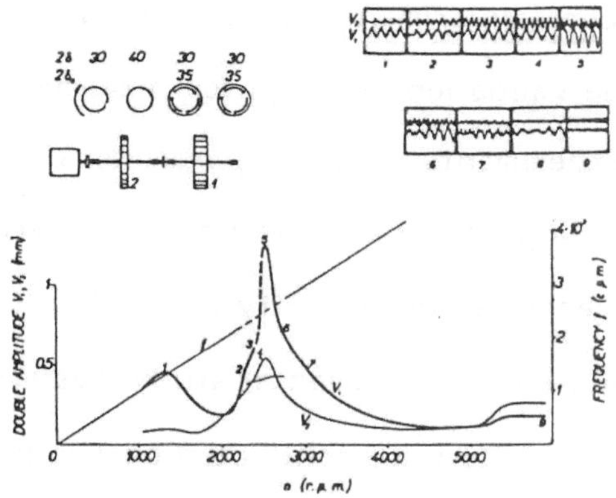

Figure 10 Same Case as Figure 9, but with Greater Rotor Unbalance [10]

Figure 11 Loose-Bushing Bearing; Loose Bushing is of the Flexible Element Type [10]

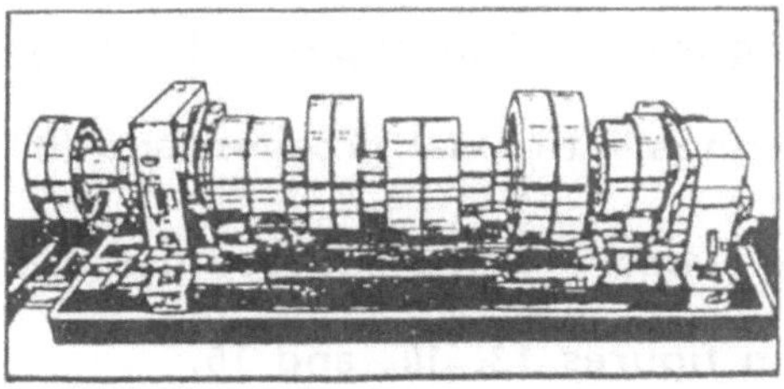

Figure 12 General View of Test Rotor No. 2 [12]

Most of the recent experiments on rotor stability have been performed in connection with the validation of computer programs for stability analysis. In these experiments separate parameters which influence the onset of unstable whirling have been examined. Several studies have also investigated the influence of non linearity of the fluid-film on threshold and whirl radius. One such experimental study was made by Lund and Tonnesen [12]. Experiments were conducted on two rotor systems having the following properties:

	Rotor 1	Rotor 2
Weight	40 kg	187.5 kg
Length	1040 mm	1190 mm
Bearing span	880 mm	880 mm
Shaft diameter	80 mm	80 mm
Journal diameter	62.7 mm	62.7 mm

Rotor 2 is shown in figure 12. Two types of bearings and supports were tested, (a) rigid bearings with axial grooves, and (b) cylindrical bearings mounted on a flexible support with a squeeze film damper. These rotor supports are shown in figures 13, 14, and 15.

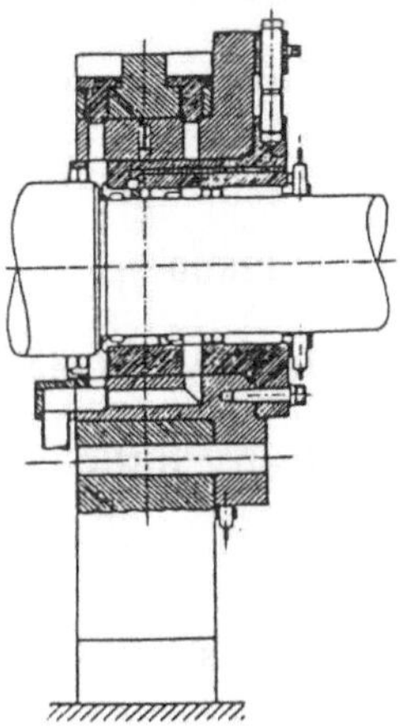

Figure 13

Cross Section View of Bearing Arrangement with Rigid Support [12]

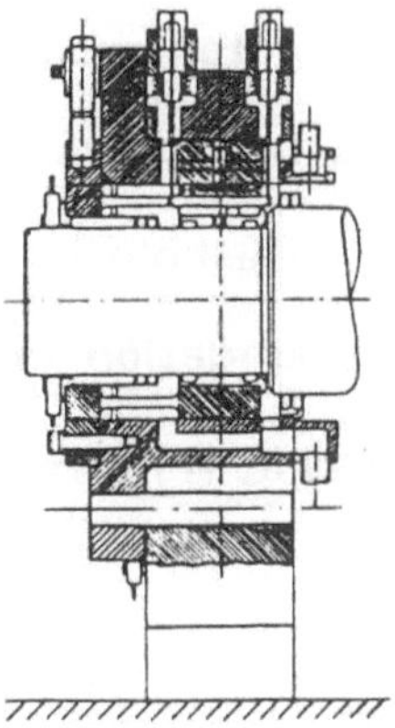

Figure 14

Cross Section View of Bearing Arrangement with Damped Flexible Support [12]

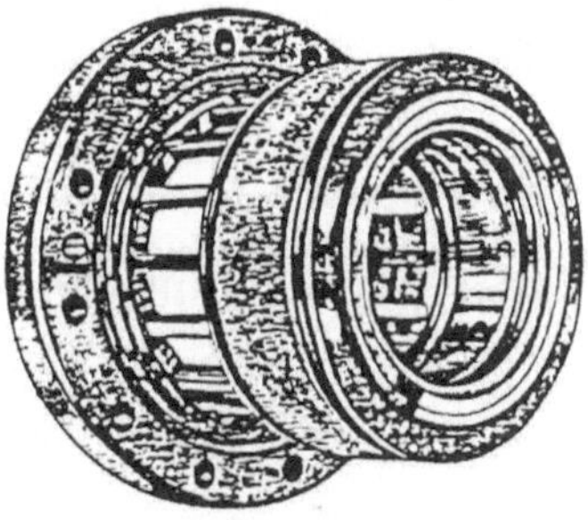

Figure 15

View of Flexible Support Structure for Journal Bearing with Squeeze Film Damper [12]

Lund and Tonnesen calculated stability threshold and whirl frequency ratio by a computer program and verified the predicted results experimentally with the two rotors described above. For the heavier rotor, figure 13, the stability threshold in rigid supports was found to be 12600 rpm. Use of a flexible foundation with a squeeze film damper enabled this same rotor to be operated at its maximum speed, 20,000 rpm without any indication of instability.

Lund and Tonnesen [12] obtained the following conclusions from their test program:

- The experiments confirmed the general validity of using an analytical model to predict the threshold speed and whirl frequency of a rotor-bearing system. The linear model predicted instability threshold in good agreement with the experimental findings. Discrepancies can more readily be ascribed to causes other than deficiencies in the analytical model.
- Unbalance vibrations were found to initiate self-excited whirl, with the result that the instability threshold speed is lowered.
- Unstable whirling was itself found to excite a spectrum of frequencies in the rotor system, but the whirl orbit is a stationary limit cycle, associated with a single whirl frequency. If a large limit cycle amplitude can be accepted, it has been shown to be feasible to operate the test rotors past the onset of instability. In practice, however, the threshold speed should be considered the maximum acceptable speed.

- The addition of flexibility and viscous damping to the foundation is the crucial point in the suppression of instability for a rotor bearing system.

Theoretical Investigations

About the same time that oil whip was identified by Newkirk in 1924 [7] and [13], Stodola [14] undertook a theoretical investigation of the influence of the journal bearings on the critical speeds of a rotor. In the course of their investigation, he and Hummel [15] arrived at the conclusion that the fluid-film forces in a bearing may induce rotor instability in cases where the journal eccentricity in the bearing falls below 65 per cent of the radial clearance.

Stodola's analysis is based on a linearization of the fluid-film forces which allows the stability of the equilibrium position to be calculated. This linearization procedure has subsequently been greatly developed. Examples of such calculations appear later in this chapter. To illustrate, assume the journal to have a mass M and let the journal center position be defined by Cartesian coordinates (X,Y). The fluid-film force components F_x and F_y, depend on the instantaneous position and velocity of the journal center such that the linearized equations of motion become:

$$M\ddot{\bar{X}} = F_x(\bar{X}, \bar{Y}, \dot{\bar{X}}, \dot{\bar{Y}}) = \frac{\partial Fx}{\partial \bar{X}} \bar{X} + \frac{\partial Fx}{\partial \dot{\bar{X}}} \dot{\bar{X}} + \frac{\partial Fx}{\partial \bar{Y}} \bar{Y} + \frac{\partial Fx}{\partial \dot{\bar{Y}}} \dot{\bar{Y}} \quad (1)$$

$$M\ddot{\bar{Y}} = Fy(\bar{X}, \bar{Y}, \dot{\bar{X}}, \dot{\bar{Y}}) = \frac{\partial Fy}{\partial \bar{X}} \bar{X} + \frac{\partial Fy}{\partial \dot{\bar{X}}} \dot{\bar{X}} + \frac{\partial Fy}{\partial \bar{Y}} \bar{Y} + \frac{\partial Fy}{\partial \dot{\bar{Y}}} \dot{\bar{Y}} \quad (2)$$

where the partial derivatives are evaluated at the equilibrium position

$\bar{X} = 0$, $\bar{Y} = 0$, $d\bar{X}/dt = 0$, $d\bar{Y}/dt = 0$. There are two simultaneous, second-order ordinary differential equations with constant coefficients which are readily tested for stability once the eight bearing coefficients are known.

Although Stodola's method is correct and simple, it presents the problem of obtaining accurate values for the bearing coefficients. Before 1950, the only available solution of Reynolds' equation was Sommerfeld's solution [16], or modifications thereof, for the infinitely long 360-degree journal bearing. Thus, Stodola was forced to neglect the four damping coefficients and to arrive at the values for the four spring coefficients from an estimate of the functional relationship between the statically imposed bearing load and the journal center position.

The first attempt to investigate the motion of a rigid journal within a bearing using hydrodynamic theory was made by Harrison [17], who derived expressions for the radial and tangential components of the fluid-film forces due to the journal displacement. These expressions are based on Reynolds' assumptions and apply to an infinitely long, full (no cavitation) bearing using an incompressible lubricant.

Robertson [18] reconsidered Harrison's analysis and showed the dependence of radial force effects on the tangential components of surface velocity. In his analysis, he considered the damping coefficients neglected by Stodola, but ignored the subambient pressures in the fluid film. This led to the incorrect conclusion that the journals are inherently unstable, and will whirl with a frequency equal to one-half the speed of rotation at all rotor speeds. This finding is only correct for a flooded vertical bearing.

Poritsky [19] showed that the inclusion of a radial force component in the equations of motion for a rigid journal predicts stability at speeds below the rotor first critical speed. It also predicts that, at speeds above twice the first critical speed, the rotor will become unstable and whirl at the rotor critical frequency. These predictions agreed with observed performance and led to the well-known stability criterion:

$$M\omega^2\left[\frac{1}{K_r} + \frac{1}{K_s}\right] < 4 \qquad (3)$$

where K_r and K_s are the rotor stiffness and bearing support stiffness respectively, M is the mass of the rotor at the journal, and ω is the rotational speed in rad/sec. Poritsky's analysis neglected the influence of fluid-film damping, and no attempt was made to determine the value of the fluid film stiffnesses, except to postulate that these would be linear with displacement for small amplitude motions. Later investigations into the elastic and damping properties of the cavitated fluid-film [20], [21], verified the existence of the radial force component, and also provided values for four-spring and damping coefficients. When damping is included, the predicted whirl frequency is less than 0.5 ω. This agrees with practical experience.

Flooded vertical bearings were investigated by Boecker and Sternlicht [20], who found that plain bearings were inherently unstable at all speeds, whereas grooved bearings had a small stability margin. Bowman, Collingwood, and Midgley [22] also studied the performance of flooded bearings under laboratory conditions.

To test the linearized equations for stability, most authors have used Routh's stability criterion Inclusion of the flexibility of the rotor in the equations of motion leads to a characteristic equation of sixth order. The problem, however, is easily reduced to one of fourth order and a solution can be obtained directly in closed form in terms of the eight bearing coefficients, as shown in references [23] and [24]. This method makes it easy to recognize that the contribution of rotor flexibility to the oil whip problem is to lower the instability threshold speed, and not to influence the basic character of the onset of instability. This was first shown in reference [25].

The interest in the problem of oil whip increased markedly after World War II, beginning with Hagg [26] in 1946, and has been growing since that time, primarily because of the trend towards high-speed machinery and the use of low kinematic viscosity lubricants. A considerable amount of additional work, both experimental and analytical, has been devoted to determining the eight Taylor coefficients (the spring and damping coefficients) for a wide variety of bearing types [21], [27], [28], primarily to determine the bearing's influence on the critical speeds of the rotor. This work, however, has made it feasible to broaden the scope of the oil whip investigation to cover other and more practical bearing types than the plain cylindrical bearings. Hagg [29] evaluated the damping capacity of a tilting-pad bearing, and later gave curves for spring and damping constants for 120 degree partial-arc bearings, together with a simple stability chart.

Later work by Hagg and Sankey [30], [31] gave more complete data on spring and damping constants for partial-arc and tilting-pad bearings for small-amplitude whirl, determined experimentally. At both low and high Sommerfeld numbers these curves are in error, as in these zones they had been obtained by extrapolation. Accurate values for the cylindrical bearing have been obtained [21] for the 150 degree partial bearing [23], and for the tilting-pad bearing [32]. Each of these analyses was for incompressible lubricants, but the results apply for compressible lubricants at low $\Lambda < 1.0$ compressibility numbers. Hagg and Warner [25] further examined the stability threshold speed, using an analog computer to study the stability limit. This work gave good qualitative correlation with both test results, and with data obtained from an industrial turbine set. Typical oscillograms obtained are shown in figure 16. The upper curve shows a well developed whirl, and the lower curve shows transition from whirl to stable running, with decrease in speed.

Badgley and Booker [33] made a theoretical study of the rigid body dynamics of rotors supported in plain cylindrical, cavitated, fluid-film journal bearings. Expressions for journal force due to the fluid film are developed using the short bearing (Ocvirk), the long bearing (Sommerfeld), and Warner's finite length bearing approximate solution to the Reynolds equation. The nomenclature used by Badgley and Booker in reference [33] is:

C	= bearing radial clearance	p	= local fluid film pressure
D	= bearing diameter	t	= time
L	= bearing length	T	= dimensionless time (radians)
F^{ϕ}, F^{ε}	= fluid film force components	ε	= eccentricity ratio
g	= acceleration of gravity	ϕ	= journal center attitude angle
j_n^{lm}	= integral notation	Θ	= fluid film orientation angle
M	= rotor mass per journal	μ	= local fluid film velocity
w	= rotor weight per journal	ωg	= equal to $(g/c)^{\frac{1}{2}}$
		ωj	= journal angular velocity

Forces acting on the journal e.g. from the fluid film are shown in figure 17.

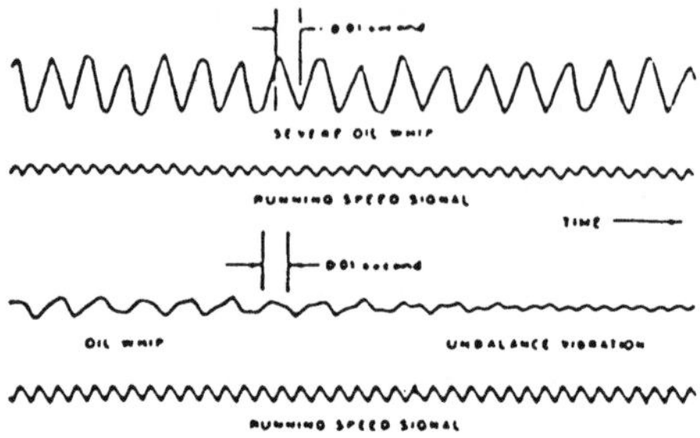

Figure 16 Whirl Oscillogram Showing (a) Well-developed Whirl and (b) Transition from Whirl to Stable Running with Decreasing Speed [25].

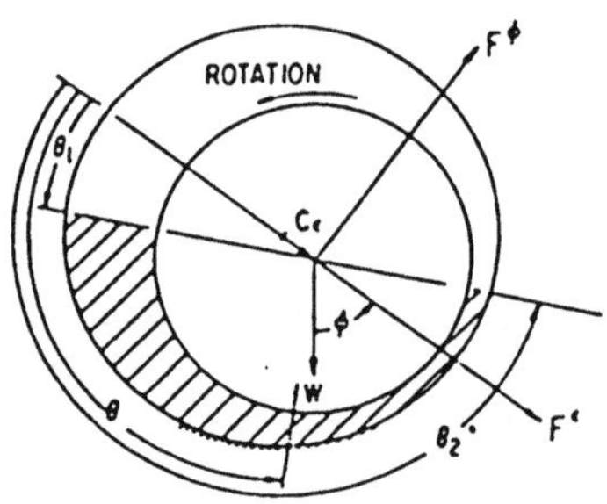

Figure 17 Geometry of Dynamically Loaded Journal Bearing [33].

With the assumption that only plane journal motion occurs, this motion (and thus that of the rigid rotor) may be described by two polar equilibrium equations. In polar coordinates, with $T = \omega_j t$

$$\frac{d^2\phi}{dT^2} = \frac{F^\phi}{\epsilon\omega_j^2 MC} - \frac{W}{\epsilon\omega_j^2 MC}\sin\phi - \frac{2}{\epsilon}\left(\frac{d\epsilon}{dT}\right)\left(\frac{d\phi}{dT}\right) \tag{4}$$

$$\frac{d^2\epsilon}{dT^2} = \frac{F^\epsilon}{\omega_j^2 MC} + \frac{W}{\omega_j^2 MC}\cos\phi + \epsilon\left(\frac{d\phi}{dT}\right)^2 \tag{5}$$

The film-force components may be obtained from the expressions:

$$F^\epsilon = \int_A p\cos\theta\, dA \tag{6}$$

$$F^\phi = \int_A p\sin\theta\, dA \tag{7}$$

which yields for the various configurations:

Short-Bearing (Ocvirk) Approximation

$$\left(\frac{F^\epsilon}{\omega_j^2 MC}\right)_s = -4\left[\frac{\omega_j\mu DL}{2W(C/R)^2}\right]\left[\frac{L}{D}\right]^2\left[\frac{\omega_e}{\omega_j}\right]^2 \times \left[\frac{d\epsilon}{dT}J_3^{02} + \epsilon\left(\frac{d\phi}{dT} - \frac{1}{2}\right)J_3^{11}\right] \tag{8}$$

$$\left(\frac{F^\phi}{\epsilon\omega_j^2 MC}\right)_s = -\frac{4}{\epsilon}\left[\frac{\omega_j\mu DL}{2W(C/R)^2}\right]\left[\frac{L}{D}\right]^2\left[\frac{\omega_e}{\omega_j}\right]^2 \times \left[\frac{d\epsilon}{dT}J_3^{11} + \epsilon\left(\frac{d\phi}{dT} - \frac{1}{2}\right)J_3^{20}\right] \tag{9}$$

Long-Bearing (Sommerfeld) Approximation

$$\left(\frac{F^\epsilon}{\omega_j^2 MC}\right)_L = -6\left[\frac{\omega_j\mu DL}{2W(C/R)^2}\right]\left[\frac{\omega_e}{\omega_j}\right]^2\left[\frac{d\epsilon}{dT}(J_1^{02} + J_3^{02}) + \left(\frac{2\epsilon}{2+\epsilon^2}\right)\left(\frac{d\phi}{dT} - \frac{1}{2}\right)(J_1^{11} + J_3^{11})\right] \tag{10}$$

$$\left(\frac{F^\phi}{\epsilon\omega_j^2 MC}\right)_L = -\frac{6}{\epsilon}\left[\frac{\omega_j\mu DL}{2W(C/R)^2}\right]\left[\frac{\omega_e}{\omega_j}\right]^2\left[\frac{d\epsilon}{dT}(J_1^{11} + J_3^{11}) + \left(\frac{2\epsilon}{2+\epsilon^2}\right)\left(\frac{d\phi}{dT} - \frac{1}{2}\right)(J_1^{20} + J_3^{20})\right] \tag{11}$$

Finite Length Bearing (Warner) Approximation

$$\left(\frac{F^{\epsilon}}{\omega_j^2 MC}\right)_F = \left(\frac{F^{\epsilon}}{\omega_j^2 MC}\right)_L \cdot \Pi \qquad (12)$$

$$\left(\frac{F^{\phi}}{\epsilon\omega_j^2 MC}\right)_F = \left(\frac{F^{\phi}}{\epsilon\omega_j^2 MC}\right)_L \cdot \Pi \qquad (13)$$

where $\Pi = \left[1 - \frac{\tanh\left(\frac{\lambda L}{D}\right)}{\left(\frac{\lambda L}{D}\right)}\right]$. In all cases the integrals $J_n^{l,m} = \int_{\theta_1}^{\theta_2} \frac{\sin^l\theta \cos^m\theta}{(1+\epsilon\cos\theta)^n}\, d\theta$

(14),(15)

are functions of θ_1, θ_2 and ε

The quantities $d^2\varepsilon/dT^2$ and $d^2\phi/dT^2$ may be evaluated using the equations 4 and 5 for given values of ε, ϕ, $d\varepsilon/dT$ and $d\phi/dT$. If the latter four quantities are specified initially, the resulting initial value problem may be solved by standard numerical "marching" techniques. In this investigation, solutions to the equations of motion were obtained by fourth order Runge-Kutta extrapolation techniques for a wide range of combinations of the ε_o and ω_j/ω_g parameters. Initial displacements of the journal from its equilibrium position were simulated by specifying starting values of ε larger than the value of ε_o with ϕ set equal to ϕ_o. Stability is determined visually and the solution is represented as a stable point on the ε_o versus ω_j/ω_g parameter plane. Figure 18 shows the results of several "experimental" simulations. The fact that these points may be separated by a single line indicates that the direction of the initial displacement does not

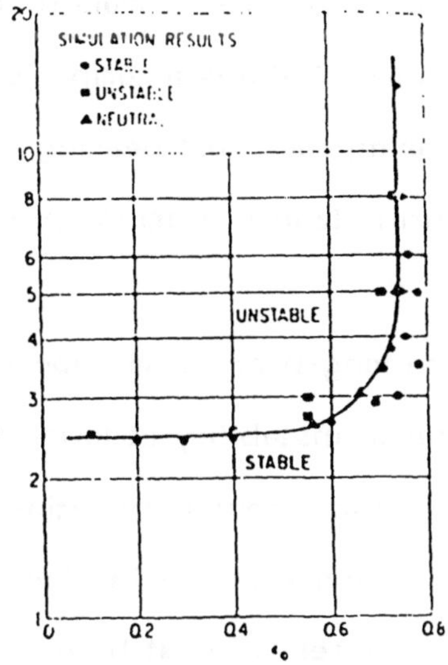

Figure 18

Short Bearing Stimulation Results after Initial Position Disturbances [33]

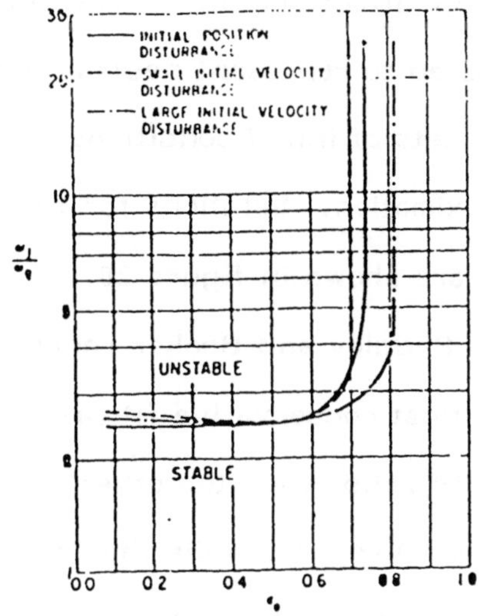

Figure 19

Instability Threshold Curves - Short Bearing Simulation after Various Initial Disturbances [33]

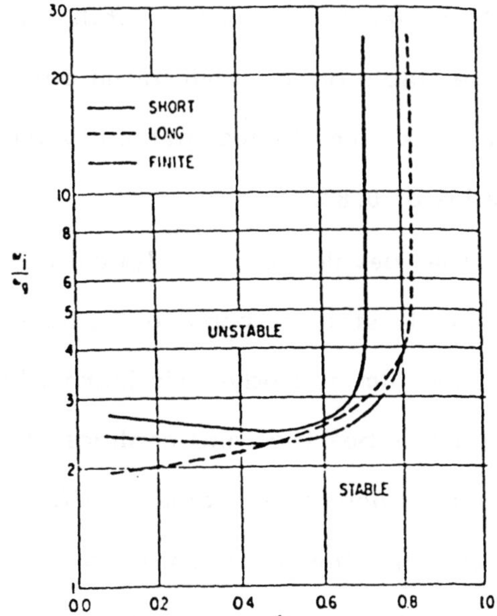

Figure 20 Instability Threshold Curves for Bearing Assumptions After Small Initial Velocity Disturbances [33]

sensibly affect stability. The stability boundary is drawn through points estimated to represent neutral stability. Figure 19 shows a comparison of three sets of initial conditions investigated using the short bearing approximation. Solutions for the long and finite length bearing approximations are shown in figure 20.

Badgley and Booker concluded that the long-bearing solution appeared to be most conservative (it predicts the onset of instability at lower angular velocity ratios than the other solutions) for static eccentricity ratios between 0 and 0.5, while the finite-bearing solution, with bearing length to diameter ratio L/D equal to 1, appeared most conservative at higher static eccentricity ratios. Variations in L/D between 0.5 and 7.0 did not affect journal path shapes appreciably. Variations in initial journal center velocity were found to be important, at least with the short-bearing solution; large initial velocities were observed to produce instability for certain parameter combinations which are stable under small initial positions or small velocity disturbances. In all cases investigated, instability was not observed above static eccentricity ratios of 0.83.

In 1975 Myrick [34] developed an analytical method for the simulation of the transient and steady state response of flexible rotors supported by "realistic" incompressible film hydrodynamic journal bearings. His method combines realistic fluid-film bearings and nonlinear flexible-rotor dynamics. In this method the coupled nonlinear ordinary differential equation of motion (which describes a flexible rotor similar to that of figure 19) is formulated as an initial-value problem, and is solved by direct integration on a digital computer with the hydrodynamic bearing forces and moments obtained for each time increment from a finite-difference formulation of the general Reynolds equation. This "realistic" bearing model includes:

- Finite-length hydrodynamic bearings
- Wedge and squeeze films
- Fluid-film cavitation
- Oil inlet geometry and pressure
- Eccentricity and tilt (gyroscopics) of the journal.

With the numerical extrapolation approach, the transient and steady-state response of the rotor may be simulated, and the stability of the nonlinear rotor-bearing system can be determined through inspection of the generated rotor trajectories. The results given in ref. [34] were obtained using this approach; the response of a symmetric flexible rotor system for speeds up to the threshold of instability is demonstrated as a function of disk unbalance (single plane), and of viscous damping (see figure 23). The symmetric rotor parameter values for this study are listed in table 2,and the hydrodynamic journal bearing parameter values are listed in table 3. Curves for inlet oil supply pressure P_{in} and effective absolute oil viscosity P_{eff} as functions of rotor speed are presented in figure 21.

The numerical simulations of the response of the symmetrical test rotor were then compared with results obtained from experimental whirl data. Speeds through the critical speed region to the onset of oil whip were studied using an experimental test rig to obtain whirl and whip data for a flexible, horizontal rotor bearing system. The rotor consisted of a disk and two journals pressed onto a shaft, supported in two cylindrical hydrodynamic journal bearings. The shaft was comparatively long, and of

Modulus of elasticity, E	20.2×10^{11} N·m^{2}
Shaft diameter, D_s	2.54×10^{2} m
Shaft moment of inertia, I	2.04×10^{8} m^{4}
Shaft segment length, L_s	1.96×10^{1} m
Rigid journal width, z_1^s	4.07×10^{2} m
Shaft length, L_{total}	5.81×10^{1} m
Oil inlet hole diameter, D_{oh}	6.35×10^{3} m
Lubricating oil number	SAE 15
Rotor support length, L_{spt}	4.74×10^{-1} m
Rotor specific weight, γ (steel)	7.68×10^{4} N/m^{3}
Total rotor weight	1.83×10^{2} N
Effective journal mass, m_1	1.68 N·s^{2}/m
Disk mass, m_d	14.0 N s^{2}/m
Effective disk mass, m_3	15.3 N·s^{2}/m
Effective journal polar mass moment of inertia, I_{z_1}	5.83×10^{4} N.m.s.^{2}
Effective journal transverse mass moment of inertia, I_{xy_1}	1.63×10^{3} N. .s.^{2}
Undamped fundamental natural frequency, ζ_n	55.7 Hz (calculated)
Rotor speed, ζ	0 110 Hz
Disk mass-eccentricity, e_R	1.54×10^{6}; $e.09 \times 10^{6}$ m
Disk damping ratio, ζ_v	0, 0.025, 0.04; 0.05

Table 2 Rotor Parameter Values for the Flexible Rotor Simulations [34].

Bearing length, L_b	2.54×10^{-2} m
Journal radius, R = R	2.75×10^{-2} m
Radial bearing clearance, C_r	7.62×10^{-5} m
Bearing grid size, $n_v \times n_z$	16 x 6
SOR relaxation factor ω_b	1.290
Oil inlet hole location, v_{oh}	180 degrees
Effective lubricant absolute viscosity, v_{eff}	(see figure 5)
Atmospheric pressure, P_{atm}	1.01×10^{-5} N/m^{2}
Oil inlet supply pressure, P_{in}	(see figure 5)
Pressure distribution error criterion, ϵ_p	6.89×10^{3} N/m^{2}

Table 3 Hydrodynamic Journal Bearing Parameter Values for the Flexible Rotor Simulation [34].

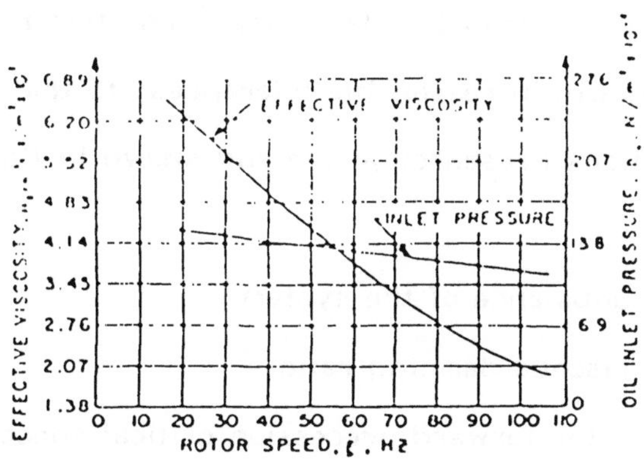

Figure 21 Effective Oil Viscosity and Bearing Inlet Pressure Versus Rotor Speed from Experimental Data [34].

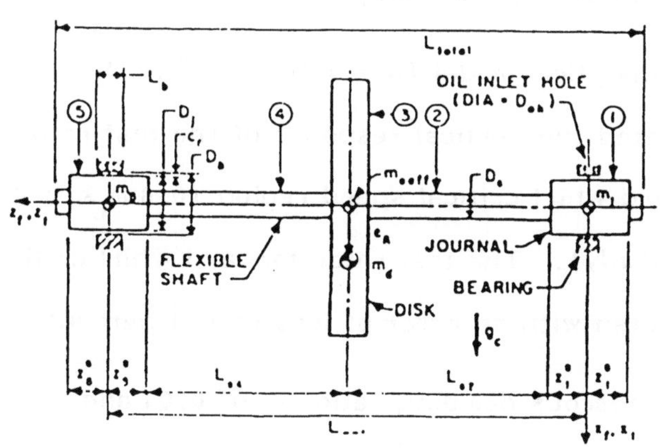

Figure 22 Three Mass Station, Elastic Shaft Rotor with Single Plane Unbalance [34].

small diameter to insure reasonable flexibility. The test rotor bearing system was identical, within reasonable tolerances, to that used for the flexible rotor simulations presented above and shown in figure 22. Using the notations:

W = unbalance of the system

ξ = viscous damping ratio

ω_c = rotor forward precession critical speed, Hz

ω = running speed,

the most significant results of this analysis can be summarized as follows:

- Oil whip has been successfully simulated by this analytical model of a flexible rotor/hydrodynamic journal bearing system;
- The oil whip could be suppressed by means of the addition of external viscous damping;
- The analytical model ($W = 4.24 \times 10^{-4}$ N.M.; $\xi = 0.04$) accurately simulated the vertical response of the test rotor through the forward whirl critical speed region ($0.8\ \omega_c \leq \omega \leq 1.15\ \omega_c$, for this study). The test rotor forward whirl critical speed was predicted with an error of less than 1 percent;
- Using viscous damping ratio ξ of 0.04 and 0.05, the analytical model ($W = 6.24 \times 10^{-4}$) simulated the onset of unstable oil whip ($W = 105.5$ Hz) of the test rotor with less than 3 percent error in this system.

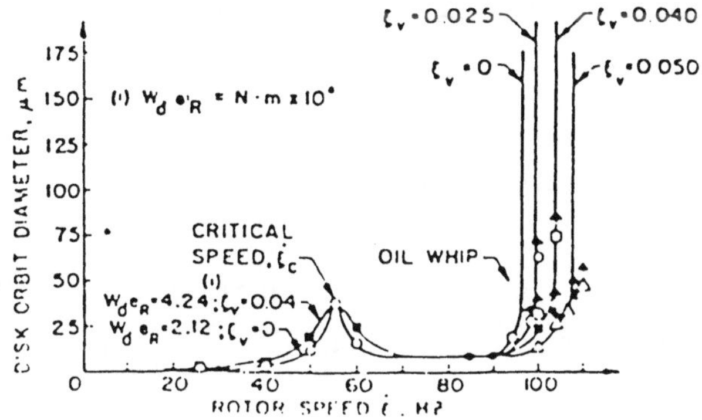

Figure 23 Disk Orbital Response as a Function of Rotor Speed, Disk Unbalance, and External Disk Damping [34].

Theory of Instability for a Rigid Rotor in Fluid Film Bearings

Consider the rigid rotor in fluid film bearings shown in figure 24. For convenience, this rotor is taken as a simple cylinder of mass M, translatory inertia I_t and polar inertia I_p. Both bearings are identical and share the rotor load (gravity) equally, so that they both have the same operating eccentricity and bearing dynamic coefficients. At the instability threshold speed ω_t the whirl frequency ν will be somewhat less than 0.5ω, as seen previously. Two possible modes of whirling exist, (a) translatory whirl (ends in in-phase), and (b) conical whirl (ends in anti-phase), figure 25. The question of the mode in which the rotor instability will occur is determined by the inequality

$$4.\frac{I_t - \frac{\nu}{\omega} I_p}{ML^2} \quad \begin{matrix} < 1.0, & \text{translatory whirl} \\ > 1.0, & \text{conical whirl,} \end{matrix} \qquad (16)$$

where $I_t = \frac{1}{12} M(3R^2 + L^2)$ and $I_p = \frac{1}{2} MR^2$ for a solid cylindrical rotor.

Writing $\frac{\nu}{\omega} = 0.5$ and substituting gives:

$$4. \frac{\frac{1}{12} M(3R^2 + L^2) - 0.5 \; \frac{1}{2} MR^2}{ML^2} = 4. \frac{ML^2[1+3(\frac{R}{L})^2 - 3(\frac{R}{L})^2]}{12ML^2} = \frac{1}{3} \quad (17)$$

This expression applies for a solid cylindrical rotor at all speeds.

In general, however, the ratio is $4[I_t-(\nu/\omega)I_p]/ML^2$ for a rotor symmetrical about the mid-span plane. The above result indicates

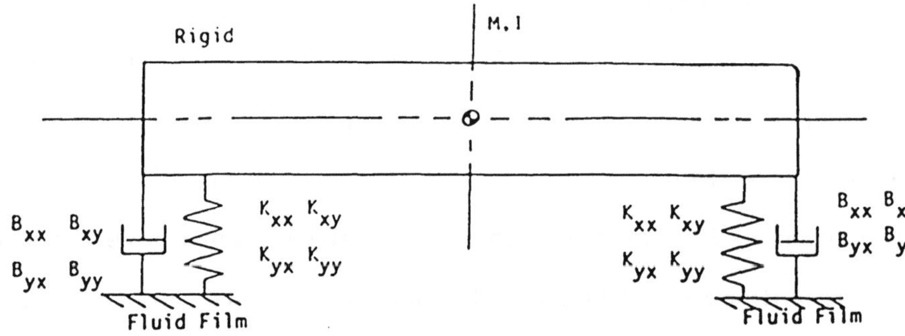

Figure 24 Rigid Rotor in Fluid Film Bearings

that the tendency is for a rigid rotor to whirl in its translatory mode.

To determine the half-frequency whirl threshold condition, consider the rotor to be perfectly balanced, as in figure 26, and symmetrical in all respects about the mid-span plane. Assuming that the rotor will whirl first in its translatory mode, the threshold conditions may be obtained by the procedure first given by Lund [35]. The equations of motion are

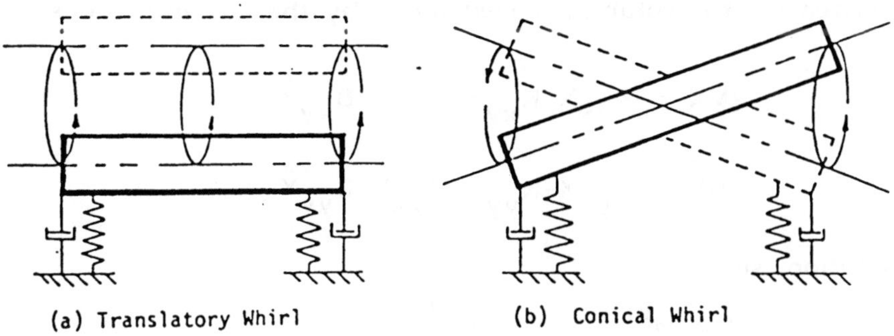

Figure 25 Whirl Modes for Rigid Rotor in Flexible Bearings

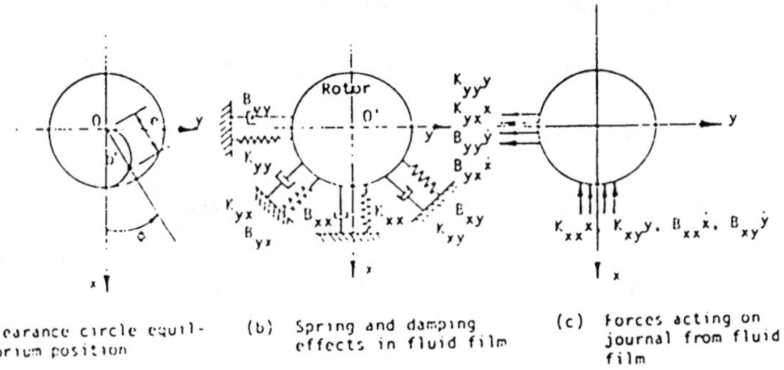

Figure 26 Film Force Effects

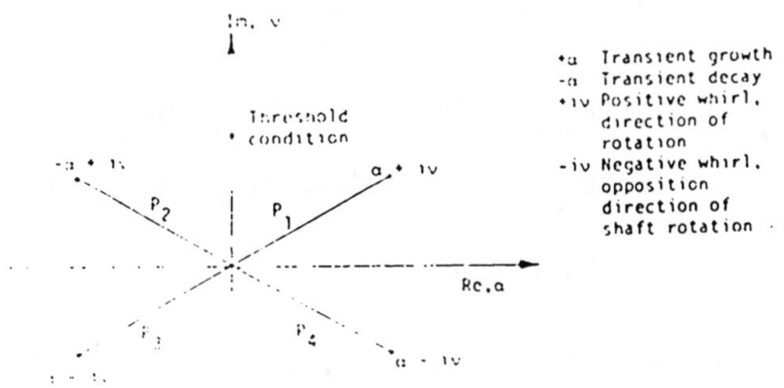

Figure 27 Complex Plane Plot of Eigenvalue P Conditions

first written for the rotor c.g. restrained by the bearing forces:

$$\tfrac{1}{2}M\ddot{X} = -K_{xx}X-K_{xy}Y-B_{xx}\dot{X}-B_{xy}\dot{Y} \tag{18}$$

$$\tfrac{1}{2}M\ddot{Y} = -K_{yx}X-K_{yy}Y-B_{yx}\dot{X}-B_{yy}\dot{Y} \tag{19}$$

For a solution, set:

$$X = xe^{pt}; \qquad Y = ye^{pt} \tag{20}$$

where $p = \alpha + i\nu$. In this expression ν is the rotor whirl frequency and α is the system damping coefficient. At the onset of whirl $\alpha = 0$ and $p = \alpha + i\nu$ occurs at the boundary of the real axis. In this condition α is changing from negative to positive, as ω increases (see figure 27). Substituting $\alpha = 0$ and $p = i\nu$ gives:

$$\begin{bmatrix} 2K_{xx}-M\nu^2 + i\nu\, 2B_{xx} & 2K_{xy} + i\nu\, 2B_{xy} \\ 2K_{yx} + i\nu 2B_{yx} & 2K_{yy} - M\nu^2 + i\nu 2B_{yy} \end{bmatrix} \begin{Bmatrix} X \\ Y \end{Bmatrix} = \begin{Bmatrix} 0 \\ 0 \end{Bmatrix} \tag{21}$$

Expanding this yields:

$$(2K_{xx}-M\nu^2+i\nu 2B_{xx})(2K_{yy}-M\nu^2+i\nu 2B_{yy})-(2K_{xy}+i\nu 2B_{xy})\cdot$$
$$\cdot(2K_{yx}+i\nu 2B_{yx}) = 0 \tag{22}$$

$$(2K_{xx}-M\nu^2)(2K_{yy}-M\nu^2)-\nu^2 2B_{xx}2B_{yy}-2K_{xy}2K_{yx}-\nu^2 2B_{xy}2B_{yx}+$$
$$i\nu 2B_{xx}(2K_{yy}-M\nu^2)+i\nu 2B_{yy}(2K_{xx}-M\nu^2)-i\nu 2K_{xy}2B_{yx}-$$
$$i\nu 2K_{yx}2B_{xy} = 0 \tag{23}$$

Equating the real and the imaginary expressions to zero gives:

$$2K_{xx}2K_{yy}+(M\nu^2)^2-M\nu^2(2K_{xx}+2K_{yy})-2K_{xy}2K_{yx}$$
$$-\nu^2(2B_{xx}2B_{yy}-2B_{xy}2B_{yx}) = 0 \tag{24}$$

and

$$\nu[2B_{xx}2K_{yy}+2B_{yy}2K_{xx}-M\nu^2(2B_{xx}2B_{yy})-(2K_{xy}2B_{yx}+2K_{yx}2B_{xy})] = 0 \quad (25)$$

writing $\nu = \gamma\omega$ where γ is the whirl frequency ratio (ν/ω) gives:

$$\frac{(K_{xx}\omega B_{yy}+K_{yy}\omega B_{xx})-(K_{xy}\omega B_{yx}+K_{yx}\omega B_{xy})}{(\omega B_{xx}+\omega B_{yy})} = \frac{1}{2}M\omega^2\gamma^2 = \kappa \quad (26)$$

$$\frac{(K_{xx}-\frac{1}{2}M\nu^2)(K_{yy}-\frac{1}{2}M\nu^2)-K_{xy}K_{yx}}{\omega B_{xx}\omega B_{yy}-\omega B_{xy}\omega B_{yx}} = (\frac{\nu}{\omega})^2 = \gamma^2 \quad (27)$$

The bearing stiffness and damping coefficients are now written in terms of the speed of rotation. They may thus be selected directly for any given operating condition.

To determine the whirl threshold speed, the procedure is as follows:

- Select a trial threshold speed ω
- Determine the corresponding bearing coefficients Kxx, Bxx, etc.
- Substitute in equation 26, determine $M\nu^2=\alpha K$
- Substitute in equation 27, determine γ^2
- Calculate ω^2 from $2K/M\gamma^2=\omega_{cal}^2$ hence ω_{cal}
- Compare ω_{cal} with ω_{trial}. When $\omega_c=\omega_t$ this is the threshold speed.

Several trial calculations may be necessary before agreement is reached. The cross plot shown in figure 28 facilitates convergence.

These equations may be easily programmed for direct solutions. Values of the bearing coefficients must be supplied for each trial speed assumed.

Rigid-rotor conical whirl stability threshold speeds may also be found by replacing the mass term in the above equations by the "effective mass" M acting at the bearings. As indicated previously, this is given by:

$$M_e\left(\frac{L}{2}\right)^2 = I_t - \frac{\nu}{\omega} I_p \approx I_t - 0.5\, I_p = \frac{ML^2}{12} \text{ (cylinder)} \tag{28}$$

$$M_e = \frac{1}{3} M \tag{29}$$

i.e., the effective mass is one-third of the actual mass, for a rigid cylindrical rotor. Thus the term M becomes M_e in equations 28 and 29, the value of which is

$$M_e = \begin{cases} \left(I_t - \frac{\nu}{\omega} I_p\right) / \left(\frac{L}{2}\right)^2 \approx 4(I_t - 0.5\, I_p) / L^2 & \text{[General Rotor Symmetry]} \\ \frac{1}{3}M & \text{[Cylindrical Rotor].} \end{cases} \tag{30}$$

Instability of a Flexible Rotor in Fluid-Film Bearings

The above analysis may be readily extended to the case of a flexible rotor in fluid-film bearings. Consider the flexible rotor shown in figure 29 which carries two disks each of mass $\frac{1}{2}M$ a distance ξL apart, where L is the rotor span between bearings. The system has mid-span symmetry, as previously. Two disks are used in this example so that either the

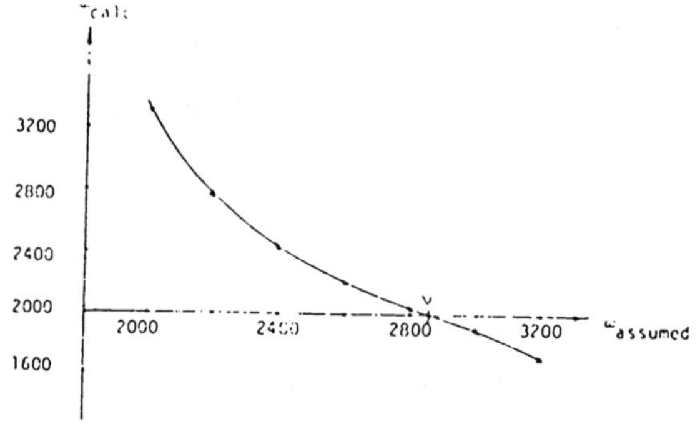

Figure 28 Plot of $\omega_{calc.}$ vs. $\omega_{assumed}$

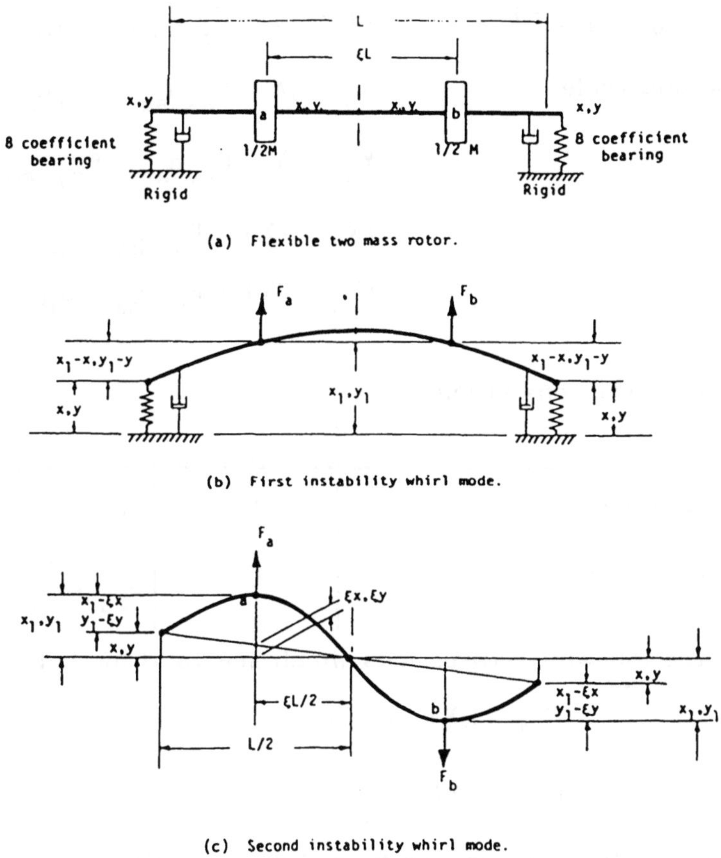

Figure 29 Flexible Rotor and Whirl Modes

translatory instability whirl mode or the conical instability whirl mode may be obtained directly from the following analysis given by Lund [35].

Shaft bending deflection between disk and bearing due to forces F:

First mode:

$$(X_1 - X) = F_{ax}\alpha_{aa} + F_{bx}\alpha_{ab} \tag{31}$$

$$(Y_1 - Y) = F_{ay}\alpha_{aa} + F_{by}\alpha_{ab} \tag{32}$$

as $F_{ax} = F_{bx} = F_x$, $$(X_1 - X) = F_x\,(\alpha_{aa} + \alpha_{ab}) \qquad (33)(34)$$

and $F_{ay} = F_{by} = F_y$, $$(Y_1 - Y) = F_y\,(\alpha_{aa} + \alpha_{ab}) \qquad (35)(36)$$

Second mode: $$X_1 - \xi X = F_{ax}\alpha_{aa} - F_{bx}\alpha_{ab} \qquad (37)$$

$$Y_1 - \xi Y = F_{ay}\alpha_{aa} - F_{by}\alpha_{ab} \qquad (38)$$

i.e., as above, $$(X_1 - \xi X) = F_x(\alpha_{aa} - \alpha_{ab}) \qquad (39)$$

$$(Y_1 - \xi Y) = F_y(\alpha_{aa} - \alpha_{ab}) \qquad (40)$$

Introducing the conventions:

	ξ	α
First mode	1	$\alpha_{aa} + \alpha_{ab}$
Second mode	ξ	$\alpha_{aa} + \alpha_{ab}$

we may write the following expressions which are valid for both modes:

$$(X_1 - \xi X) = F_x . \alpha \qquad (41)$$

$$(Y_1 - \xi Y) = F_y . \alpha \qquad (42)$$

The equations of motion may now be formed by recognizing that:

$$F_x = -\frac{1}{2} M\ddot{X}_1 \qquad (43)$$

$$F_y = -\frac{1}{2} M\ddot{Y}_1 \qquad (44)$$

Hence, $$-\frac{1}{2}\alpha M\ddot{X}_1 = (X_1 - \xi X) \qquad (45)$$

$$-\frac{1}{2}\alpha M\ddot{Y}_1 = (Y_1 - \xi Y) \qquad (46)$$

The force balance at the bearings is shown in figure 30.

First mode, (forces)

$$\frac{1}{\alpha}(X_1 - \xi X) = K_{xx}X + K_{xy}Y + B_{xx}\dot{X} + B_{xy}\dot{Y} \quad (47)$$

$$\frac{1}{\alpha}(Y_1 - Y) = K_{yx}X + K_{yy}Y + B_{yx}\dot{X} + B_{yy}\dot{Y} \quad (48)$$

Second mode, (moments)

$$F_x \cdot \xi L = (K_{xx}X + K_{xy}Y + B_{xx}\dot{X} + B_{xy}\dot{Y}) \cdot L = \frac{\xi L}{\alpha}(X_1 - \xi X) \quad (49)$$

$$F_y \cdot \xi L = (K_{xy}X + K_{yy}Y + B_{xy}\dot{X} + B_{yy}\dot{Y}) \cdot L = \frac{\xi L}{\alpha}(Y_1 - \xi Y) \quad (50)$$

Again using the previous conventions, the above equations for both modes may be written as:

$$\frac{\xi}{\alpha}(X_1 - X) = K_{xx}X + K_{xy}Y + B_{xx}\dot{X} + B_{xy}\dot{Y} \quad (51)$$

$$\frac{\xi}{\alpha}(Y_1 - Y) = K_{yx}X + K_{yy}Y + B_{yx}\dot{X} + B_{yy}\dot{Y} \quad (52)$$

It is now necessary to eliminate X_1, Y_1 from the above equations. To do this, substitute

$$X_1 = x_1 e^{i\nu t} \qquad X = xe^{i\nu t} \quad (53)(54)$$

$$Y_1 = y_1 e^{i\nu t} \qquad Y = ye^{i\nu t} \quad (55)(56)$$

into equations (45) and (46) and solve for x_1, y_1. This gives:

$$X_1 = \frac{\xi x}{1 - \frac{1}{2}\alpha M \nu^2} = \frac{\xi x}{1 - \frac{1}{2}(\frac{\nu}{\omega n})^2} \quad (57)$$

$$Y_1 = \frac{\xi y}{1 - \frac{1}{2}\alpha M \nu^2} = \frac{\xi y}{1 - \frac{1}{2}(\frac{\nu}{\omega n})^2} \quad (58)$$

where $\omega_n^2 = 1/\alpha M$. Substituting equations 53 through 58 into equations 51 and 52 and cancelling the $e^{i\nu t}$ on both sides gives:

$$\frac{\xi}{\alpha}\left\{\frac{\xi x}{1-\frac{1}{2}(\frac{\nu}{\omega})^2} - \xi x\right\} = K_{xx}X + K_{xy}Y + i\nu B_{xx}X + i\nu B_{xy}Y \tag{59}$$

$$\frac{\xi}{\alpha}\left\{\frac{\xi y}{1-\frac{1}{2}(\frac{\nu}{\omega n})^2} - \xi y\right\} = K_{yx}X + K_{yy}Y + i\nu B_{yx}X + i\nu B_{yy}Y \tag{(6)(60)}$$

Multiplying through by C/W and writing:

$$\kappa = \frac{C\xi^2}{W\alpha} \cdot \frac{\frac{1}{2}(\frac{\nu}{\omega n})^2}{1-\frac{1}{2}(\frac{\nu}{\omega n})^2} = \frac{C\xi^2}{W\alpha} \cdot \frac{\frac{1}{2}(\frac{\nu}{\omega n})^2(\frac{\omega th}{\omega n})^2}{1-\frac{1}{2}(\frac{1}{\omega n})^2(\frac{\omega th}{\omega n})^2} = \frac{C\xi^2}{W\alpha} \cdot \frac{\frac{1}{2}(\gamma S)^2}{1-\frac{1}{2}(\gamma S)^2} \tag{61}$$

allows equations 59 and 60 to be written in terms of dimensionless stiffness and damping ratios, as follows:

$$\begin{bmatrix} \bar{K}_{xx} - \kappa + i\gamma\bar{B}_{xy} & \bar{K}_{xy} + i\gamma\bar{B}_{xy} \\ \bar{K}_{yx} + i\gamma\bar{B}_{yx} & \bar{K}_{yy} - \kappa + i\gamma\bar{B}_{yy} \end{bmatrix} \begin{Bmatrix} x \\ y \end{Bmatrix} = \begin{Bmatrix} 0 \\ 0 \end{Bmatrix} \tag{62}$$

where

$$\bar{K}_{xx} = \frac{CK_{xx}}{W} \qquad \frac{\nu}{\omega}\bar{B}_{xx} = \frac{\nu}{\omega}\frac{C\omega B_{xx}}{W} = \gamma\bar{B}_{xx}. \tag{63}$$

$$\bar{K}_{xy} = \frac{CK_{xy}}{W} \qquad \frac{\nu}{\omega}\bar{B}_{xy} = \frac{\nu}{\omega}\frac{C\omega B_{xy}}{W} = \gamma\bar{B}_{xy} \tag{64}$$

$$\bar{K}_{yx} = \frac{CK_{yx}}{W} \qquad \frac{\nu}{\omega}\bar{B}_{yx} = \frac{\nu}{\omega}\frac{C\omega B_{yx}}{W} = \gamma\bar{B}_{yx} \tag{65}$$

$$\bar{K}_{yy} = \frac{CK_{yy}}{W} \qquad \frac{\nu}{\omega}\bar{B}_{yy} = \frac{\nu}{\omega}\frac{C\omega B_{yy}}{W} = \gamma\bar{B}_{yy} \tag{66}$$

and $\gamma = \nu/\omega_{th}$, the whirl frequency ratio, usually somewhat less than 0.5, and $s = (\frac{\omega_{th}}{\omega n})$, the threshold speed ratio.

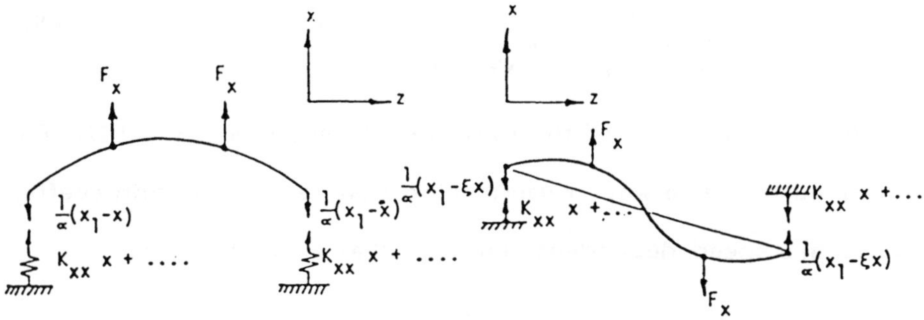

Figure 30 Forces and Moments on Rotor

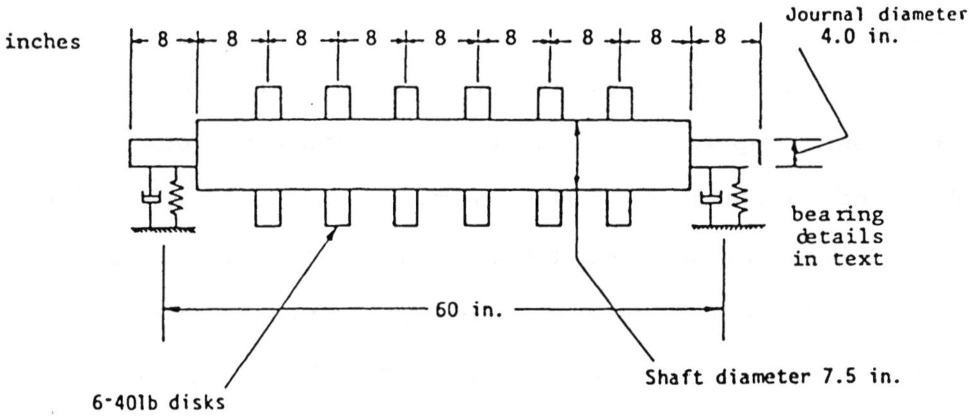

Figure 31 Compressor Rotor

To find the eigenvalues of these equations, multiply out the determinant of the coefficients and separate into real and imaginary parts, to obtain:

$$\frac{(\bar{K}_{xx}\omega\bar{B}_{yy} + \bar{K}_{yy}\omega\bar{B}_{xx}) - (\bar{K}_{xy}\omega\bar{B}_{yx} + \bar{K}_{yx}\omega\bar{B}_{xy})}{\bar{K}_{xx}\omega\bar{B}_{yy} + \bar{K}_{yy}\omega\bar{B}_{xx}} = \kappa \qquad (67)$$

$$\frac{(\bar{K}_{xx} - \kappa)(\bar{K}_{yy} - \kappa) - \bar{K}_{xy}\bar{K}_{yx}}{\omega\bar{B}_{xx}\omega\bar{B}_{yy} + \omega\bar{B}_{xy}\omega\bar{B}_{yx}} = \gamma^2 \tag{68}$$

These equations are of the same form as equations 26 and 27. They may be solved the same way, remembering that (a) the bearing coefficients $\bar{K}_{xx}$...etc. are speed-dependent, and (b) that κ is given by:

$$\kappa = \frac{C\xi^2}{W\alpha} \cdot \frac{\frac{1}{2}(S\gamma)^2}{1 - \frac{1}{2}(S\gamma)^2} \tag{69}$$

Notice that $C/W\alpha$ acts as the dimensionless rotor stiffness in either mode ($\bar{K}$).

As an example of the use of this method, consider the rotor system shown in figure 31. A horizontal compressor rotor weighs 1000 lb. and operates in 2 - 4.00 in. diam. x 2.00 in. long plain cylindrical bearings. Bearing clearance is 0.0040 in. on the diameter. Oil of viscosity 22 cp. at 110°F and 4.84 cp. at 200°F is supplied to each bearing at the rate of 1.25 gal/min at operating speed under 20 lb/in^2 gage inlet pressure, at 120°F. For these operating conditions the variation of the bearing dynamic coefficients is as shown in figure 32. What is the whirl threshold speed for this rotor in its bearings? At 10,000 rpm the bearing coefficients are, for S =2.2, ε = 0.17:

$K_{xx} = 0.33 \times 10^6$ lb / in. $\qquad \omega B_{xx} = 2.89 \times 10^6$ lb/in. (70)(71)

$K_{xy} = 1.49 \times 10^6$ lb/in. $\qquad \omega B_{xy} = 0.61 \times 10^6$ lb/in. (72)(73)

$$K_{yx} = -1.120 \times 10^6 \text{ lb/in.} \qquad \omega B_{yx} = 0.58 \times 10^6 \text{ lb/in.} \quad (74)(75)$$

$$K_{yy} = 0.59 \times 10^6 \text{ lb/in.} \qquad \omega B_{yy} = 2.52 \times 10^6 \text{ lb/in.} \quad (76)(77)$$

By substitution in equation 67

$$\kappa = 0.445 \times 10^6 \text{ lb/in.} \qquad (78)$$

From equation 68 $\quad \gamma^2 = 0.256, \quad \gamma = 0.506.$

Now

$$\kappa = \frac{\xi^2}{\alpha} \cdot \frac{\frac{1}{2}(S\gamma)^2}{1-\frac{1}{2}(S\gamma)} \qquad (\xi = 1, \quad \alpha = \frac{1}{M\omega^2}) \qquad (79)$$

A separate calculation for ω_n gives N_n = 8600 rpm. Note that this is the critical speed for the rotor in rigid bearings, $\omega_n^2 = 1/M\alpha$. Then:

$$\alpha = \left\{ \left(\frac{500}{386}\right) \times (900)^2 \right\}^{-1} = 0.955 \times 10^{-6} \text{ in/lb (half-weight).} \quad (80)$$

$$\kappa = 1.045 \times 10^6 \cdot \frac{\frac{1}{2}(0.256)s^2}{1 - \frac{1}{2}(9.256)s^2} = 0.445 \times 10^6 \text{ lb/in.} \quad (81)$$

Therefore $\quad s = \dfrac{\omega_{th}}{\omega_n} = 1.165 \qquad (82)$

$$9.55\ \omega_{th} = (1.165)(8600) = 10.100 \text{ rpm} \qquad (83)$$

$$\underline{N_{th}} = \underline{10{,}100 \text{ rpm}}. \qquad (84)$$

This is the whirl threshold speed. It could be raised, if required, by (a) increased clearance (0.004 to 0.006 in.), (b) decreased bearing length (2.0 in. to 1.5 in.), (c) increased oil inlet temp (120° to 140°) or by changing the bearing type (tilting pad).

Comparisons of Theoretical and Experimental Values for Threshold Speed

A computer program by Rieger and Thomas [36] to calculate the threshold speed associated with the lowest instability mode gave good correlation with results of previous work, e.g., Newkirk [1], Pinkus [2], and Tondl [9]. Variation of the eight bearing coefficients was incorporated into the program by curve-fitting the bearing coefficient charts.

Table 4 gives details of several experimental rotor-bearing systems used by previous investigators. Using the program mentioned above, the threshold speed was calculated for each of these systems. Where details

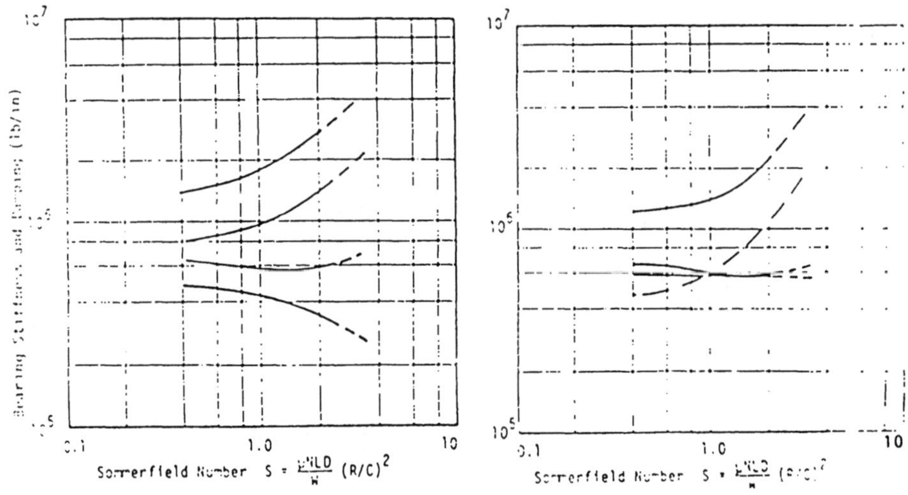

Figure 32 Plain Cylindrical Bearing Coefficients

were insufficient, e.g., descriptions of the bearing surface, oil supply pressure, outlet temperature, etc., reasonable assumed values were taken. Correlation of predicted threshold speeds with the experimentally observed values was quite close in most instances, and within the usual range of practical threshold speed variation. It can be concluded that, where valid data on bearing geometry and operating conditions are available, the threshold speed of smooth-running rotors in fluid bearings can be predicted to within design accuracy by using linear bearing theory. Common sources of prediction errors are: (i) viscosity variation, (ii) bearing clearance, (iii) misalignment, and (iv) inlet pressure effects on the fluid

	ROTOR SYSTEM PARAMETERS									THRESHOLD SPEED	
	D	L	C	R	W	μ	L1	F1	D1	Experiment	Cal.
Source	IN.	IN.	IN.	IN.	LBS	$X10^{-6}$	IN.	RPM	IN.	RPM	RPM
Pinkus-1956	2	1	.0025	1	84	9.425	32	2500	2	5000-7500	6200
Pinkus-1956	2	2	.0025	1	82	9.425	46	6100	2	12,000	13,200
Newkirk-1956	2	1	.002	1	96	3.45	36	1600	2	3375	3500
Newkirk-1956	2	1	.004	1	95	5.24	36	1580	2	4200	4100
Newkirk-1956	2	2	.004	1	48	2.57	42	1580	2	3275	2950
Tondl-1964	1.55	1.55	.008	.75	62.7	3.25	40	100	1.38	3800	3950
Tondl-1964	1.55	1.55	.008	.75	30.3	3.25	40	2500	1.38	4900	4900
Rao-1970	1	1	.002	.5	4.87	7.7	20	1500	.5	3000-4000	4000

Table 4 Details of Experimental Rotor Systems
Comparison of Threshold Speeds Observed and Calculated

film. Similar conclusions were obtained by Lund and Tonnesen [12].

Other computer programs have been written by Reddi, Lund, Giberson and Chang to investigate the stability of rotors in fluid-film bearings. These programs are discussed in more detail in Chapter 1 of this series.

Conclusions

a) The conditions under which fluid-film bearing and shaft gas seals may lead to unstable whirling of a rotor in its bearings have been identified.

b) Rigid rotors have been observed to whirl at speeds above 1.5 2.0 times the lowest system critical speed. The whirl frequency for this class of rotors occurs at approximately half rotational frequency. Whirl frequency increases with increase in rotor speed.

c) Flexible rotors have been observed to whirl at speeds above 1.5 ~ 2.0 times the lowest bending critical speed of the rotor in its bearings. The whirl frequency for flexible rotors occurs at the bending critical frequency of the rotor. This whirl frequency remains constant with increase in rotor speed.

d) The whirl threshold speed depends on the type of journal bearing used. References are included to several investigations which have ranked the different bearing types in terms of their susceptibility to whirling.

e) A variety of techniques have been developed and demonstrated for the calculation of whirl threshold speed, and whirl precession frequency. The features of several proven computer programs for stability analysis have been reviewed.

f) Small-amplitude instability in which the whirl motions occur about the journal steady-state equilibrium position has been found to be accurately predicted by analysis in which the bearing first-order displacement and velocity terms of the forces alone are used.

g) Large amplitude unstable whirling of the journal within and around the bearing circumference has been found to require some procedure in which the Reynolds equation is integrated step-by-step around the jouranl orbit, or some similar nonlinear approach.

h) Small amplitude unstable whirl orbits cannot be predicted by linear analysis. Orbit studies involve some form of nonlinear analysis.

i) The major problem in prediction of the actual whirl threshold speed lies in accurately representing bearing conditions in analytical terms, in obtaining reliable bearing dynamic coefficients for the acutual bearing geometry and oil film conditions.

References

1. Newkirk and Lewis, "Oil Film Whirl, An Investigation of Disturbances due to Oil Films in Journal Bearings," Trans. ASME, Vol. 78, p. 21 (1954).

2. Pinkus, O.J., "Experimental Investigation of Resonant Whip," Trans. ASME, Vol. 87, p. 975 (1956).

3. Hori, Y., "A Theory of Oil Whip," Journal of Applied Mech., p. 189 (June, 1959).

4. Tondl, A., "Experimental Investigation of Self-Excited Vibrations of Rotors due to the Action of Whirling Oil Film in Journal Bearings," Monographs and Memoranda No. 1 (1961).

5. Sternlicht, B., "Stability and Dynamics of Rotors Supported on Fluid-Film Bearings," ASME Paper 62-WA-190 (1962).

6. Smith, D.M., "Journal Bearings in Turbomachinery," Chapman and Hall, Ltd. (1970).

7. Newkirk, B.L., "Shaft Whipping," General Electric Review (1924).

8. Newkirk, B.L., "Varieties of Shaft Disturbances due to Fluid Films in Journal Bearings," Trans. ASME Vol. 78, p. 985 (1956).

9. Tondl, A., "Notes on the Problem of Self-Excited Vibrations and Non-Linear Resonances of Rotors Supported in Several Journal Bearings," Wear, 8 (1965).

10. Tondl, A., "The Effect of the Out-of-Roundness of Journals

on Rotor and Bearings Dynamics," ACTA Technica Grav. No. 1, p. 62 (1967).

11. Tondl, A., "Some Problems of Rotor Dynamics," Publishing House of the Czechoslovakian Academy of Sciences, Prague, (1965).

12. Tonnessen, J. and Lund, J.W., "Some Experiments on Instability of Rotors Supported in Fluid-Film Bearings," ASME Paper No. 77-DET-23 (1977).

13. Newkirk, B.L., and Taylor, H.D., "Shaft Whipping due to Oil Action in Journal Bearings," General Electric Rev. (1925).

14. Stodola, A., "Kritische Wellenstorung Infolge der Nachgiebigkeit des Oelpolsters im Lager," Schweizerische Bauzeitung, Vol. 85, p. 265 (1925).

15. Hummel, C., "Kritische Drehzahlen als Folger der Nachgiebigkeit des Schmiermittels im Lager," VDI-Forschehft, Vol. 287 (1926).

16. Sommerfeld, A., "Zur Hydrodynamischen Theorie der Schmiermittelreibung," A. Math. Phys. Vol. 50, p. 124 (1904).

17. Harrison, W.J., "The Hydrodynamical Theory of the Lubrication of a Cylindrical Bearing under Variable Load, and of a Pivot Bearing," Trans Edinburgh Philosophical Soc., Edinburgh, Scotland, Vol. 22, p. 373 (1919).

18. Robertson, D., "Whirling of a Journal in a Sleeve Bearing," Phil. Mag. Series 7, Vol. 15, p. 113 (1933).

19. Poritsky, H., "Contribution to the Theory of Oil Whip," Trans. ASME Vol. 75, p. 1153 (1953).

20. Boeker, G.F. and Sternlicht, B., "Investigation of Translatory Fluid Whirl in Vertical Machines," Trans. ASME Vol. 78, p. 13 (1956).

21. Sternlicht, B., "Elastic and Damping Properties of Cylindrical Journal Bearings," Trans. ASME, Jnl. Basic Eng., Series D, Vol. 81, p. 101 (1959).

22. Bowman, R.M., Collingwood, L.D., and Midgley, J.W., "Some Factors Affecting the Whirl Instability of a Journal Bearing," Part I and Part II, Proc. First and Second Lubrication and Wear Conventions, Inst. Mech. Engrs., London (1963-1964).

23. Lund, J.W., "The Stability of an Elastic Rotor in Journal Bearings with Flexible, Damped Support," Trans. ASME Jnl. Appl. Mech., Series E, Vol. 87, p. 911.

24. Sternlicht, B., Poritsky, H., and Arwas, E.B., "Dynamic Stability of Cylindrical Journal Bearings Using Compressible and Incompressible Fluids," First International Symposium on Gas-Lubricated Bearings, Washington, D.C., ONR/ACR-49 p. 119 (October, 1959).

25. Hagg, A.C. and Warner, P.C., "Oil Whip of Flexible Rotors," Trans. ASME, Vol. 75 (No. 7), p. 1339 (1953).

26. Hagg, A.C., "The Influence of Oil-Film Journal Bearings on the Stability of Rotating Machines," Trans. ASME, Jnl. Appl. Mech., Vol. 68 , p. A-211. (1946).

27. Sternlicht, B., "Gas-Lubricated Cylindrical Journal Bearings of Finite Length - Part II-Dynamic Loading," General Electric Tech. Rep. ONR Contract No. Nonr 2844 (00), Task No. NR 097-348 (September, 1960).

28. Warner, P.C. and Thoman, R.J., "The Effect of the 150 Degree Partial Bearing on Rotor-Unbalance Vibration," Trans. ASME, Jnl. Basic Eng., Series D, Vol. 86, p. 337 (1964).

29. Hagg, A.C., "Some Vibration Aspects of Lubrication," Lubric. Engng., p. 166 (August, 1948).

30. Hagg, A.C. and Sankey, G.O., "Some Dynamic Properties of Oil-Film Journal Bearings with Reference to the Unbalance Vibration of Rotors," Trans. ASME, Jnl. Appl. Mech., Vol. 78, p. 302 (1956).

31. Hagg, A.C. and Sankey, G.O., "Elastic and Damping Properties of Oil-Film Journal Bearings for Application to Unbalance Vibration Calculation," Trans. ASME, Jnl. Appl. Mech., Vol. 80, p. 141 (1958).

32. Lund, J.W., "Spring and Damping Coefficients for the Tilting Pad Journal Bearing," Trans. Am. Soc. Lubric. Engrs., Vol. 7, p. 342 (1964).

33. Badgley, R.H. and Booker, J.F., "Turborotor Instability: Effect of Initial Transients on Plane Motion," Journal of Lubrication Technology, Trans ASME,Vol.91,p.625 (Oct.1969).

34. Myrick, S.T. and Rylander, H.G., "Analysis of Flexible Rotor Whirl and Whip Using a Realistic Hydrodynamic Journal Bearing Model," ASME Paper No. 75-DET-68.

35. Lund, J.W. Written Discussion of Paper by Warner, B.C. and Thoman, R.K. Trans ASME, Jnl. Basic Engineering, Series D, Vol. 86, p. 337 (1964): Ref. [28] above.

36. Rieger, N.F., Thomas, C.B., "Some Recent Computer Studies on Theoretical Stability of Rotors in Fluid-Film Bearings," International Union of Theories and Applied Mechanics, Prc. Dynamics of Rotor Symposium, Lyngby, Denmark, p. 436 (1976).

CHAPTER 2.9

EFFECT OF SUPPORTING STRUCTURE ON ROTOR DYNAMICS

Z.A. Parszewski

1 Introduction

Receptance. It is advantageous in many scientific as well as industrial situations to treat considered systems (mechanism, machine or its subassemblies as parts (subsystems) of a larger system. It may contain, besides the considered assembly, also other assemblies of the machine, bearings, supporting structure, floor and building and the environment in general.

Two types of problems are important:

1. In the first type of problem - the influence of the rest of the system on dynamics of the given subsystems, is considered. These are in particular the problems of flexible support influence on machine dynamics, and also isolation problems of (passive) objects from environmental (kinematic) excitation.

2. In the second type of problems spreading of dynamic effects form a subsystem (from machine or its subassembly) to another part of the system is considered (to another subassembly of the machine, to other machine or structure, etc.).

These are hence problems of dynamic isolation of active objects. In both type of problems at least two subsystems interacting dynamically always exist.

A method of subsystems composition will be presented for finding dynamic characteristics and vibration of more complicated systems composed of simpler subsystems [1], [4], [7], [8].

Receptance methods is advantageous in those problems. Responses of the subsystem along connecting co-ordinates are hence necessary to harmonic excitation. These can be found experimentally if necessary and hence the method can be also applied to systems difficult for sufficiently accurate analytical description.

The receptance matrix of the rank equal to the number of connecting co-ordinates is sufficient for the composition of the subsystems considered. It gives not however a complete dynamic characteristics of the system. For some other problem another matrix may be required but corresponding dynamic model may be different as well.

The advantage of this method is simple possibility of finding (measuring) these quantities without any simplifications and omitments for complicated systems (eg. for the whole turbo-generator supporting subsystem including casings, supporting structure, foundation and ground).

To introduce the receptance concept let us consider a mechanical system. It may be any continuous or discrete however linear system. It may represent a whole structure or machine or any of its elements or assemblies (Fig. 1).

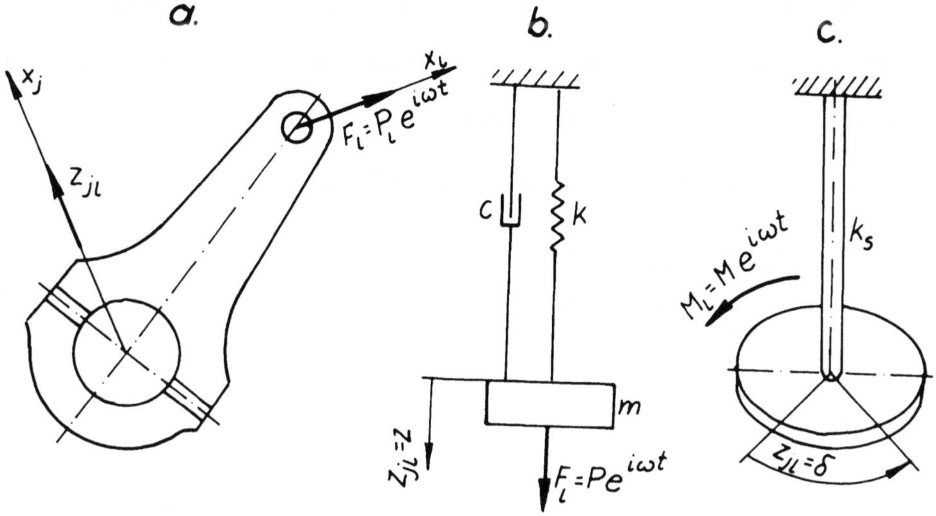

Fig. 1

Let numbers j, ℓ, denote the x_j, x_ℓ co-ordinates ie. directions of displacements linear or angular in given points of the system.

Let an harmonic complex force act in the direction of ℓ co-ordinate (Fig. 1).

$$F_\ell = P_\ell e^{i\omega t} \tag{1}$$

Vibration excited (in steady state) along the j co-ordinate (ie. dynamic displacement at a point and direction designed with number j) can be represented in the form

$$z_{j\ell} = a_{j\ell} e^{i(\omega t - \beta_{j\ell})} \quad \text{where } |z_{j\ell}| = a_{j\ell} \tag{2}$$

is amplitude of the vibrations, and argument $\beta_{j\ell}$ is the phase lag of the displacement with respect to the force.

$$\text{The ratio } c_{j\ell} = \frac{z_{j\ell}}{F_\ell} = \frac{a_{j\ell}}{P_\ell} e^{-i\beta_{j\ell}} \tag{3}$$

does not depend on time. This ratio is a function of circural frequency ω of exciting force F_ℓ depending on the system parameters only. Is hence characteristic for the system.

For a system on one degree of freedom (Fig. 1 b or c) is

$$\alpha_{j\ell} = a = x_{st} \frac{1}{\sqrt{\left(1-\frac{\omega^2}{\alpha^2}\right)^2 + 4\frac{h^2}{\alpha^2}\frac{\omega^2}{\alpha^2}}} \tag{4}$$

$$\mathrm{tg}\beta_{j\ell} = \mathrm{tg}\beta = 2\,\frac{\frac{h}{\alpha}\,\frac{\omega}{\alpha}}{1-\frac{\omega^2}{\alpha^2}} \tag{5}$$

where $x_{xt} = \frac{P}{k}$ or $x_{st} = \frac{M}{k_s}$

and $\alpha = \sqrt{\frac{k}{m}}$ or $\alpha = \sqrt{\frac{k_s}{B}} = \sqrt{\frac{GI}{B\ell}}$ $2h = \frac{c}{m}$

For any linear subsystem continuous or discrete interacting with some other subsystem along n connecting co-ordinates, the displacement z_j can be expressed by the forces F_ℓ as follows

$$\begin{Bmatrix} z_1 \\ z_2 \\ \cdot \\ \cdot \\ z_n \end{Bmatrix} = \begin{Bmatrix} c_{11} & c_{12} & \cdots & c_{1n} \\ c_{21} & c_{22} & \cdots & c_{2n} \\ \cdot & & & \\ \cdot & & & \\ c_{n1} & c_{n2} & \cdots & c_{nn} \end{Bmatrix} \cdot \begin{Bmatrix} F_1 \\ F_2 \\ \cdot \\ \cdot \\ F_n \end{Bmatrix} \tag{6}$$

or $\{z_j\} = [c_{j\ell}]\ \{F_\ell\}$ (7)

Here

$$[c] = [c_{j\ell}] = \begin{Bmatrix} c_{11} & c_{12} & \cdots & c_{1n} \\ c_{21} & c_{22} & \cdots & c_{2n} \\ \cdot & & & \\ \cdot & & & \\ \cdot & & & \\ c_{n1} & c_{n2} & \cdots & c_{nn} \end{Bmatrix} \tag{8}$$

is the dynamic flexibility matrix or receptance matrix.

Solving (6) for forces F_ℓ we have

$$F_\ell = [c_{j\ell}]^{-1}\{z_j\} = [k_{j\ell}]\{z_j\} \tag{9}$$

here

$$[k] = [k_{j\ell}] = \begin{bmatrix} k_{11} & k_{12} & \cdots & k_{1n} \\ k_{21} & k_{22} & \cdots & k_{2n} \\ \cdot & & & \\ \cdot & & & \\ \cdot & & & \\ k_{n1} & k_{n2} & \cdots & k_{nn} \end{bmatrix} \tag{10}$$

is the dynamic stiffness matrix of the system for its interaction along j, ℓ co-ordinates of number n.

As an example let us consider a rotating machine Fig. 2c composed of a rotor (Fig. 2a) and casing with supporting structure included (Fig. 2b).

The connecting co-ordinates are four; directions q_1, q_2, q_3, q_4 perpendicular to the rotor axis in the middle of the bearings 0 and ℓ. They can be best taken orthogonal (corresponding to support symmetry) and equally oriented at each end (Fig. 2).

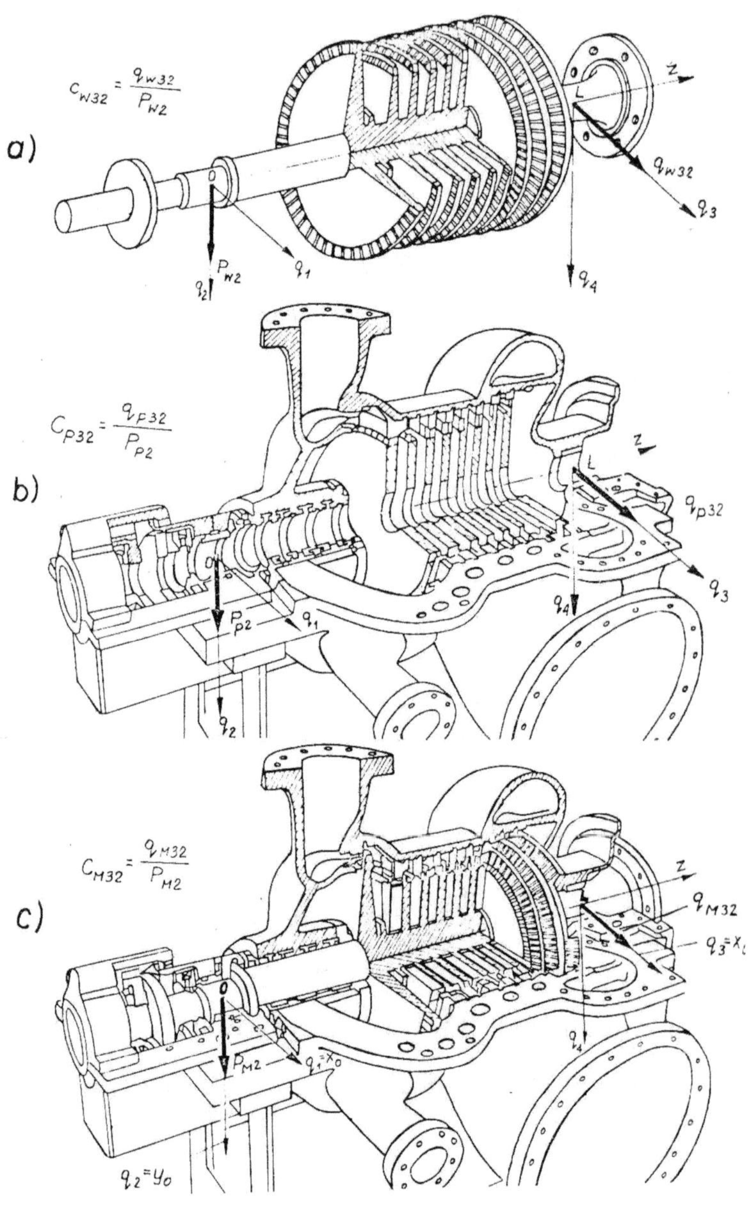
$c_{W32} = \frac{q_{W32}}{P_{W2}}$
a)
z
L
q_{W32}
q_3
0
P_{W2}
q_2
q_1
q_4
$C_{P32} = \frac{q_{P32}}{P_{P2}}$
b)
z
L
q_{P32}
q_3
0
q_4
P_{P2}
q_1
q_2
$C_{M32} = \frac{q_{M32}}{P_{M2}}$
c)
z
L
q_{M32}
$q_3 = x_L$
0
q_4
P_{M2}
$q_1 = x_0$
$q_2 = y_0$

Fig. 2

Receptance matrices will be:

for the rotor

$$[c_w] = [c_{wrs}] = \begin{bmatrix} c_{w11} & c_{w12} & c_{w13} & c_{w14} \\ c_{w21} & c_{w22} & c_{w23} & c_{w24} \\ c_{w31} & c_{w32} & c_{w33} & c_{w34} \\ c_{w41} & c_{w42} & c_{w43} & c_{w44} \end{bmatrix} \tag{11}$$

$r,s = 1,2,3,4$

For the supporting structure.

$$[c_p] = [c_{prs}] \ , \ r, s = 1, 2, 3, 4. \tag{12}$$

Examples of the matrices elements are defined in Fig. 2

The receptance matrix elements are functions of excitation frequency (and rotor speed in some cases) and are complex quantities when damping is introduced.

They can be easily measured if not calculated.

In the considered case of machine casing with all supporting structures (Fig. 2b) a rotary exciter (Fig. 3) can for example be installed consecutively in each bearing with the exciting force directed once vertically and next horizontally. For each exciter position (eg. P_{p2}. Fig. 2c) dynamic displacements are measured along all the connecting co-ordinates. The receptances are then defined as in Fig. 2.

Doing it at various excitation circular frequencies ω and measuring amplitude and phase, diagrams of the type shown in Fig. 4 are obtained.

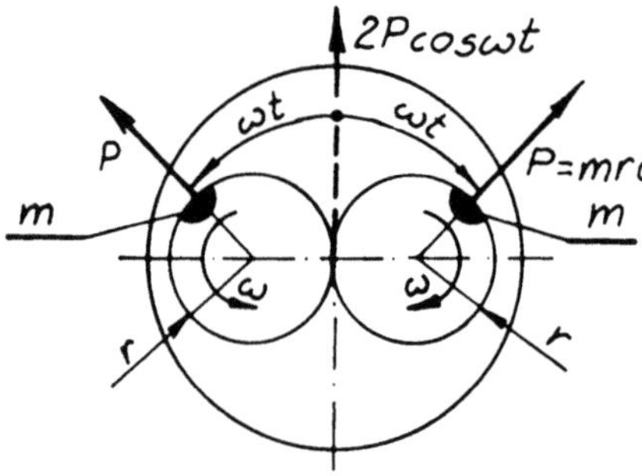

Fig. 3

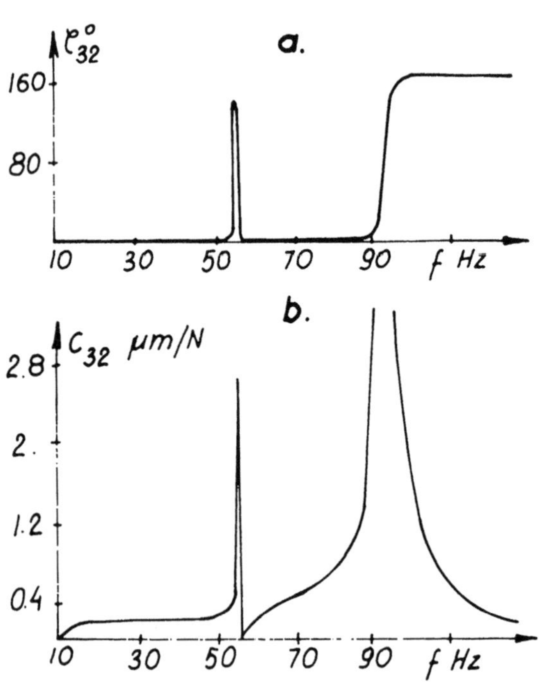

Fig. 4a,b

Rotor receptances (Fig.2) should be computed or measured for the rotor free in space. In case of measurements the rotor may be suspended eg. on ropes, so that its vibrations are practically uninfluenced by this constraint of very low frequency. This is necessary for the following subsystem composition with the receptance method.

Alternatively, rotor on rigid supports can be considered. This has the advantage that first approximation for the real system is in this way at once obtained (as compared with at first meaningless free rotor parameters). Subsystem composition is however with that method somewhat less straightforward (called mixed

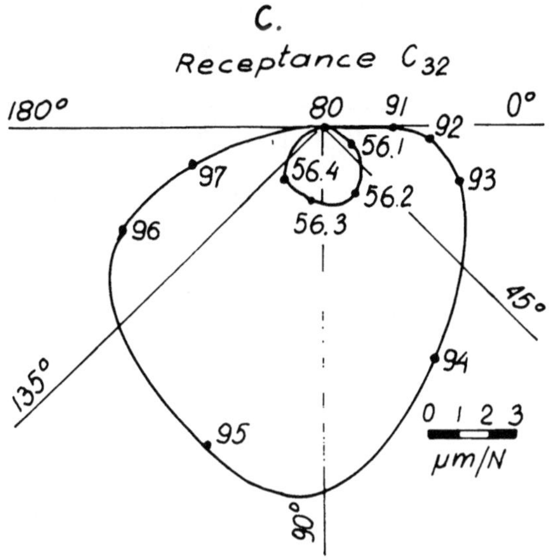

Fig. 4c

method).

2 Receptance Method.

Receptance matrices are used for both subsystems being composed. [10],[11]

Displacements of the subsystems a and b (Fig. 5a,c) connected (interacted) along n co-ordinates (Fig. 5c) are in the form

$$\{q_a\} = [c_a]\ \{P_a\}$$
$$\{q_b\} = [c_b]\ \{P_b\} \qquad (13)$$

Corresponding displacements (ie. at the same points in the same directions) for the system c composed of both subsystems (Fig. 5) are

$$\{q_c\} = [c_c]\ \{P_c\}$$

The system c may as well be a part of a larger system including subsystem d etc.

Composition conditions are

1. Equilibrium conditions for the forces acting on the whole system along the co-ordinates q connecting the both subsystems (Fig. 5)

$$\{P_a\} + \{P_b\} = \{P_c\} \qquad (14)$$

2. Continuity conditions

$$\{q_a\} = \{q_b\} = \{q_c\} \qquad (15)$$

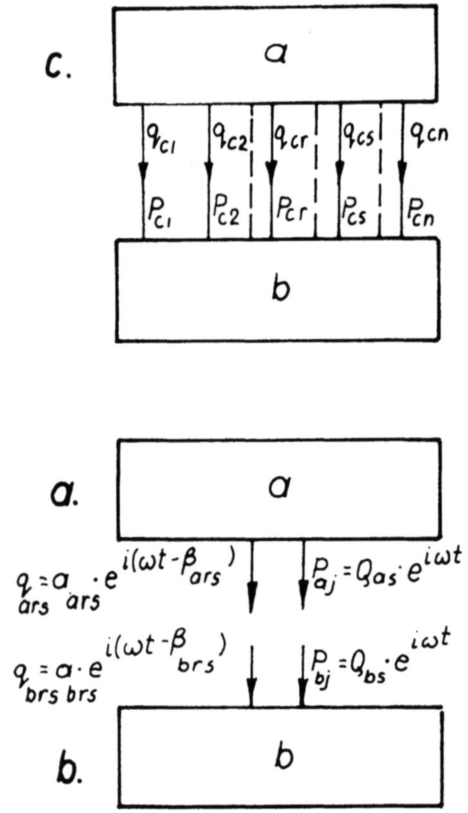

Fig. 5

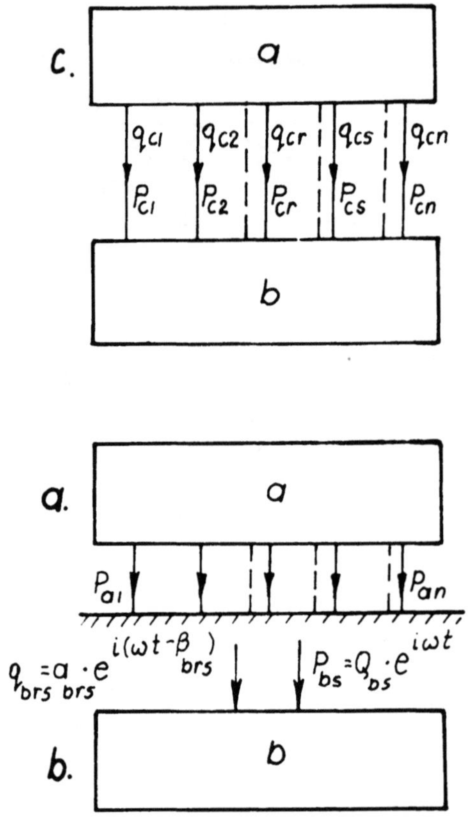

Fig. 6

Conditions (14) and (15) with (13) give the expression of the displacements $\{q_c\}$ of the whole system by the forces $\{P_c\}$ acting on it. This hence defines the receptance matrix for the whole system.

$$[c_c] = [c_a] \left[[1] + [c_b]^{-1} [c_a]\right]^{-1} \tag{16}$$

In the case of a rotor, let us repeat, the receptances $[c_{wjk}]$ have to be in this method found for the rotor free in space (Fig. 2a) This applies always to subsystems which have no external constraints (from outside of the system considered), which, for example, gives the foundation for the supporting structure.

3 Mixed Method.

In most practical cases it is the influence of the rest of the system on the dynamics of one distinguished subsystem, that is considered. In the example given previously this distinguished subsystem is the rotor and the influence of the flexibility of supporting structure on its dynamics is considered.

It is necessary already on design stage of that subsystem (rotor in this example) to find its dynamic properties even if some idealisation of the rest of the system is necessary, eg. assuming its rigidity. These properties give dynamic characteristics of the subsystem and at the same time an approximation of the characteristics of the actual whole system. In many practical cases, this approximation is finally sufficient, when support flexibility influence is negligible.

This approximation may be unavoidable because of no information at that stage of the rest of the system, eg. bearings often and supporting structures always are designed and manufactured by another producer. It is hence necessary to use this type of characteristics for the subsystem composition (Fig.6).

At the same time there is no necessity for finding an additional characteristic of the subsystem as free in space.

The mixed method assumes the description of the considered subsystem a (Fig.6) with idealisation (assumption of rigidity) of the rest b of the system. Of necessity hence not in receptance form but in modal form.

This is an agreement with general method of dynamic model extension.

On the other hand this also corresponds to the practical cycles of design and manufacture, using the characteristics of simplified system for the system extension and if necessary for introduction of further influences. [5], [6], [7], [8], [9].

Assessment is at the same time obtained of this influence on the basic dynamic quantitites i.e. principal modes and natural frequencies of the main subsystem (the simplified system with rigid supports). Block diagram is for this method shown in Fig. 6a.

The block diagram for the subsystem b is as before. It is better now to give its dynamic description in the form of dynamic stiffness matrix [k].

$$[k_b] = [c_b]^{-1} \qquad c_{brs} = \frac{a_{brs}}{Q_{bs}} e^{-i\beta_{rs}} \tag{17}$$

The subsystem a distinguished in this consideration (Fig. 6a), the rotor for example, is described with its modal functions f_j fulfilling end conditions of the subsystem a (Fig. 6a) corresponding to ideal rigid support and with corresponding natural circural frequencies α_j.

Then the modal functions F_k of the same rotor supported flexibly are found by superposition of motion of the shaft as rigid body on its vibrations with rigid support.

For one span rotor is

$$F_k = F_{k1} \frac{\ell - z}{\ell} + F_{kz} \frac{z}{\ell} + \sum_{j=1}^{\infty} B_{kj} f_j \tag{18}$$

The composition conditions are now

1. $\{P_b\} = \pm\{T_a\}$ (19)

Where T_a are shearing forces for the subsystem a.

2. The continuity conditions requires again the same displacements.

An example of application of the mixed method is given in chapter 2.4 for fluid film influence on critical speeds. Direct method is applied in following examples.

For multi-supported multi-discs or continuous rotor systems with flexible supports approximate methods are more effected than those based on the solution of vibration equation. Rotor is divided into a number of segments and the mass of each segment is concentrated at its ends (Myklestad - Prohl's Method) or in its middle (Finite elements method). Deflections and internal forces for the segments far-end are expressed by their values at its other end and a numerical procedure is applied to fulfil given end conditions.

4 Myklestad-Prohl's Method.

A method based on Holzer's approach can be used for calculating transverse natural frequencies and principal modes for any shaft or beam.

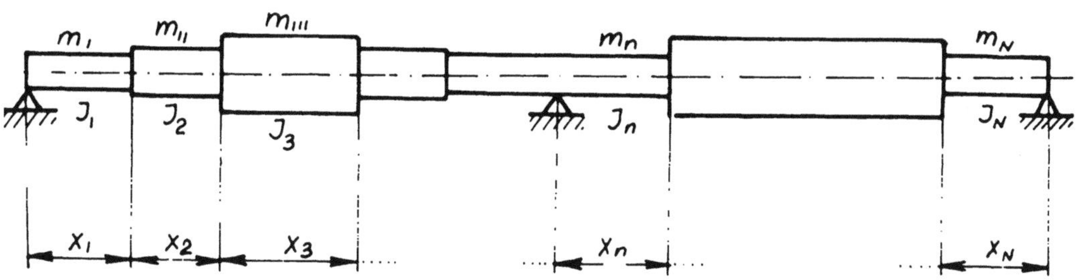

Fig. 7a

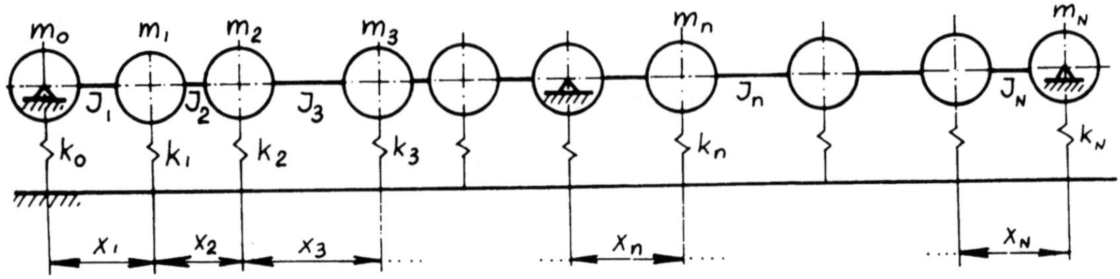

Fig. 7b

The shaft is divided into N segments each of possible constant radius (Fig. 7) and concentrated masses are located at the ends of the segments. Flexible supports are introduced at each of the masses, only those actually existing will however have stiffness coefficient k_n different from zero. This gives the possibility of introducing any number of flexible supports (bearings) and simplifies programming.

The deflections and forces at the n + 1 cross-section are expressed by their values at the n section (Fig. 8)

$$Y_{n+1} = Y_n + X_{n+1}\,\Phi_n + \tfrac{1}{2}\,\frac{X^2_{n+1}}{EJ_{n+1}}\,M_n + \frac{1}{6}\,\frac{X^3_{n+1}}{EJ_{n+1}}\,T_n - \frac{K_{n+1}Y^4_{n+1}}{3EJ_{n+1}}$$

$$\begin{aligned}
\Phi_{n+1} &= \Phi_n + \frac{X_{n+1}}{EJ_{n+1}}\,M_n + \tfrac{1}{2}\,\frac{X^2_{n+1}}{EJ_{n+1}}\,T_n \\
M_{n+1} &= M_n + X_{n+1}\,T_n \\
T_{n+1} &= T_n + m_{n+1}\alpha^2 Y_{n+1} - K_{n+1}Y_{n+1}
\end{aligned} \qquad (20)$$

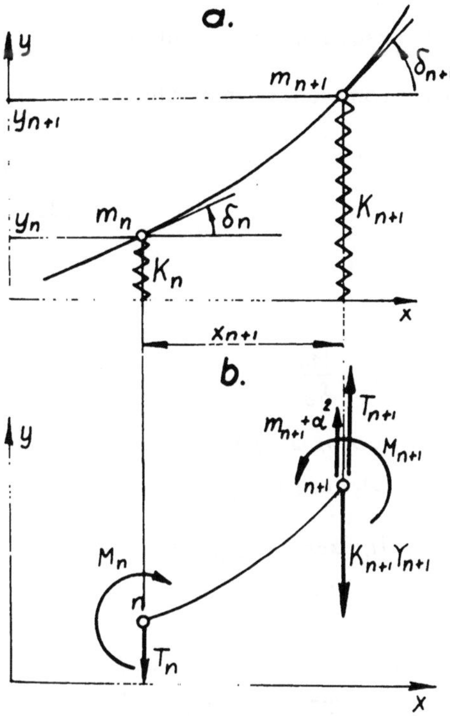

Fig. 8

Bending moments M_o, M_N and shearing forces T_o, T_N are known at the shaft ends. Using equations 8 and moving from segment to segment, everything is expressed by the deflection Y_o and deflection angle δ_o at the starting shaft end. Eventually also T_N and M_N are expressed but these are vanishing, hence, the following equations are obtained.

$$T_N = A\delta_o + BY_o = 0$$
$$M_N = C\delta_o + DY_o = 0 \qquad (21)$$

Two cycles of computations are required for finding the coefficients A, B, C, D values corresponding to each frequency α

The end conditions for x = o are for the first cycle.

1) $Y_o = 1 \quad T_o = \alpha^2 m_o - k_o 1 , \quad M_o = 0 , \delta_o = 0$

For the second cycle they are

2) $Y_o = 0, \quad T_o = 0, \quad M_o = 0, \quad \delta_o = 1$

The coefficients are hence

$$A = (T_N) \text{ at } \delta_o = 1, \; Y_o = 0 \qquad C = (M_N) \text{ at } \delta_o = 1, \; Y_o = 0$$
$$B = (T_N) \text{ at } \delta_o = 0, \; Y_o = 1 \qquad D = (M_N) \text{ at } \delta_o = 0, \; Y_o = 1 \qquad (22)$$

The above coefficients are functions of frequency α. Equations (21) can give for δ_o and Y_o non zero solutions only when their determinant vanishes.

$$\Delta(\alpha) = AD - BC = 0 \tag{23}$$

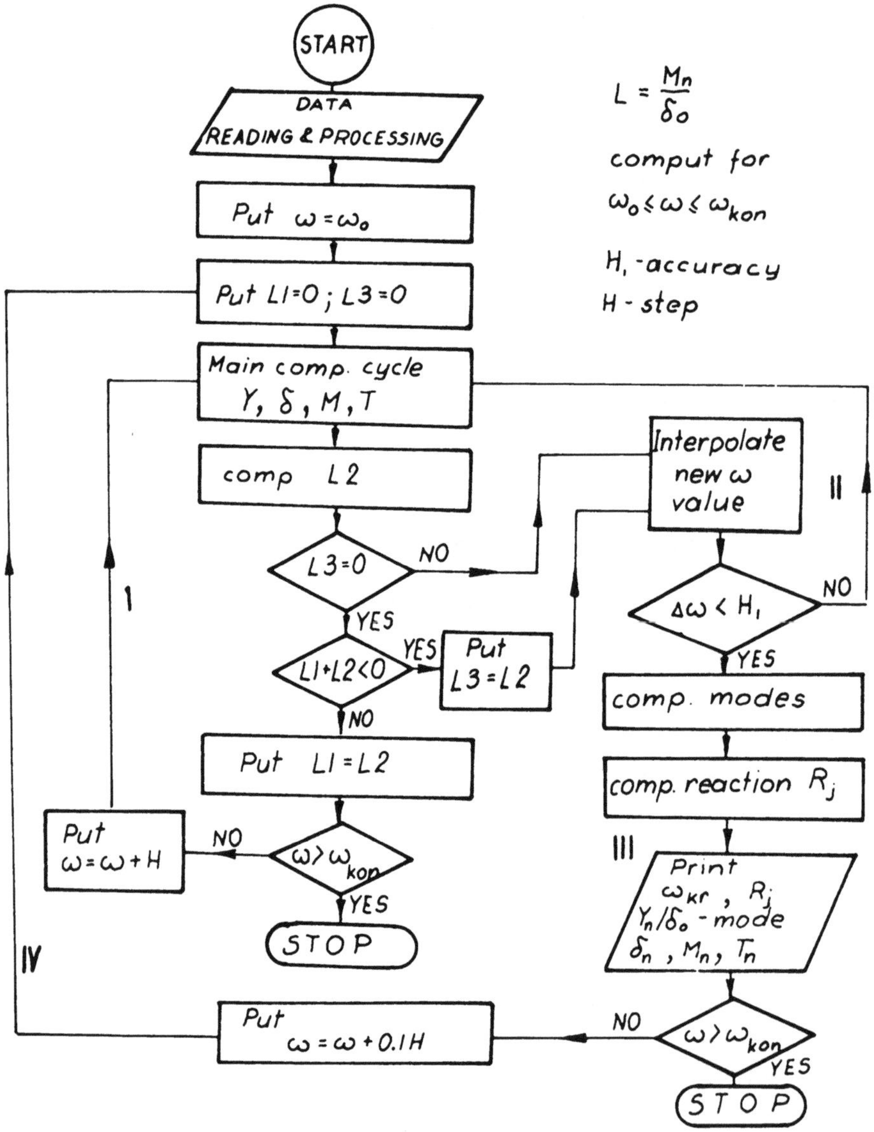

Fig. 9

Its roots α_i give the natural circular frequencies of the rotor. Its principal modes are described by corresponding deflection Y.

The method is convenient for programming for computer.

Flow diagram is given in Fig. 9.

An example showing support flexibility influence on consecutive four modes and corresponding natural frequencies, for a shaft shown in Fig. 10 is given in Fig. 11 for first mode, Fig. 12 for second mode, Fig. 13 for third mode and in Fig. 14 for the fourth mode.

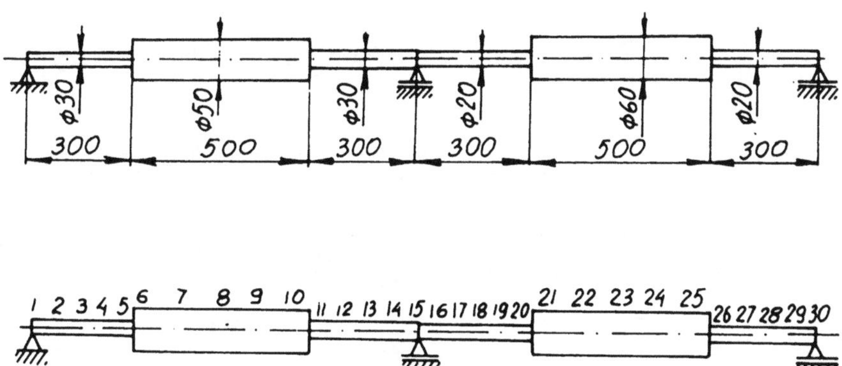

Fig. 10

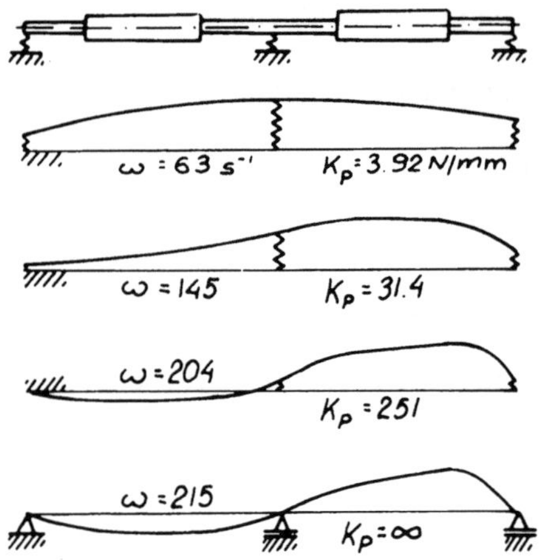

Fig. 11

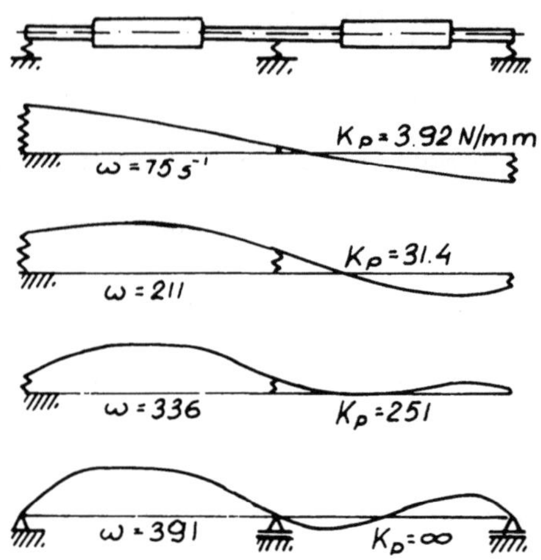

Fig. 12

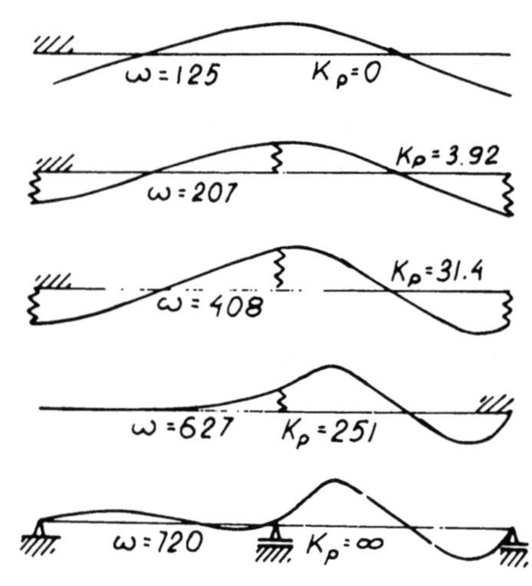

Fig. 13

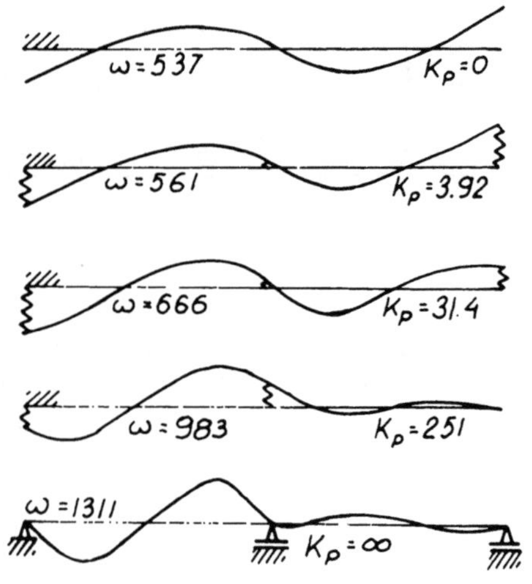

Fig. 14

The consecutive modes for the shaft supported rigidly are repeated in Fig. 15.

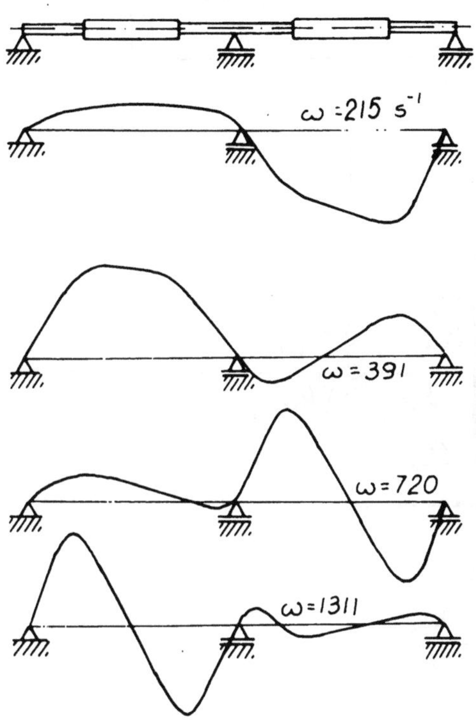

Fig. 15

5 Rigid Finite Elements Method.

Specially convenient for programming and applications for interconnected rotor systems (with couplings) on flexible supports is the following approach developed by J. Skoraczynski [11] and based on Kruszewski's considerations [12].

The given rotor is divided into sufficient for necessary accuracy, number of segments of constant cross-section (Fig. (16a). Each section is then modelled as composed of two rigid parts connected by spring elements modelling bending and shearing contributions (stiffnesses) to deflection (Fig. 16b). Bending and shearing connecting springs stiffnesses are defined in Fig. 17. Equivalence of both actual element and its model requires equal displacements at equal forces. This gives for the bending spring (Fig. 17 a and b):

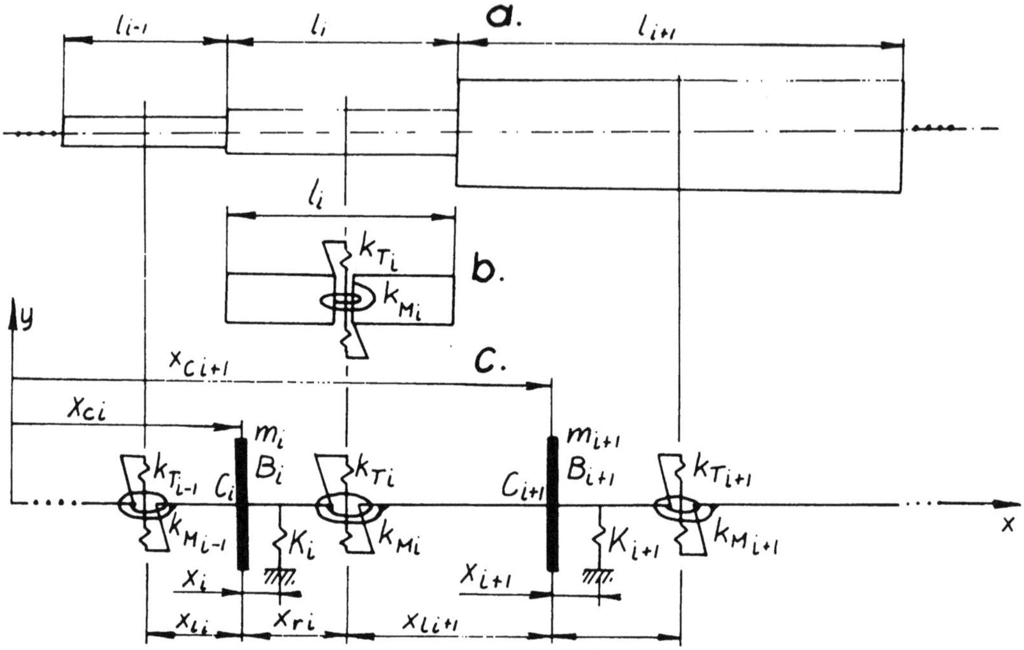

Fig. 16

$$\delta_{ri} = \frac{M_i \ell_i}{2EJ_i} = \delta_{zi} = \tfrac{1}{2}\frac{M_i}{k_{Mi}} \quad \text{hence} \quad k_{Mi} = \frac{EJ_i}{\ell_i} \tag{24}$$

and for the shear spring (Fig. 3.9.17 c, d).

$$\gamma_{ri} = \frac{T_i \ell_i}{GA_i} = \gamma_{zi} = \tfrac{1}{2}\frac{T_i}{k_{Ti}} \quad \text{hence} \quad k_{Ti} = \frac{GA_i}{\ell_i} \tag{25}$$

The rotor element between two connecting elastic elements is considered rigid and its mass is represented as a disc of mass m_i and moment of inertia B_i with respect to its diameter. A system presented in Fig. 16c is hence obtained.

Each of the rigid elements may be supported flexibly (Fig.16c) at a given position, with only the actual supports having stiffness K_K different from zero. In this way complete symmetry is obtained

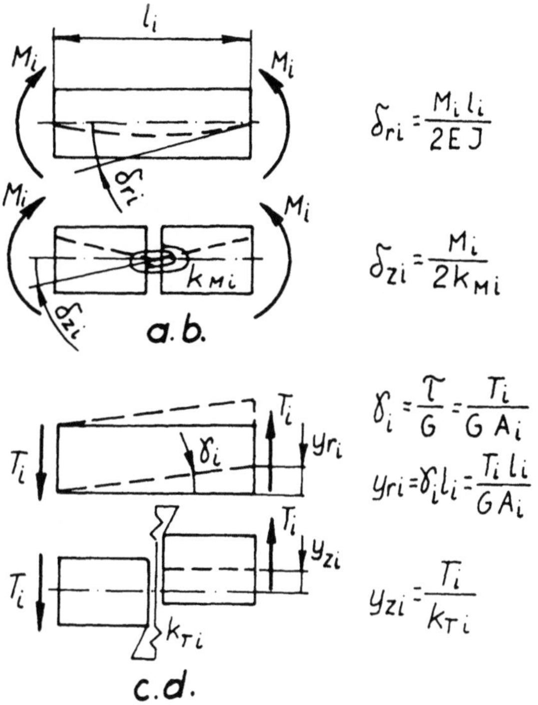

Fig. 17

that gives simplest programming for computer analysis.

Lagrange's equations

$$\frac{\partial}{\partial t}\left(\frac{\partial E}{\partial \dot{y}_i}\right) + \frac{\partial V}{\partial y_i} = 0$$

$$\frac{\partial}{\partial t}\left(\frac{\partial E}{\partial \dot{\delta}_i}\right) + \frac{\partial V}{\partial \delta_i} = 0 \tag{26}$$

with the energy expressions for the described system of rotor elements,

$$E = \sum_{i=1}^{I} \frac{m_i \dot{y}_i^2}{2} + \frac{B_i \dot{\delta}_i^2}{2}$$

$$V = \sum_{i=1}^{I-1} \left[\frac{k_{Mi} (\delta_{i+1} - \delta_i)^2}{2} + \frac{k_{Ti} (y_{i+1} - x_{\ell i+1}\delta_{i+1} - y_i - x_{ri}\delta_i)^2}{2} \right] + \sum_{k=1}^{K} \frac{K_k (y_k + \delta_k x_k)^2}{2} \tag{27}$$

give the equations of motion in matrix form

$$\mathbf{M}\ddot{\mathbf{z}} + \mathbf{K}\mathbf{z} = \mathbf{0} \tag{28}$$

with the elements of the matrices of the form

$$\mathbf{M} = \begin{bmatrix} \mathbf{M}_{1,1} & 0 & 0 & 0 & 0 \\ 0 & \ddots & 0 & 0 & 0 \\ 0 & 0 & \mathbf{M}_{i,i} & 0 & 0 \\ 0 & 0 & 0 & \ddots & 0 \\ 0 & 0 & 0 & 0 & \mathbf{M}_{I,I} \end{bmatrix} \qquad \mathbf{z} = \begin{bmatrix} y_1 \\ \delta_1 \\ \vdots \\ y_i \\ \delta_i \\ \vdots \\ y_I \\ \delta_I \end{bmatrix}$$

$$\mathbf{K} = \begin{bmatrix} \ddots & \ddots & \ddots & 0 & 0 & 0 & 0 \\ 0 & \square & \square & \square & 0 & 0 & 0 \\ 0 & 0 & K_{i,i-1} & K_{i,i} & K_{i,i+1} & 0 & 0 \\ 0 & 0 & 0 & \square & \square & \square & 0 \\ 0 & 0 & 0 & 0 & \ddots & \ddots & \ddots \end{bmatrix}$$

$$\mathbf{M}_{i,i} = \begin{bmatrix} m_i & 0 \\ 0 & B_i \end{bmatrix}$$

$$\mathbf{K}_{i,i-1} = \begin{bmatrix} -k_{Ti-1} & -k_{Ti-1}\, x_{ri-1} \\ k_{Ti-1} x_{\ell i} & -k_{Mi-1} + k_{Ti-1} x_{ri-1} x_{\ell i} \end{bmatrix}$$

$$\mathbf{K}_{i,i} = \begin{bmatrix} k_{Ti-1} + k_{Ti} + K_{Pi} & -k_{Ti-1} x_{\ell i} + k_{Ti} x_{ri} + k_{Pi} x_i \\ -k_{Ti-1} x_{\ell i} + k_{ri} x_{ri} + K_{Pi} x_i & k_{Mi-1} + k_{Mi} + k_{Ti-1} x_{\ell i}^2 + k_{Ti} x_{ri}^2 + K_{Pi} x_i^2 \end{bmatrix}$$

$$\mathbf{K}_{i,i+1} = \begin{bmatrix} -k_{Ti} & +k_{Ti}\, x_{\ell i+1} \\ -k_{Ti} x_{ri} & -k_{Mi} + k_{Ti}\, x_{\ell i+1}\, x_{ri} \end{bmatrix} \tag{29}$$

Vibrations in principal modes are harmonic

$$\mathbf{z} = \mathbf{A} \sin \alpha t \tag{30}$$

Hence principal modes and natural frequencies are given by the equation

$$(\mathbf{1}\alpha^2 - \mathbf{M}^{-1}\mathbf{K}) = \mathbf{0} \tag{31}$$

There are handy methods and computer programs for finding the eigen - values (natural frequencies) of equation (31).

6 Interaction with Real Supporting Structure.

To introduce the influence of real casing and any supporting structure of any mass distribution (Fig. 18) dynamic stiffness $k_{kp}(\omega)$ is necessary.

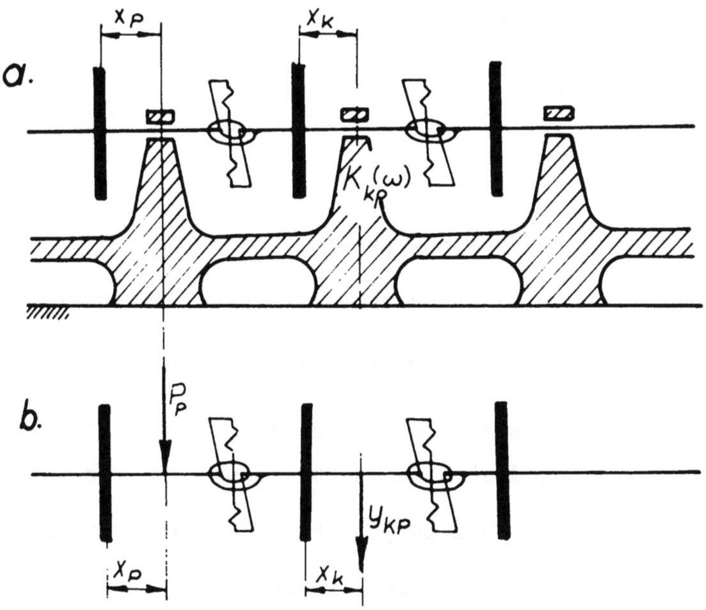

Fig. 18

The stiffness matrix along the connecting co-ordinates (at the bearings) is inverse with respect to the receptance matrix (see 1)

$$[K]_r = [C]_r^{-1}$$

The last one can always be measured if not computed as it was explained in 1. The elements $k_{kp}(\omega)$ of the stiffness matrix are hence known as function of circular frequency ω.

Receptance for the rotor free in space (Fig. 18c) can be computed with all the $K_k = 0$.

With harmonic excitation P_p applied to the element p at x_p giving displacement y_{kp} of element k at x_k will be the receptance

$$c_{kp} = \frac{y_{kp}}{P_p}$$

Then k_{kp} is the kp element of the matrix $[k_{kp}] = [c_{kp}]^{-1}$

Connecting equations are hence

$$[k_{kp}]_r \{y_k\}_r + [K_{kp}]_s \{y_k\}_s = 0 \qquad (32)$$

with $\{y_k\}_r = \{y_k\}_s$

Equations (32) replace now previous equations (31) . Their eigenvalues depend however on the assumed frequency and/or rotor speed ω. Critical speed can hence be found by repetitions leading to $\omega \to \alpha$.

7 Water Feed Pumps Vibrations and Corrections Analysis:

Final element method, described in the point 5 was used for follow-up analysis of water feed pumps vibrations elimination problem.

Not all the system parameters were known, hence it is the effectiveness of the applied remedy that is the analysis object rather than the actual speeds or frequencies.

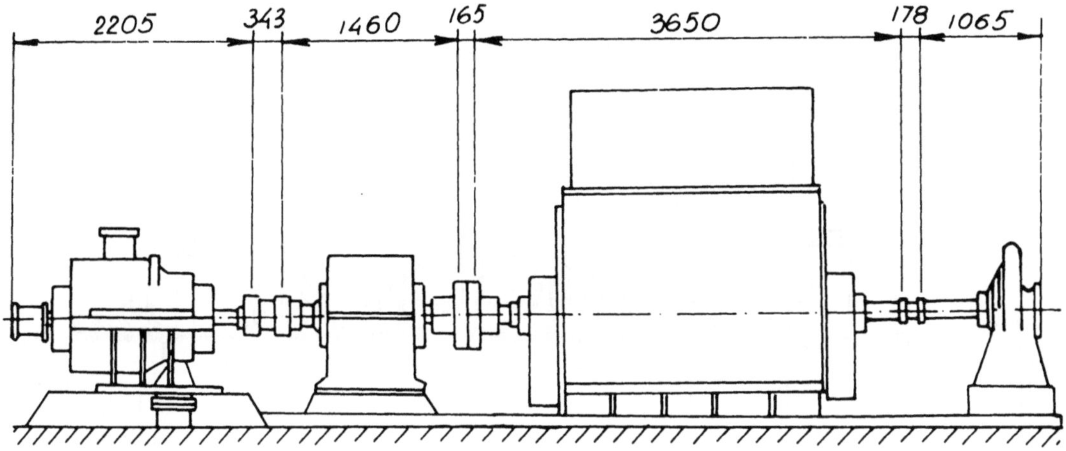

Fig. 19

The pumps, (eg. Yalourn W2 (Fig. 19)), developed severe vibrations of over 20mm/sec at working speed of around ω = 630 rad/sec and all went out of commission. External sealing were damaged. Second critical was responsible. Exchange of the flexible coupling (Fig. 20) between the pump and gear box, for another of about half the mass, lowered the vibrations, in all cases, to the level of 3.5mm/sec. This was achieved however after long periods of test trials and errors by many home and overseas experts.

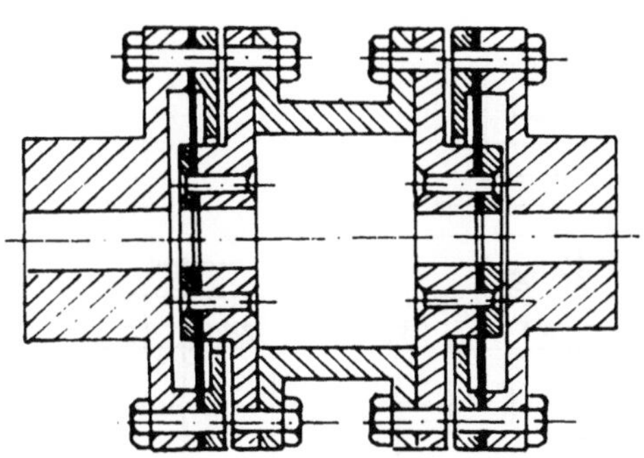

Fig. 20

The following shows how easily the analysis applied in advance could have saved considerable numbers of menhours and could have given increased numbers of mega-watts-hours.

The pump cross section is shown in Fig. 21 . Dimensions

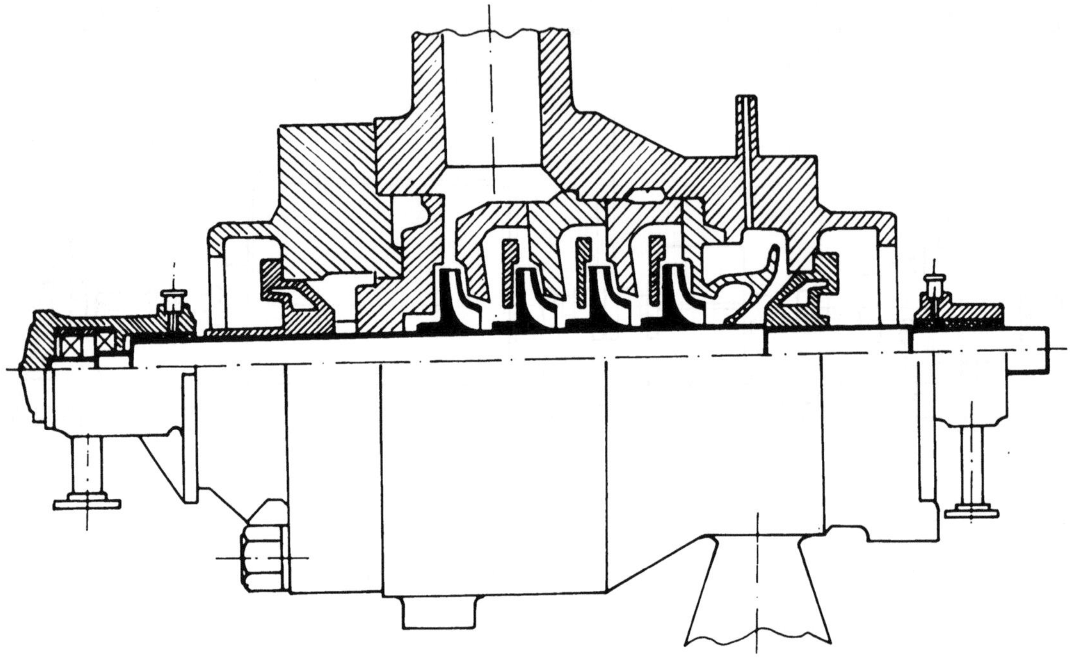

Fig. 21

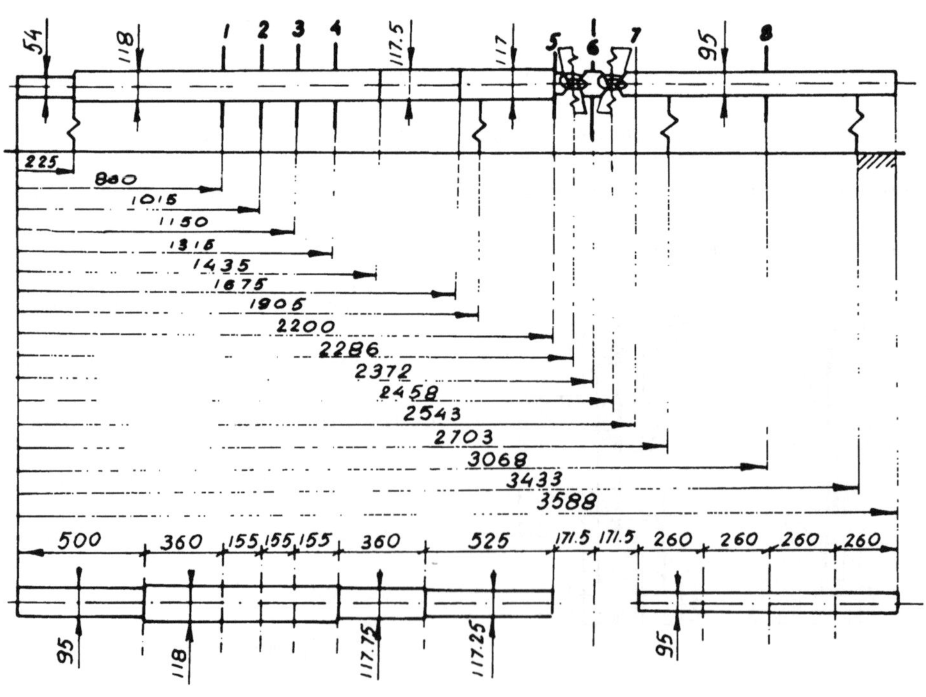
54
118
1
2
3
4
117.5
117
5
6
7
95
8
225
880
1015
1150
1315
1435
1675
1905
2200
2286
2372
2458
2543
2703
3068
3433
3588
500
360
155
155
155
360
525
171.5
171.5
260
260
260
260
95
118
117.75
117.25
95

Fig. 22

TABLE

ASSESSED			
	№	mass kg	moment of inertia kgm^2
pump	1	60	0.5
	2	60	0.5
	3	60	0.5
	4	60	0.5
coupling	5	25	0.1
	6	100	0.4
	7	25	0.1
gear	8	40	0.2

ASSUMED	
Stiffness of the coupling membranes	$k_M = 1 \cdot 10^6$ Nm/rd
	$k_T = 1 \cdot 10^9$ N/m
Bearing stiffness	$k_1 = k_2 = k_3 = k_4 = 2 \cdot 10^8$ N/m
Working speed	$\omega = 630\ s^{-1}$
E	$0.206 \cdot 10^{12}$ N/m^2
G	$0.81423 \cdot 10^{11}$ N/m^2
γ	$7.75 \cdot 10^3$ kg/m^3

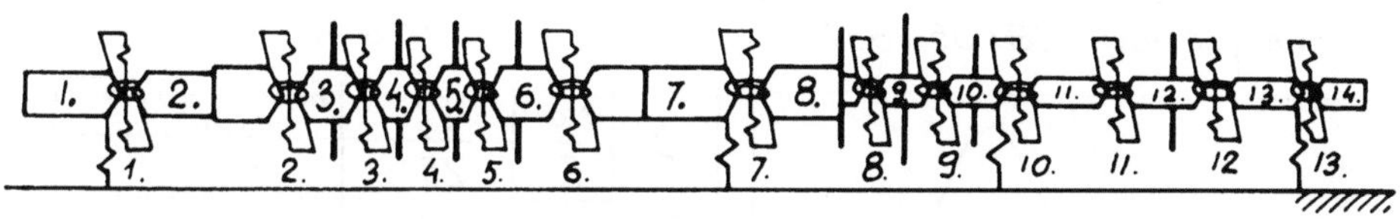

No	mass kg	moment of inertia kgm^2	bending spring stiffness Nm/rd	shear spring stiffness N/m
1	13.7	0.0715	$0.1647 \cdot 10^7$	$0.1154 \cdot 10^{10}$
2	28.9	0.4467	$0.5446 \cdot 10^7$	$0.2473 \cdot 10^{10}$
3	81.8	0.6626	$0.1264 \cdot 10^8$	$0.5744 \cdot 10^{10}$
4	73.1	0.5263	$0.1264 \cdot 10^8$	$0.5744 \cdot 10^{10}$
5	73.1	0.5273	$0.1264 \cdot 10^8$	$0.5744 \cdot 10^{10}$
6	81.7	0.6798	$0.5408 \cdot 10^7$	$0.2465 \cdot 10^{10}$
7	37.1	0.6062	$0.3639 \cdot 10^7$	$0.1673 \cdot 10^{10}$
8	46.9	0.3972	$1 \cdot 10^6$	$1 \cdot 10^9$
9	100.	0.4	$1 \cdot 10^6$	$1 \cdot 10^9$
10	32.1	0.1413	$0.3167 \cdot 10^7$	$0.2219 \cdot 10^{10}$
11	14.2	0.0804	$0.3167 \cdot 10^7$	$0.2219 \cdot 10^{10}$
12	54.2	0.2807	$0.3167 \cdot 10^7$	$0.2219 \cdot 10^{10}$
13	14.2	0.0845	$0.3167 \cdot 10^7$	$0.2219 \cdot 10^{10}$
14	7.1	0.0100	–	–

Fig. 23

and masses assessed from that drawing are given in Fig. 22 and the table. The table gives also assumed values.

The final rigid element model is shown in Fig. 23 together with the table of values.

Results: Computed first three principal modes and natural frequencies, for the system with heavy coupling (i.e. before correction) are shown in Fig. 24 together with approximate mass distribution of the coupling.

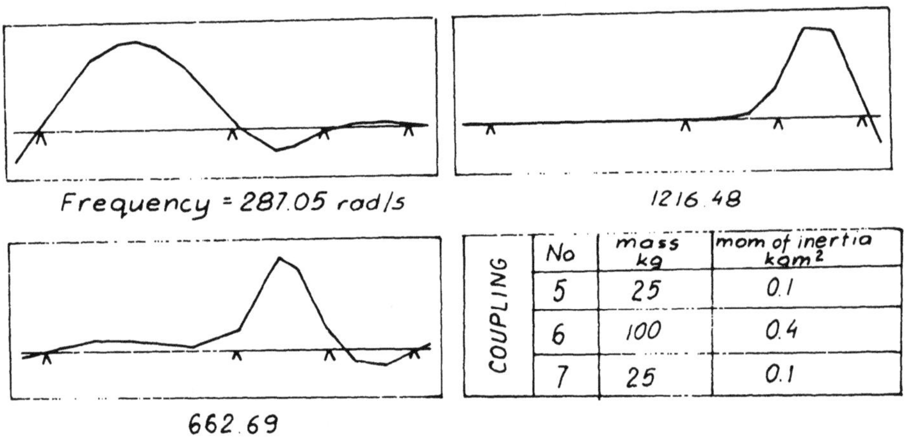

COUPLING No	mass kg	mom of inertia kgm^2
5	25	0.1
6	100	0.4
7	25	0.1

Fig. 24

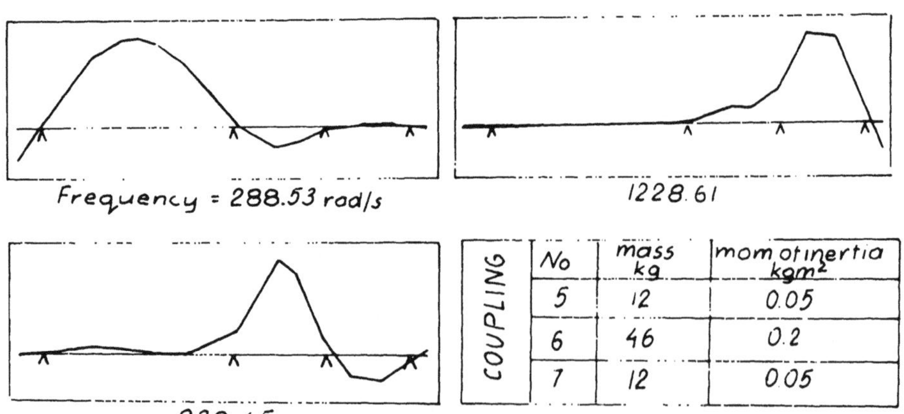

COUPLING No	mass kg	mom of inertia kgm^2
5	12	0.05
6	46	0.2
7	12	0.05

Fig. 25

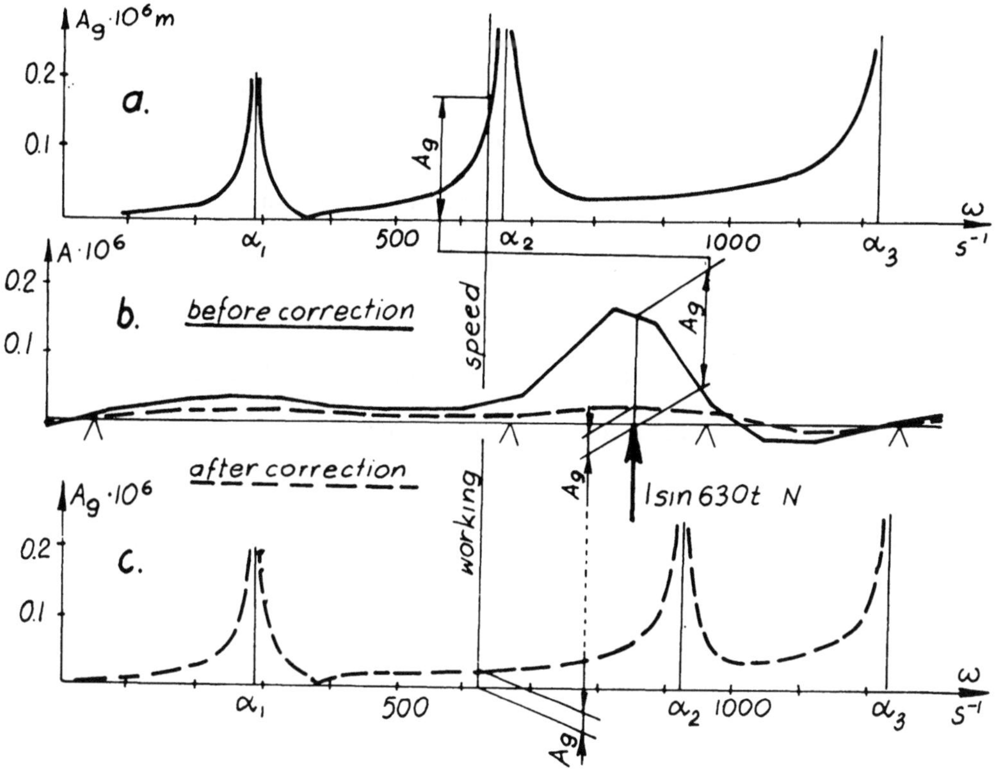

Fig. 26

Second critical is close to working speed. Corresponding modes and frequencies for the system with the light coupling (i.e. after correction) are shown in Fig. 25 with the assumed mass distribution of the coupling.

Second critical was increased by 30% and lies at a comfortable distance over working speed. First and third criticals were hardly changed. Resonance diagram for the coupling (element No.9 in Fig.23) before (diagram a) and after correction (diagram c) are given in Fig. 26. Rotor response to unit imbalance of the coupling at working speed is shown in Fig. 26b before (continuous line) and after correction (broken line).

REFERENCES

(1) Parszewski, Z.A.: Metoda wywazania maszyn w łożyskach własnych, Proc. Polish Conf. TMM, Rogów 1956.

(2) Parszewski, Z.A.: Pewna metoda dynamicznego wyrównoważania maszyn wirnikowych, Arch. Bud. Masz, Vol. V, No.2, Warszawa 1958, p 199-212.

(3) Parszewski, Z.A., Grootenhuis, P.: Balancing Multi-Bearing Machines, The Engineer London February 1962.

(4) Bishop, R.E.D.: Johnson, D.C.: The Mechanics of Vibrations, Cambridge Univ. Press, 1960.

(5) Parszewski, Z.A.: Critical Speeds of Shaft-Bearing-Foundation Systems, Proc. I.Mech.E. Conf. Vibr. in Rotating Mach., London 1972.

(6) Parszewski, Z.A.: Dynamic Interaction Between Machine and Support, Proc. I Tribology Congress, New Delhi, 1972.

(7) Parszewski, Z.A.: Hincz, M.: Schwingungen von Elastisch gelagerten Wellen, Proc. III Congr. TMM, Dubrownik, 1971.

(8) Parszewski, Z.A.: Drgania i Dynamika Maszyn, WNT, Warszawa 1982, 420p.

(9) Parszewski, Z.A.: Introduction to Dynamic Analysis, Publ. School in Power Generation, Univ. of Melbourne, Dept. of Mech. Eng. Melbourne, 1983, Vol. I, pp 5.1-5.25.

(10) Parszewski, Z.A.: Vibrations of Rotating Machines, ibid p.5.83-5.109.

(11) Skoraczynski, J.: Subsystem Synthesis Approach for Dynamic Analysis of Machinery, part. Ph.D. thesis, Dept. of Mech. Eng. The University of Melbourne, 1982.

(12) Kruszewski, J.: Gawronski, W.: Application of Finite Elements Method to the Analysis of Vibration of Complex Linear Systems, Rozprawy Inzynierskie, No.20, 4, Warszawa 1972.

CHAPTER 2.10

DYNAMICS OF MACHINE FOUNDATIONS INTERACTING WITH SOIL

L. Gaul
O. Mahrenholtz

1 Introduction

A method for calculating the global three-dimensional dynamic response of machine foundations interacting with subsoil and excited by rotor unbalances is presented. Fig. 1 shows a low-tuned steel foundation with a concrete raft. The light upper steel plate on flexible columns has the advantage to minimize the amplitudes of shaft whirling relative to the bearings.

Formerly, high-tuned concrete foundations were dominating. Now, all foundations - except some special cases - are low-tuned. This development is influenced by the need of space for auxiliary aggregates below the foundation plate.

Fig. 1 Low-tuned steel foundation with concrete raft [1]

Dynamic response results primarily from rotor unbalances, short circuit moments, shaft misalignment and seismic excitation. A dynamic analysis shall predict the amplitudes of vibration. The contents of the present contribution on this subject is listed below:

o Analysis by
  - direct method
  - substructure method

o Substructure behavior of
  - frame foundation
  - subsoil

o Lumped parameter model of soil

o Response of frame foundation
  - on subsoil
  - on piles and subsoil
  - interaction between adjacent structures

2 Analysis

Soil-structure interaction analysis based on the direct method treats structure and soil in a combined analysis. Discretization by finite elements is used throughout (Fig. 2). Due to the halfspace geometry of soil the analysis is restricted to plane and axisymmetric problems. At the lower rigid boundary undesired reflexions of waves generated by the structures occur trapping the energy in the finite model domain. This leads to errors in the response. The so-called geometrical damping due to three-dimensional wave propagation can only be approximated by viscous dashpots in a plane model.

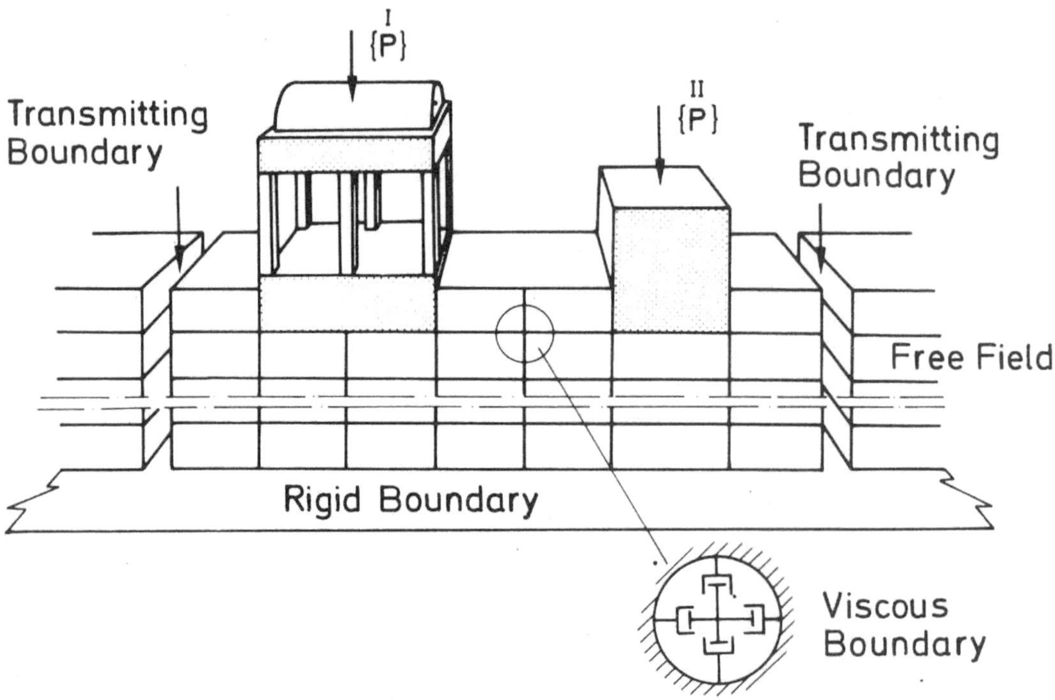

Fig. 2 Soil-structure model of direct method

The substructure technique presented here provides solutions for three-dimensional displacements of structures which are of importance concerning machine foundations [3]. The substructures are machine foundations and soil (Fig. 3). Dynamic stiffness matrices describing elastic, damping and inertia properties of the substructures are evaluated separately, thus reducing the number of degrees of freedom. In a second step the substructure matrices are coupled at the interfaces between soil and base plates shown in Fig. 3 by continuity conditions of displacements and by the reaction principle of generalized forces.

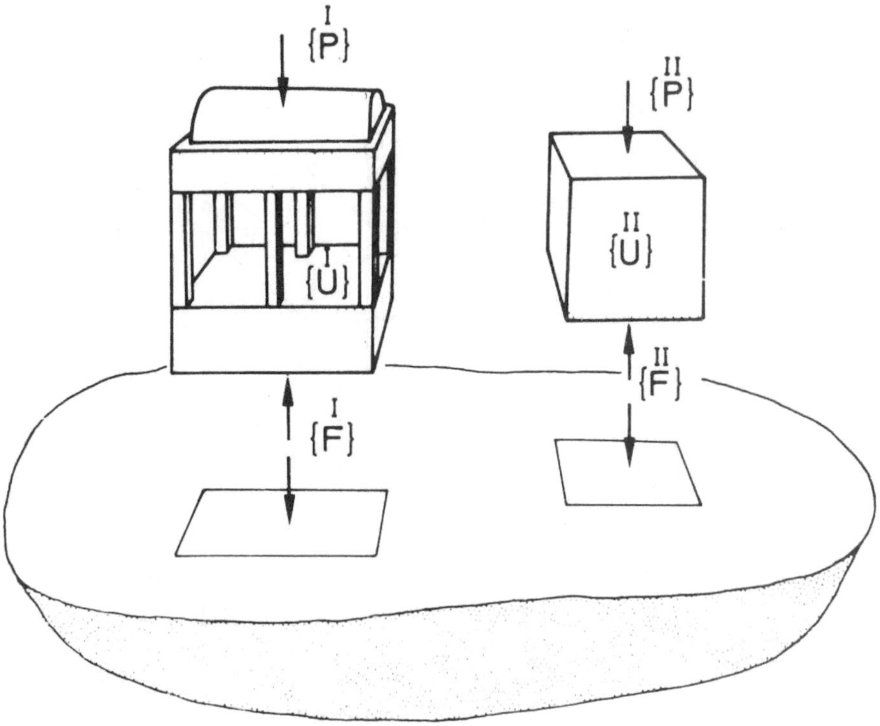

Fig. 3 Soil-structure model of substructure method

3 Substructure behavior

To evaluate the global vibration behavior of a frame foundation a simplified model (Fig. 4) is treated. This substructure consists of a machine on the upper plate excited by inertia forces of the unbalanced rotor. Upper plate and base plate are connected by viscoelastic columns. The halfspace reactions are reduced to point B of the interface. The motion can be described by six displacement coordinates v_i, u_i and the six angles describing small rotations Φ_i and φ_i of upper plate and base plate, respectively.

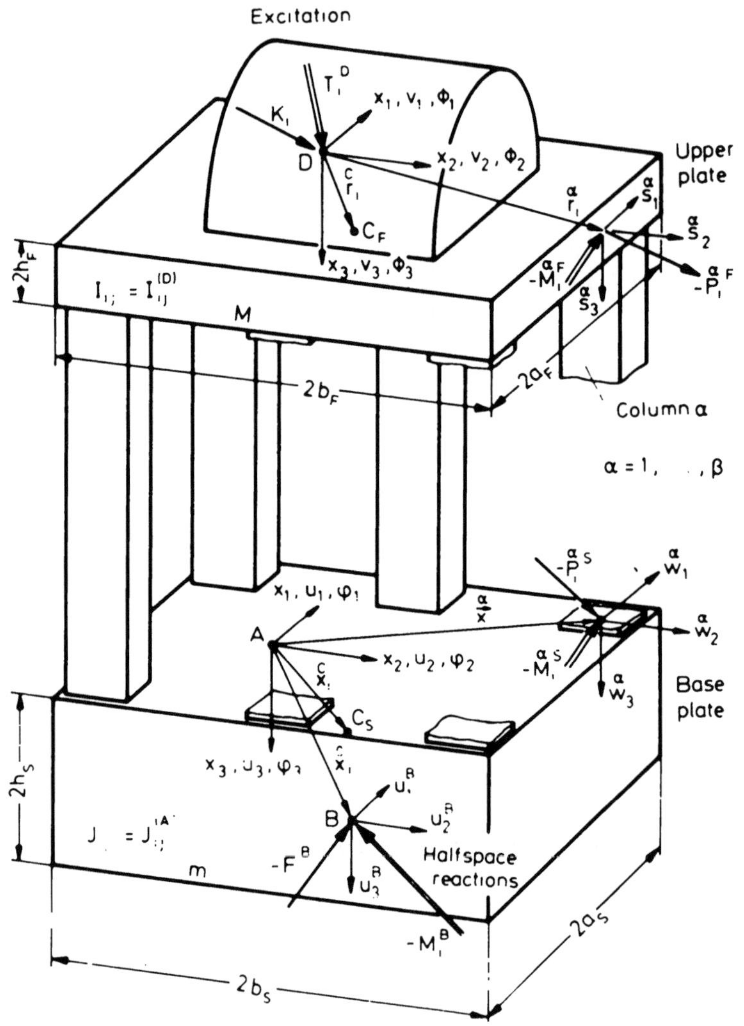

Fig. 4 Frame foundation on halfspace

The linearized equations of motion for small rotations are formulated by the Newton-Euler approach (Table 1, eqs. (1) to (5)). For example, Euler's law for the upper plate contains inertia terms on the left-hand side. The coordinates of the inertia tensor I_{ij}^{D} correspond to point D. M is the mass of upper plate and machine. The right-hand side contains, as indicated in eq. (1) the moment of excitation and the moments of the column reactions. Eq. (2) gives Newton's second law: Mass times acceleration of the center of gravity equals the resultant of the force of excitation and column reactions.

The column reactions are related to the generalized displacements by the dynamic stiffness matrix $[K]^\alpha$ of column α, eq. (3).

Upper plate

$$I^D_{ij}\ddot{\Phi}_j + M e_{ijk} r^c_j \ddot{v}_k = -\sum_{\alpha=1}^{\beta}\left(e_{ijk} r^\alpha_j P^{F}_k + M^{\alpha F}_i\right) + T^D_i \exp(i\omega t) \qquad (1)$$

$$M(\ddot{v}_i - e_{ijk} r^c_j \ddot{\Phi}_k) = -\sum_{\alpha=1}^{\beta} P^{\alpha F}_i + K_i \exp(i\omega t) \qquad (2)$$

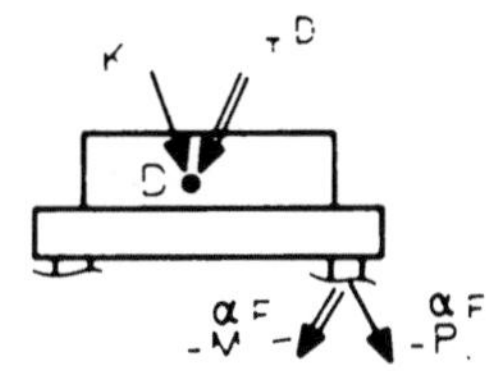

Columns

$$\begin{Bmatrix} M^S_i \\ P^S_i \\ M^F_i \\ P^F_i \end{Bmatrix}^{\alpha} = [K_{il}]^{\alpha} \begin{Bmatrix} \varphi_l \\ u_l - e_{lmn} x^\alpha_m \varphi_n \\ \Phi_l \\ v_l - e_{lmn} r^\alpha_m \Phi_n \end{Bmatrix} \qquad (3)$$

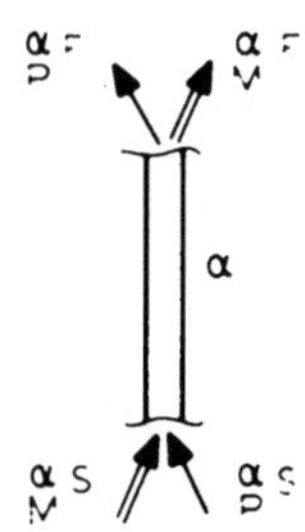

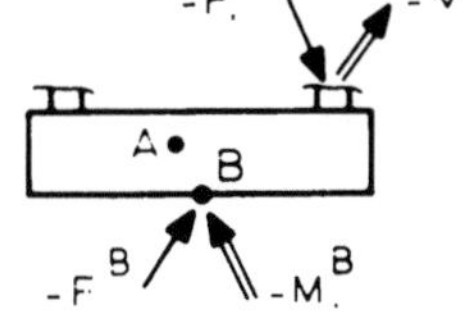

Base plate

$$m(\ddot{u}_i - e_{ijk} x^c_j \ddot{\varphi}_k) = -\sum_{\alpha=1}^{\beta} P^{\alpha S}_i - F^B_i \qquad (4)$$

$$J^A_{ij}\ddot{\varphi}_j + m e_{ijk} x^c_j \ddot{u}_k = -\sum_{\alpha=1}^{\beta}\left(e_{ijk} x^\alpha_j P^{\alpha S}_k + M^{\alpha S}_i\right) - \left(M^B_i + e_{ijk} x^0_j F^B_k\right) \qquad (5)$$

Table 1 Frame foundation, equations of motion

Analogous to the equations of motion of the upper plate, those of the base plate are given in eqs. (4) , (5) . The inertia, damping and elastic forces are combined in a dynamic stiffness matrix corresponding to time-harmonic motion. This describes the substructure behavior of the frame foundation.

Now, the substructure soil is treated. First, rigid base plates are modeled, later on the generalization for flexible plates is explained. Fig. 5 shows the free body diagram of halfspace surface. The plane interfaces are excited by forces and moments generated by the superstructures. Compared to the direct method, the halfspace is not discretized in depth.

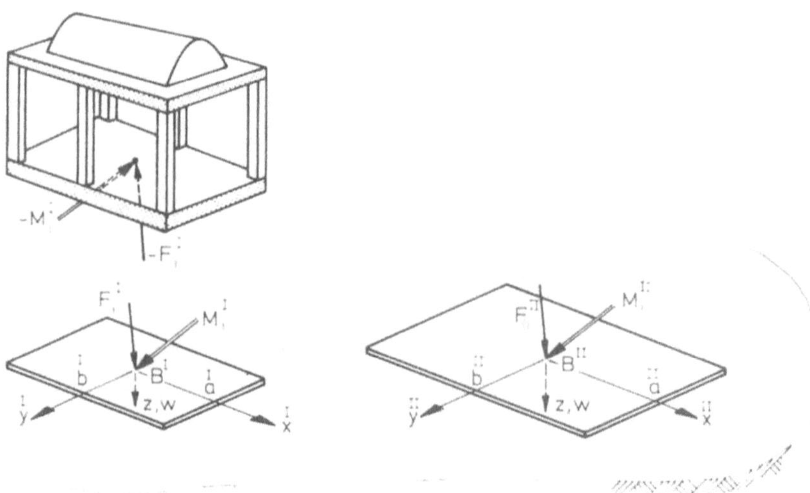

Fig. 5 Mixed boundary value problem for soil

The interaction is formulated as a mixed boundary value problem of continuum mechanics. The field equations of soil are integrated:

- displacement boundary conditions are prescribed at the interfaces

- a stress-free surface is required elsewhere.

Rigorous formulations of mixed boundary value problems by dual integral equations restricted to simple base geometricies can be given [2,3]. Here, a superposition method provides solutions for arbitrary shapes of base plates and allows for taking flexible base plates into account.

4 Lumped parameter model of soil

Arbitrary shapes can be modeled by subdividing the interfaces into rectangular surface elements. The continuous stress distribution in the interface is discretized into constant pressures in one element, acting harmonically in time (Fig. 6). Each loaded element defines a stress boundary value problem of the halfspace. To bound the influence of shear stresses in the interface

- a perfectly smooth contact where the shear stresses vanish
- a welded contact where the inplane displacements u, v vanish

are assumed.

Analytical solutions of both boundary value problems leading to displacement influence matrices are obtained then. One element of this matrix gives the displacement in the middle of element k due to the loading of element ℓ (Fig. 6). The total displacements are found by superposition. The displacement boundary condition is locally fullfilled in the center of each element. It requires plane interfaces of both bases. This superposition converges to the rigorous formulation with decreasing size of elements. Additionally, the resultants of the interface stress distributions must be equivalent to the halfspace reactions (forces F_z and moments M_y, M_x).

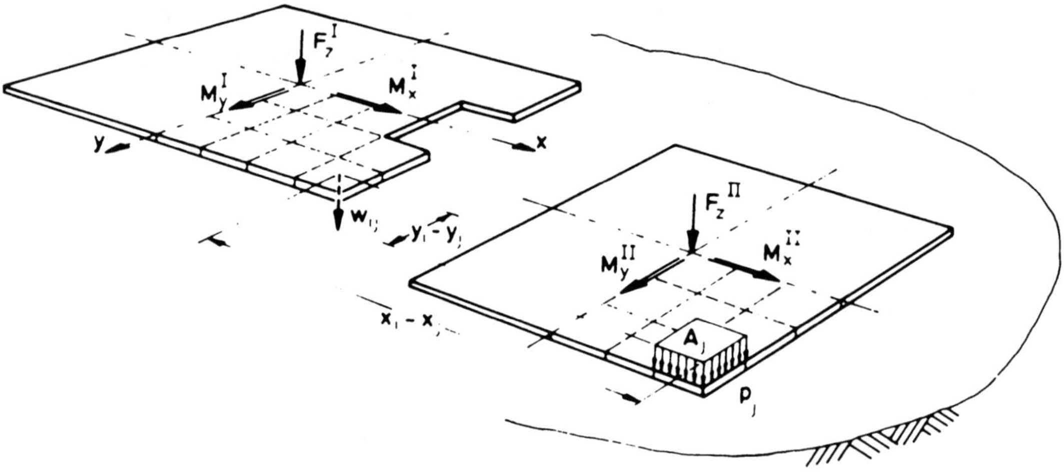

Fig. 6 Dynamic interaction of base I - halfspace - base II

The dynamic interface stress distribution as well as the relations between halfspace reactions and generalized displacements at the interfaces are evaluated, leading to the dynamic stiffness matrices of the substructure soil.

Flexibility of the base plate can be taken into account by subdividing the plate into finite elements and coupling nodal forces and displacements with the halfspace.

The solution of the stress boundary value problem of one surface element of the interface is given below (Tables 2-4, eqs. (6) to (20)). Compared to elastic halfspace theories a better approximation of the rheological properties of the soil is given by using the theory of viscoelasticity. I turns out that energy dissipation in the soil by material damping is of considerable influence when the geometrical damping by wave radiation is small, especially in the case of rocking motion.

The constitutive stress-strain relations of hereditary integral type express the dependency of the stresses in viscoelastic media on total history of strains with fading memory property of material (Table 2). The fading memory properties can be measured by relaxation functions of shear G(t) and compression K(t) corresponding to the deviatoric and the hydrostatic states of stresses and strains. Newton's and Boltzmann's laws lead to the coupled viscoelastic equations of motion in terms of displacements u_i. These can be decoupled by a decomposition of the displacement field into two wave equations describing dilatational and rotational motion. A completeness theorem is given in [3].

Steady state harmonic motions lead to reduced wave equations describing complex amplitudes of dilatation and rotation with the frequency dependent complex moduli of shear G^* and plane dilatation E^*_D instead of the relaxation functions (Table 3). The solutions of the wave equations describe dilatation and shear waves. The displacement field and the stress field are superimposed by these solutions. The complex functions of integration A, B_k have to be determined by introducing boundary conditions.

The stress boundary value problem is solved by superposition of basic harmonic solutions (Table 4). These basic solutions correspond to an exciting stress wave at the halfspace surface. The halfspace response is a vertical displacement wave. The amplitude and the phase behavior are expressed in the complex wave compliance equating the complex displacement amplitude to the complex stress amplitude.

The compliance corresponding to the elastic halfspace ($\eta_S=\eta_D=0$) as far as welded contact at the surface is concerned has a pole when the phase velocity of the stress wave v equals the shear wave velocity v_S.

Constitutive equations

$$s_{ij}(t) = 2\int_{-\infty}^{t} G(t-\tau)\,\frac{\partial e_{ij}(\tau)}{\partial \tau}\,d\tau \quad \sigma_{kk}(t) = 3\int_{-\infty}^{t} K(t-\tau)\,\frac{\partial \varepsilon_{kk}(\tau)}{\partial \tau}\,d\tau \tag{6}$$

Equations of motion for viscoelastic continuum

$$\int_{-\infty}^{t} \underbrace{[K(t-\tau) + \tfrac{4}{3}\,G(t-\tau)]}_{E_D(t-\tau)}\,\frac{\partial u_{j,ji}}{\partial \tau}\,d\tau - e_{ijk}\,e_{klm}\int_{-\infty}^{t} G(t-\tau)\,\frac{\partial u_{m,lj}}{\partial \tau}\,d\tau + \rho f_i(t) =$$

$$= \rho\,\frac{\partial u_i(t)}{\partial t^2} \tag{7}$$

Completeness theorem: Decomposition

$$u_i(t) = \varphi_{,i}(t) + e_{ijk}\,\psi_{k,j}(t) \tag{8}$$

decouples equations of motions in wave equations

$$\int_{-\infty}^{t} \frac{E_D(t-\tau)}{\rho}\,\frac{\partial \varphi_{,ll}(\tau)}{\partial \tau}\,d\tau + F(t) = \frac{\partial^2 \varphi(t)}{\partial t^2}\,, \quad \text{Dilatation} \quad \varepsilon(t) = \varphi_{,ll}(t) \tag{9}$$

$$\int_{-\infty}^{t} \frac{G(t-\tau)}{\rho}\,\frac{\partial \psi_{k,ll}(\tau)}{\partial \tau}\,d\tau + L_k(t) = \frac{\partial^2 \psi_k(t)}{\partial \tau^2}\,, \quad \text{Rotation} \quad 2\,\omega_k(t) = -\psi_{k,ll}(t) \tag{10}$$

if constraint condition $\psi_{k,k}(t) = 0$ is satisfied

Table 2 Field equations of halfspace

Harmonic motion $$u_i(x_j,t) = \bar{u}_i(x_j)\exp(i\omega t) \tag{11}$$

Reduced wave equations

$$\bar{\varepsilon}_{,11} + \frac{\rho\omega^2}{E_D^*(i\omega)}\,\bar{\varepsilon} = 0 \qquad \bar{\omega}_{k,11} + \frac{\rho\omega^2}{G^*(i\omega)}\,\bar{\omega}_k = 0 \tag{12}$$

Solutions $\mathrm{Re}(\alpha_{S,D}) \geq 0\ ,\ \bar{\omega}_{k,k} = 0$

$$\bar{\varepsilon} = A\exp[-\alpha_D z + i(\beta x + \gamma y)] \qquad \bar{\omega}_k = B_k\exp[-\alpha_S z + i(\beta x + \gamma y] \tag{13}$$

Displacement field

$$\bar{u}_i = -\bar{\varepsilon}_{,i} + 2\,e_{ijk}\,\bar{\omega}_{k,j} = \bar{u}_i(A,B_k) \tag{14}$$

Stress field

$$\bar{\sigma}_{ij} = [E_D^*(i\omega) - 2\,G^*(i\omega)]\,\bar{u}_{k,k}\,\delta_{ij} + G^*(i\omega)(\bar{u}_{i,j} + \bar{u}_{j,i}) = \bar{\sigma}_{ij}(A,B_k) \tag{15}$$

Table 3 Integration of field equations

Stress wave at surface $z = 0\ ,\ v = \frac{\omega}{(\beta^2+\gamma^2)^{1/2}}$

$$-\sigma_{zz}(x,y,t) = p(x,y,t) = \bar{p}(\beta,\gamma)\ \exp\ [i(\beta x+\gamma y+\omega t)] \tag{16}$$

excited *displacement wave*

$$w(x,y,t) = \bar{H}_w^{s,v}(\beta,\gamma,\omega)\ \bar{p}(\beta,\gamma)\ \exp\ [i(\beta x+\gamma y+\omega t)] \tag{17}$$

wave compliance (welded contact u=v=0)

$$\bar{H}_w^{\ v} = \frac{\bar{w}}{\bar{p}} = \frac{1}{Gk_s}\left\{\frac{\left(\frac{v_s}{v}\right)^2 - \left[\left(\frac{v_s}{v}\right)^2 - \frac{1}{1+i\eta_s}\right]^{1/2}\left[\left(\frac{v_s}{v}\right)^2 - \frac{n^2}{1+i\eta_D}\right]^{1/2}}{\left[\left(\frac{v_s}{v}\right)^2 - \frac{1}{1+i\eta_s}\right]^{1/2}}\right\} \tag{18}$$

Superposition: Fourier's integral

$$p(x,y) = \frac{1}{2\pi}\iint_{-\infty}^{\infty} \bar{p}(\beta,\gamma)\ \exp\ [i(\beta x+\gamma y)]\ d\beta\ \ d\gamma \tag{19}$$

$$w(x,y,t) = \frac{1}{2\pi}\iint_{-\infty}^{\infty} \bar{H}^{s,v}(\beta,\gamma,\omega)\mathcal{F}^2[p(x,y)]\ \exp\ [i(\beta x+\gamma y+\omega t)]\ d\beta\ \ d\gamma \tag{20}$$

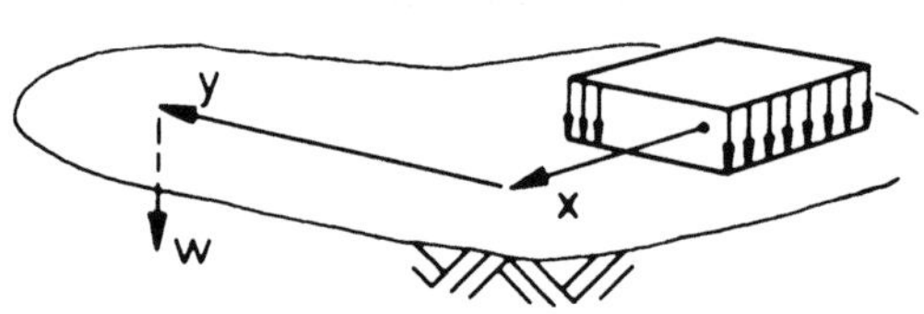

p(x,y)exp(iωt)

Table 4 Solution of stress boundary value problem

Smooth contact leads to a pole when the velocity equals the Rayleigh wave speed v_R . In order to obtain the exciting stress distribution shown in Table 4 on a rectangular element, stress waves are superimposed by double Fourier integral. The corresponding displacement field contains the Fourier transform of the stress field and the wave compliance corresponding to either smooth or welded contact.

As the elastic halfspace due to the poles leads to undetermined improper integrals, it is complicated to perform a numerical contour integration in the complex plane after choosing Cauchy's principle values of the integrals. Gaul [2, 3, 5, 6] had the idea to integrate the viscoelastic field equations in a direct manner without using an elastic-viscoelastic correspondence principle.

The reason is obvious: real and imaginary parts of the complex wave compliances corresponding to smooth and welded contact show finite resonant magnifications instead of poles (Fig. 7). The compliances are plotted versus the ratio M of exciting stress wave velocity to shear wave velocity. The integrals are no longer improper with respect to the integrant and can be integrated directly. Pointwise evaluation of the displacement field w(x,y) gives the displacement influence matrix to be used for the superposition method. It may be mentioned that welded contact leads to resonance condition at the shear wave velocity, smooth contact at the slightly slower Rayleigh wave.

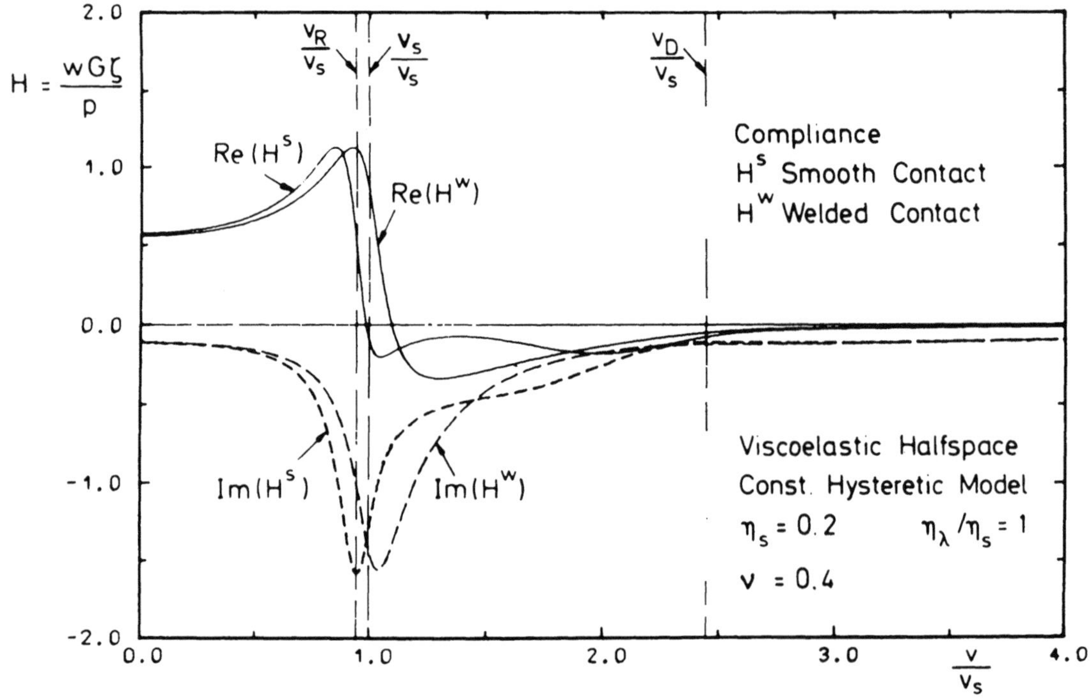

Fig. 7 Halfspace compliance. Excitation by a stress wave

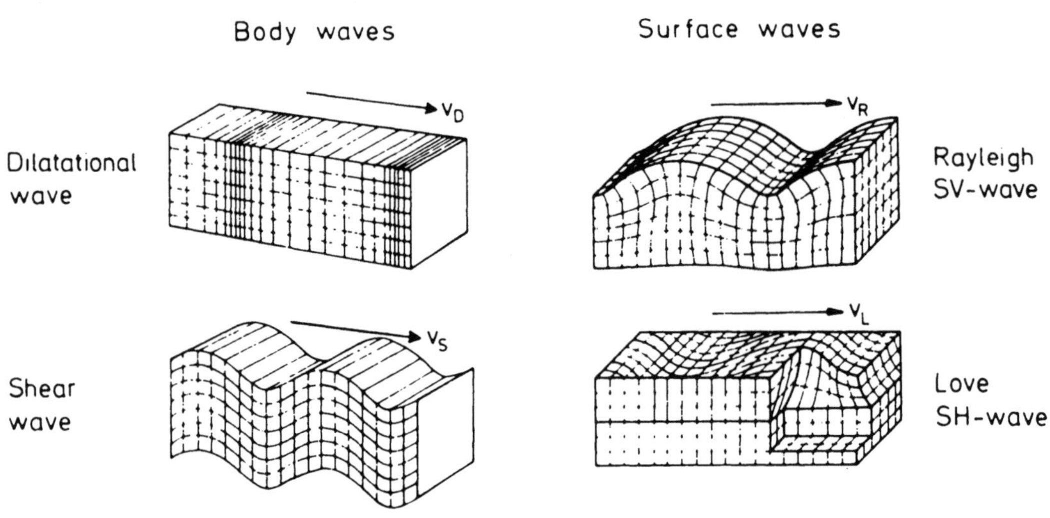

Fig. 8 Types of elastic waves

Characteristic displacement fields of waves generated in homogeneous and two-layer halfspace are shown in Fig. 8.

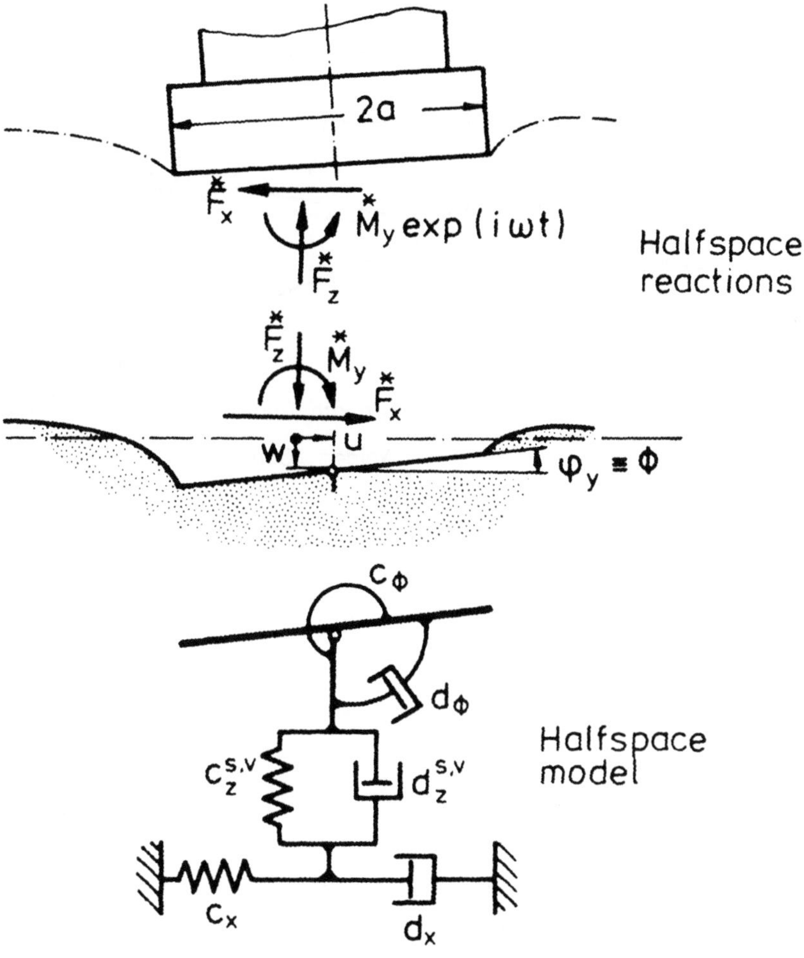

Fig. 9 Lumped parameter model

The results of the analysis can be used in practical problems without detailed knowledge of the mathematical background. The dynamic stiffness matrix of the substructure soil leads to simple lumped parameter models of soil consisting of springs and dashpots in parellel corresponding to each degree of freedom of base motion (Fig. 9). The equivalent spring and damping coefficients c_z, d_z of the soil model for vertical vibration of a rigid square base versus a dimensionless frequency a_0 are plotted in Fig. 10:

- the frequency dependent spring describes restoring as well as inertia forces,
- the damping coefficient describes geometrical damping.

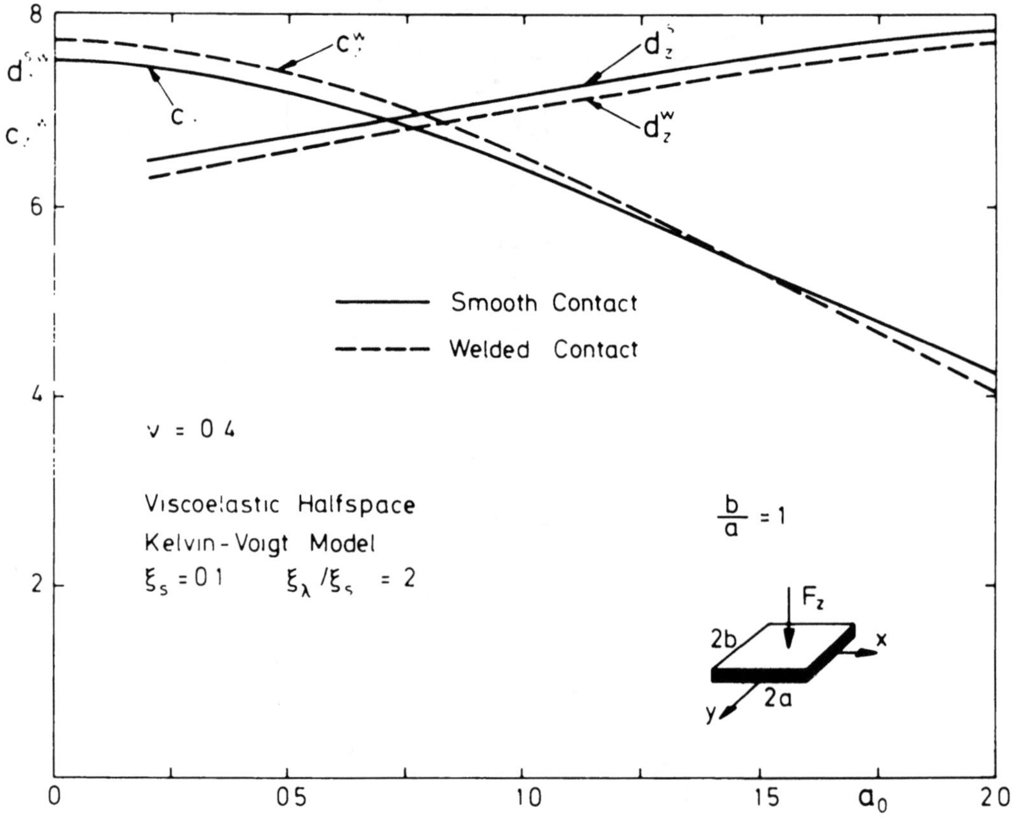

Fig. 10 Equivalent spring and damping coefficient; vertical motion of a rigid base

Both parameters depend on material damping. The influence of the contact boundary conditions (smooth and welded contact) leads only to a small difference with respect to the parameters of the halfspace models.

5 Response of frame foundation

Now, the substructure soil and frame foundation (Fig. 1) are coupled. The dynamic response of a concrete frame foundation with very stiff columns is shown in a limited frequency range (Fig. 11). It is excited by an unbalanced rotor. The amplification functions describe the amplitudes of horizontal displacements v_1 and u_1 and vertical displacements v_3 and u_3 of the upper plate and base plate respectively. The coupled rocking and sliding motions give raise to two resonant conditions indicated by the horizontal displacements.

The corresponding eigenmodes are shown in Fig. 12. The horizontal motion is strongly effected by material damping of the soil because the dominating rocking mode causes only small geometrical damping. Contrary to this, the vertical motion causes strong wave radiation shown by the small resonance magnification. Hence, the influence of material damping is negligible.

Fig. 13 displays results of the interaction between two adjacent structures through the underlying soil. Foundation I is excited by rotor unbalances. Foundation II is not loaded. The dimensionless amplitudes of vertical displacements in the middle of the foundation are plotted. The maximum amplitude of the nonexcited structure arrives at one third of the amplitude of the excited structure. The influence of small and high material damping is indicated comparing the full and dotted line corresponding to different damping factors of the constant hysteretic model of viscoelasticity.

Additional information about the dynamic loading of soil and base plates is given by the stress distributions in the interfaces. Fig. 14 shows the adjacent bases loaded at a certain frequency by forces and by moments T_x and T_y acting with a phase shift of a quarter of a period of excitation. When the exciting forces reach their maximum the pressure

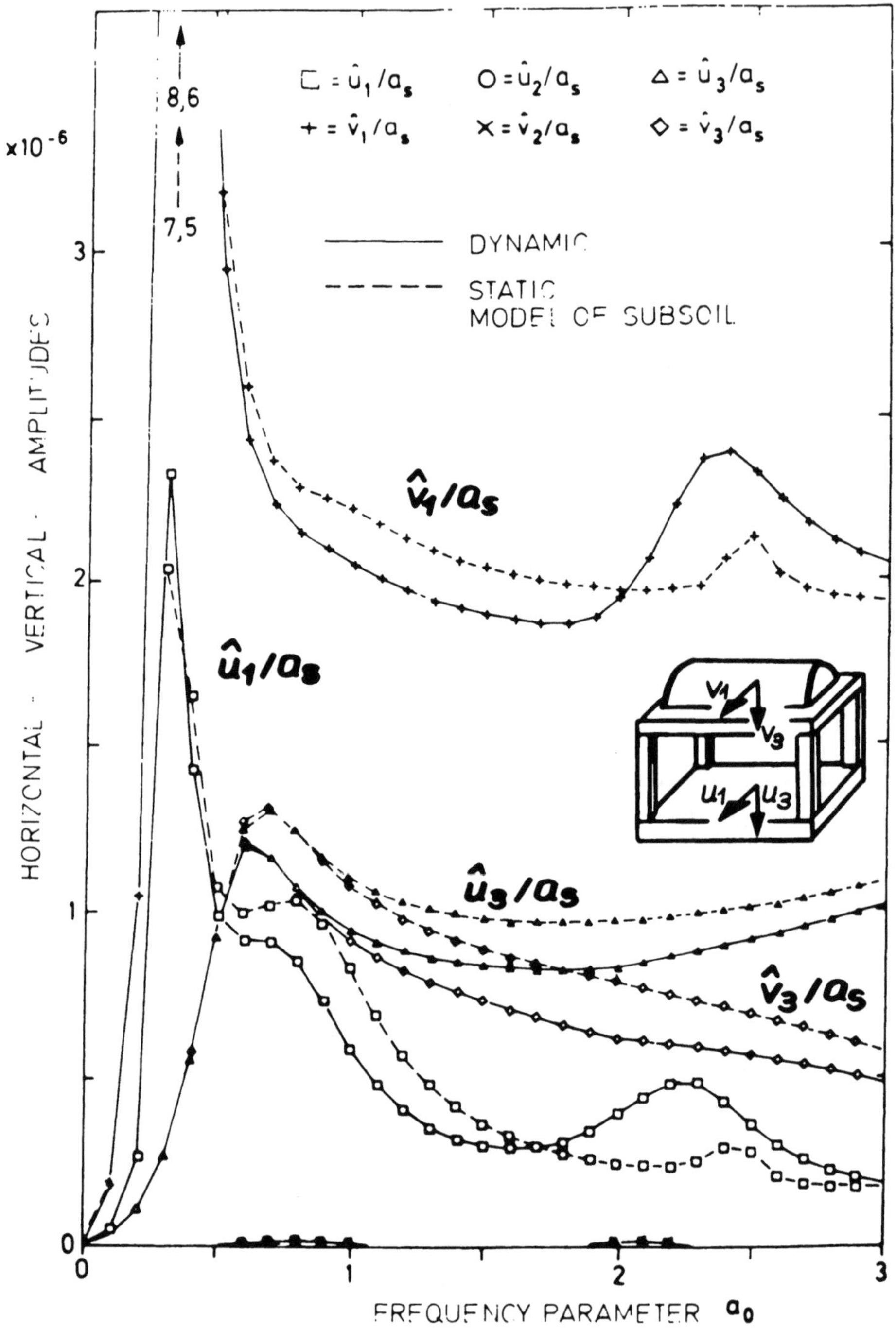

Fig. 11 Response of frame foundation on soil

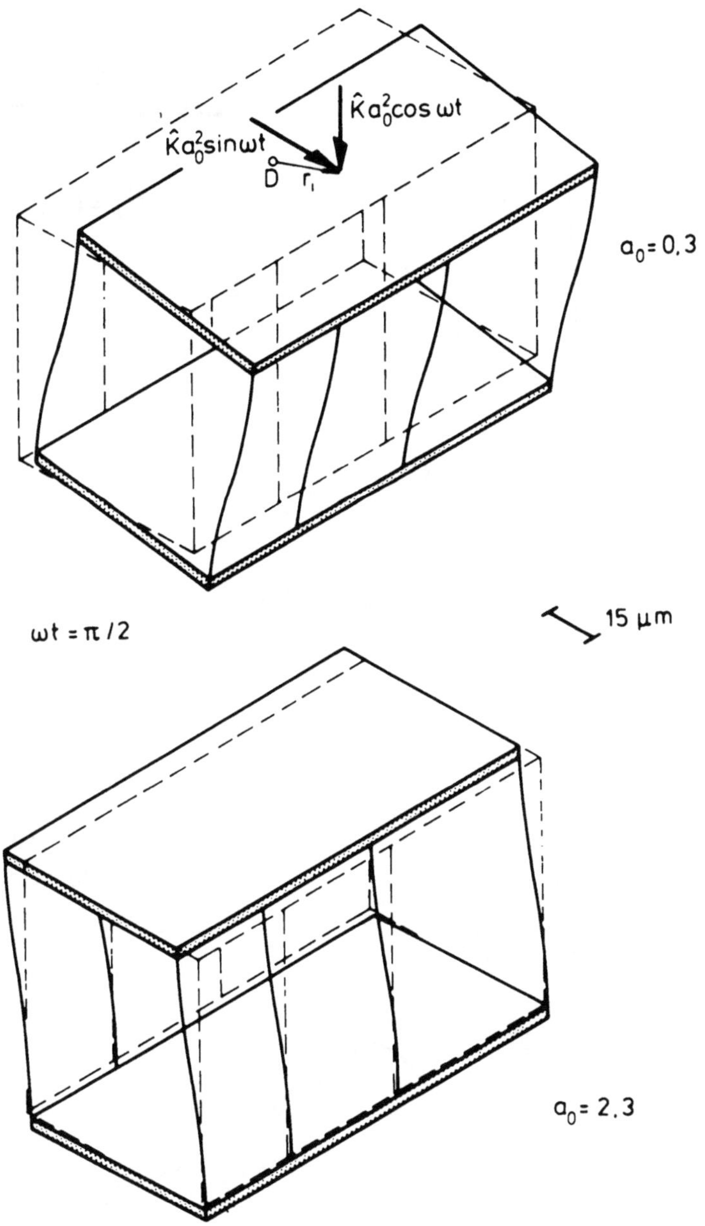

Fig. 12 Vibration modes of frame foundation on soil

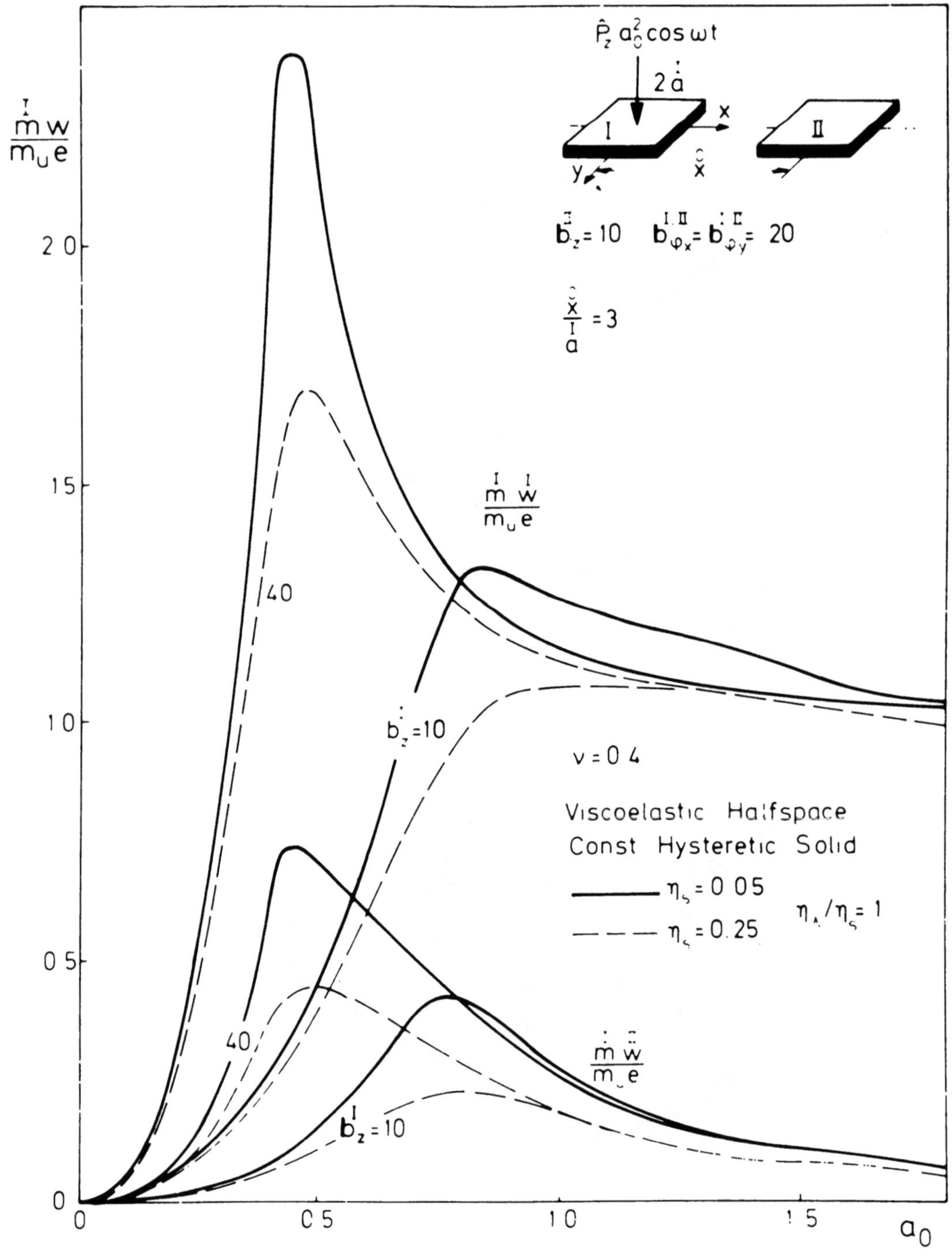

Fig. 13 Vertical response of two square footings; influence of viscoelasticity

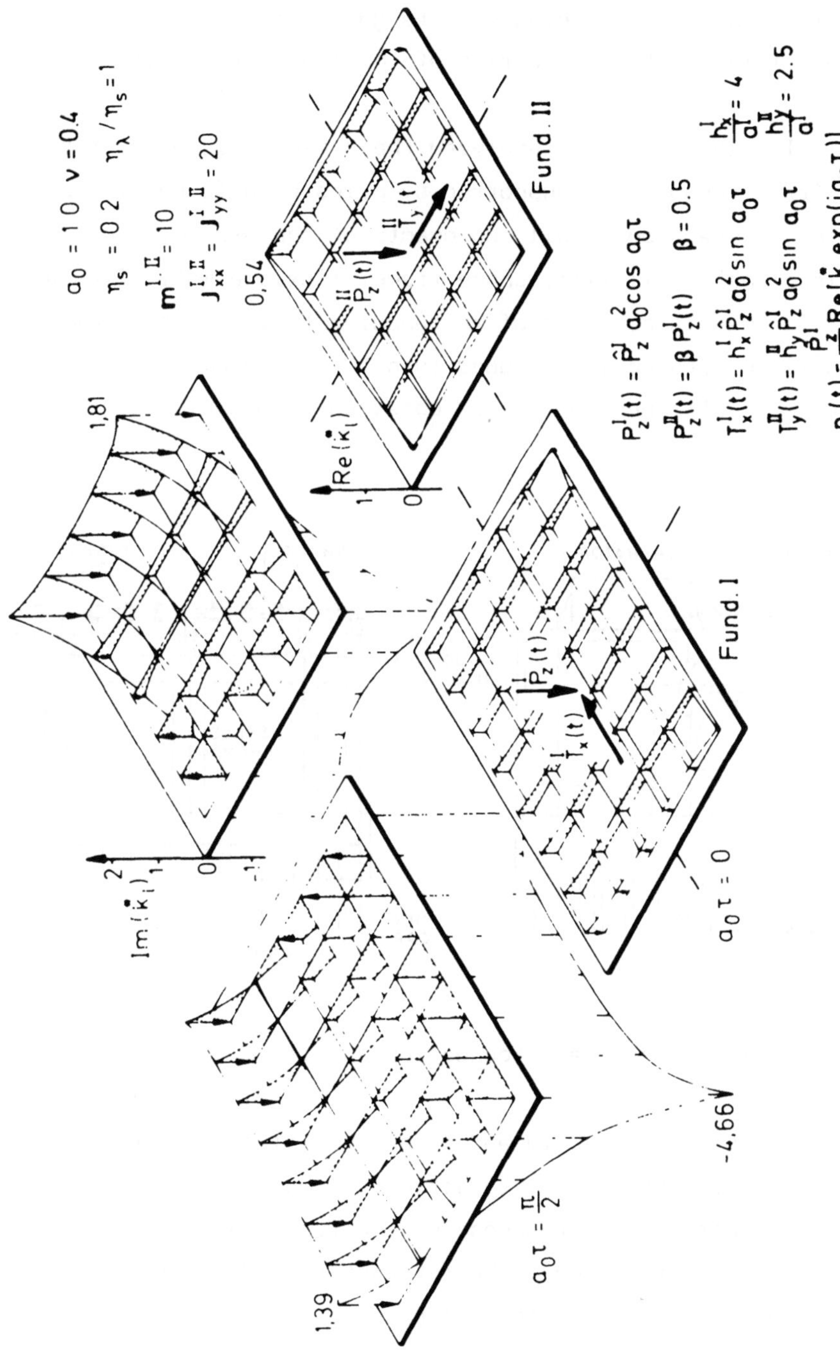

Fig. 14 Pressure distribution under adjacent excited bases

distributions are governed by the real parts of complex pressure coefficients. A quarter of a period of excitation later, the moments reach their maximum, Then, the imaginary parts govern the pressure distributions with singularities at the boundaries of the loaded rigid areas.

Aboul-Ella and Novak [4] analyzed the dynamic response of turbomachinery frame foundations supported by piles or a foundation slab. They investigated interaction of all components of the system, i.e. flexible rotors, viscoleastic oil film, space frame, flexible mat, piles and soil (Fig. 15). The mat is composed of rectangular finite plate elements. The pile and soil resistance is included into the mat element stiffness matrix. The complex soil stiffness matrix is obtained from Gaul [5].

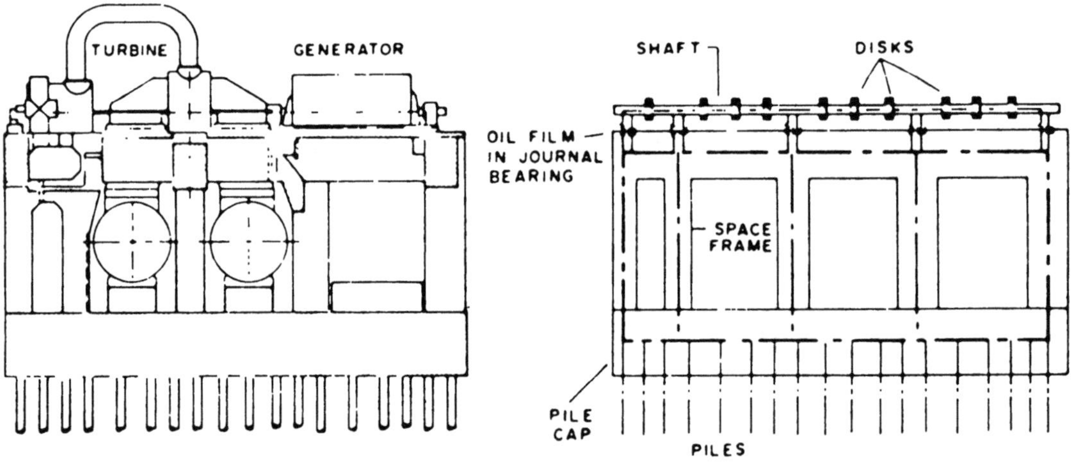

Fig. 15 Turbomachinery frame foundation and its model [4]

In the study of Aboul-Ella and Novak special attention is paid to the effects of soil-structure interaction. It was found that this interaction markedly affects the response of the frame as well as the rotors in the lowest resonant regions. The interaction reduces rotor and frame amplitudes. This results from the increase in damping due to energy radiation in the soil and viscoelastic behavior of soil and mat. The interaction reduces the frame vibration more than the shaft vibration.

E.g. Fig. 16 compares vertical response of frame under bearing pedestal corresponding to a rotor on elastic frame and rigid foundation with a rotor on elastic frame and elastic foundation.

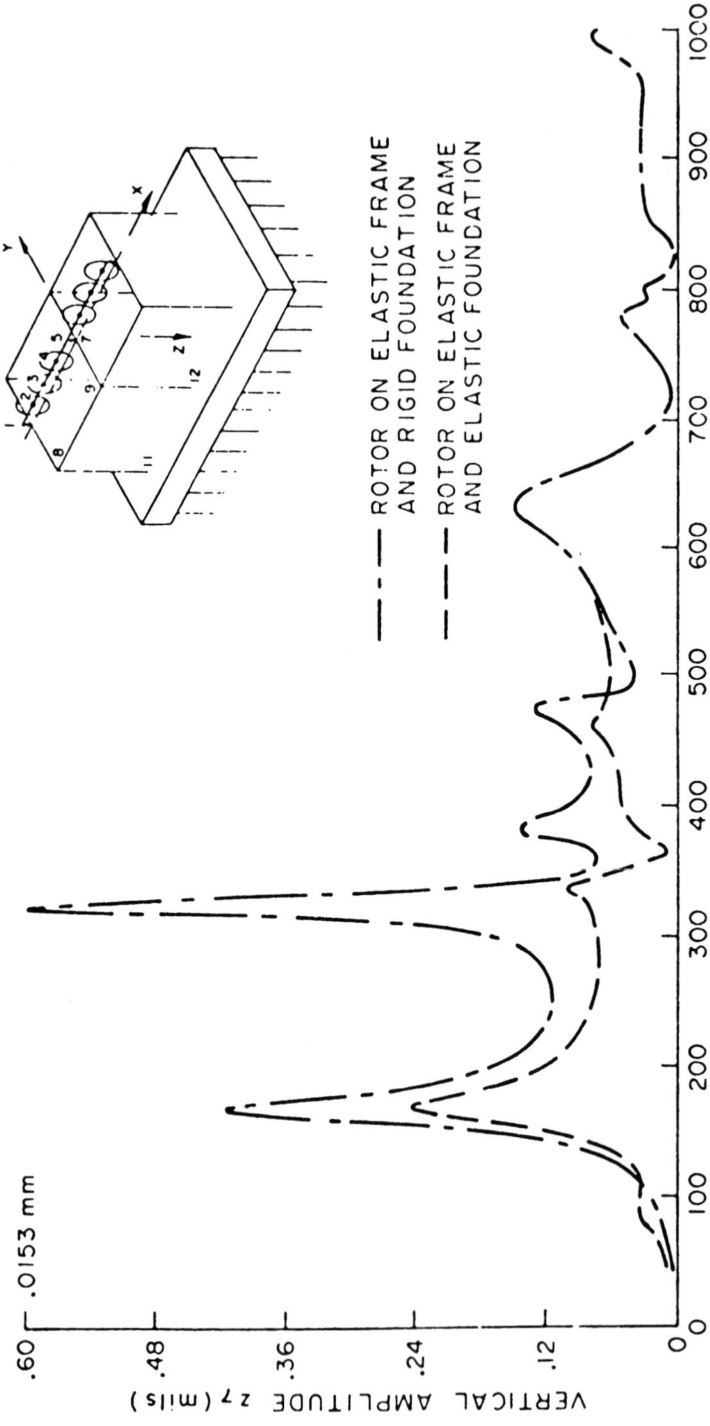

Fig. 16 Effect of foundation (piles and soil) flexibility on vertical response of frame under bearing pedestal [4]

References

[1] Dietz, H.: Steel Foundations for Turbomachines (in German). Beratungs stelle für Stahlverwendung, Düsseldorf, Merkblatt 146, 3 (1972)

[2] Gaul, L.: Dynamics of Frame Foundations Interacting with Soil. J. of Mechanical Design, 102 (1980) 303 - 310

[3] Gaul, L.: Dynamics of Soil-Structure Interaction (in German). Habilitationsschrift Universität Hannover 1980

[4] Aboul-Ella, F. and Novak, M.: Dynamic Analysis of Turbine-Generator Foundations. Fall Convention, American Concrete Institute ACI, Houston, Texas, USA (1978)

[5] Gaul, L.: Dynamic Interaction of a Footing with Viscoelastic Half-space (in German). Ingenieur-Archiv, 46 (1977) 401 - 422

[6] Mahrenholtz, O. and Gaul, L.: Soil-Structure Interaction Analysis by a Substructure Method, in: Mechanics of Inelastic Media and Structures, Mahrenholtz, O. and Sawczuk, A., Eds., Polish Scientific Publishers, Warsaw-Poznan, 1982, 119 - 132

PART III

BALANCING OF FLEXIBLE ROTORS

CHAPTER 3.1

BALANCING OF FLEXIBLE ROTORS

J. Drechsler

1 Historical Background of Flexible Rotor Balancing Techniques

When discussing flexible rotor balancing from an historical point of view, we can make out three classical development lines, mainly the modal theories developed by Bishop and Federn in England and Germany and the influence coefficient method developed in the USA. The classical papers on the modal balancing were published by Federn (1) in 1957 and by Bishop (2) in 1959. Papers on the influence coefficient approach were published by Goodman (3) in 1963, Rieger (4) in 1967 and Tessarzik, Badgley and Anderson (5) in 1971.

If we look at the modal theories first, we notice that the only important difference between these two theories lies in the number of balancing planes, that the authors claim to need in order to balance a flexible rotor correctly. Federn claimed to need (N + 2) planes, N being the number of vibration modes that the rotor experiences on the run up to operating speed. This approach is logical in an historical and practical sense, when applied to heavy rotors which are relatively stiff compared to the stiffness of the bearings. Such rotors had to be balanced in two planes even for running speeds long below the first critical speed.

For rigid rotor balancing rather sophisticated balancing machines were used. These balancing machines had a well defined bearing stiffness, so that it was possible to calculate the acting bearing forces from the measured vibration amplitudes due to the initial unbalance. Now the objective of balancing was to cancel these dynamic bearing forces P_1 and P_2 by the influence of balancing weights on these bearing forces according to the equations (1) .

$$P_1 + h_{11}\Omega^2 m_1 r_1 + h_{12}\Omega^2 m_2 r_2 = 0$$
$$P_2 + h_{21}\Omega^2 m_1 r_1 + h_{22}\Omega^2 m_2 r_2 = 0 \qquad (1)$$

In these equations the complex bearing forces have been measured as well as the rotor speed Ω , and the influence coefficients h_{ik} are functions of the rotor geometry only, thus the complex balancing weights $m_i \cdot r_i$ can be calculated by merely solving the above equation system. Electric circuits have been developed for the on line solution of this system.

As rotor constructions were reduced in weight and operating speeds increased, the operating speeds gradually approached the first critical speed. This eventually called for an additional balancing plane, because the rotors developed a new degree of freedom as they approached this critical speed, see figure 1. The practical problem was, not to upset the attained state of balance for the rigid body modes, and at the same time reduce the vibrations caused by the first flexible mode.

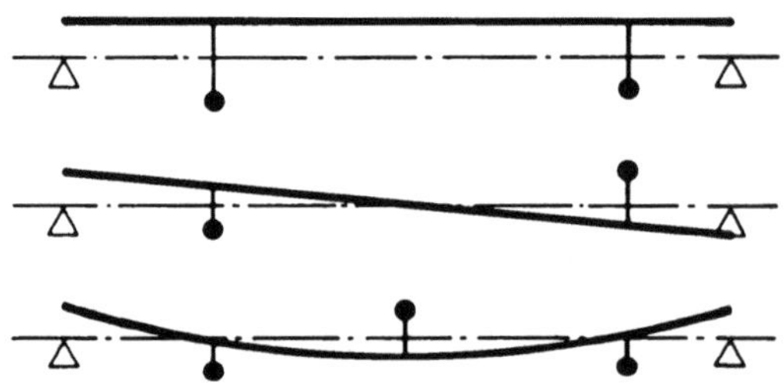

Fig. 1: Weight groups for rigid body and first flexural mode

It is intuitively obvious, that the weight distribution (1 – 2 1) exerts neither symmetric nor antisymmetric forces on the bearings at low speed. It can thus be termed a prebalanced group of balancing weights. Generalizing this approach for N critical speeds within the service speed range, this naturally leads to the N + 2 theory.

If on the other hand the unbalance response of a flexible rotor is to be studied on a scientific scale, a light weight experimental rotor in relatively stiff bearings will be designed, because this is the most economical, most practical and safest solution. Such a rotor does not have large bearing vibrations at low speed, which necessitate rigid body balancing. Only in the vicinity of the first rotor critical speed the system will display large vibration amplitudes, which call for balancing. In this case one plane will be sufficient. As even higher critical speeds are negotiated, one additional plane is necessary for each new critical speed, which naturally leads to the N-theory. But as we even in this case eventually have to take care, not to upset the state of balance attained for the low order criticals when trying to balance higher order modes, we have to work with prebalanced groups of balancing weights here as well.

As digital computers became available, the so called influence coefficient method was developed in the USA. The basic concepts for this method can be developed from the equation system (1) . Supposing we do not know the stiffness of the bearings, we can calibrate the relation between the vibration amplitudes at the bearings and the applied balancing weights by applying test weights. The basic procedure is summarized in figure 2

Run		0	1	2
Testweight	P_1	0	1	0
	P_2	0	0	1
Response	L_1	w_{10}	w_{11}	w_{12}
$\Omega = \Omega_0$	L_2	w_{20}	w_{21}	w_{22}
Influence		$-w_{i0} + w_{ik} = a_{ik}$		
Equation		$\begin{vmatrix} w_{10} \\ w_{20} \end{vmatrix} + \begin{vmatrix} a_{11} & a_{12} \\ a_{21} & a_{22} \end{vmatrix} \cdot \begin{vmatrix} u_1 \\ u_2 \end{vmatrix} = \begin{vmatrix} 0 \\ 0 \end{vmatrix}$		

w_{ik} , a_{ik} , u_i complex

p: plane

L: bearing, measuring station

Fig. 2: Basic procedure for influence coefficient balancing

This procedure can obviously be extended to an arbitrary number of balancing planes and an arbitrary number of measuring locations. As the vibration responses are dependent on the rotor speed, we can even increase the number of equations without increasing the number of measurinq stations, merely by increasing the number of balancing speeds, at which the readings are taken. Thus we can wind up with a large number of surplus equations. Nevertheless this equation system can be seriously ill conditioned or even practically singular.

Yet this eventual singularity of the equation system for a large number of balancing planes is not an unfortunate mathematical accident but can be physically explained and can even be utilized to calculate an optimum weight set, if this singularity is evaluated appropriately. For a physical understandina of this problem it is useful to review some basic results of the modal theory in more detail.

2 Modal Balancing Theory and Orthogonal Balancing Weight Sets

Strictly speaking the so called modal balancing methods are nothing but a reduced influence coefficient method. This becomes quite obvious, if we analyse the practical balancing procedure in detail which is summarized in figure 3. Comparing figure 2 and figure 3 we notice, that the only difference lies in the number of unknowns, which have to be calculated. In modal balancing we only deal with one (complex) unknown at a time.

Run	0	1
Testweight	0	1
Response	w_0	w_1
Influence	$-w_0$	$+w_1 = a_1$
Equation	w_0	$+a_1 u_1 = 0$

$$\Omega = \Omega_1 \cong \omega_1$$

Fig. 3: Basic procedure for modal balancing

At each critical speed the rotor is balanced in the same manner, but for higher critical speeds the test and balancing weights must be prebalanced groups of weights, which do not upset the balance for the previous critical speeds. If the testweight consists of a weight group the complex unknown u_i determines, how this group has to be scaled and rotated in order to obtain the desired balance.

We note, that it is not necessary to measure the responses in more than one measuring station, because the vibration state in the vicinity of a critical speed is dominated by the resonant mode shape. All the rotor responses that we might measure in different shaft locations are proportional to the resonant mode shape and do not supply other useful information.

Thus it is not the basic procedure as such which is unique for the modal balancing technique, but it is the special weight sets, which are used, to reduce the problem to a step by step procedure. These weight sets have to be discussed in more detail, because this will shed some light on the problem of singular influence coefficient matrices, which has to be discussed later.

The basic claim which is imposed on prebalanced groups of weights is as mentioned, that each new group does not upset the state of balance attained with all the previously attached groups. Denoting with $\underline{\varphi}_i$ the modal deflections of the i-th mode at the selected balancing planes and with $\underline{t}_i$ the weight set intended to balance this mode, these weight sets can be calculated from the equation system

$$\underline{\varphi}_i^t \; \underline{t}_k = 0 \quad \text{for} \quad i < k \qquad (2)$$
$$= 1 \quad \text{for} \quad i = k$$

The number of individual weights in each group is equal to the order of the critical speed to be balanced. These groups can be called backward orthogonal. They are practical as each critical speed can thus be balanced with the least amount of effort and indeed, the time which has to be invested to attach test and balancing weights to a rotor is by no means negligible.

The disadvantage connected with these groups is that they inevitably excite the next modal component, as well as all other higher modal components. This makes it more and more difficult to negotiate higher critical speeds. Moreover higher order modal components with large generalized unbalance can have a very adverse effect on the running speed. The generalized unbalance $\hat{u}_k$ due to a weight set $\underline{t}_i$ ((k > i) can be calculated from equation (2).

This problem can be overcome by employing forward orthogonal groups of balancing weights, each of which satisfying the orthogonality conditions

$$\underline{\varphi}^t_k \ \underline{t}_i = 0 \quad \text{for } i \neq k$$
$$= 1 \quad \text{for } i = k \tag{3}$$

The number of individual weights in these groups is equal to the number of planes available for the particular rotor. Although it seems to be unnecessary to install weights in all available planes in order to balance the first critical speed, the employment of such groups solves the problem of an optimum distribution of balancing weights. While the number of weight groups to be employed is naturally limited to the number of excited modes in the operating speed range, the number of balancing planes can be arbitrarily large. The more planes there are available, the better the final state of balance will be, because even the higher vibration modes have some influence on the residual vibration amplitudes at lower speed.

Thus if N is the number of available balancing planes, the mode N + 1 is the first mode, that will be adversely affected by the installed balancing weights, provided that the above recommended forward orthogonality is observed. Figure 4 illustrates the discussed backward and forward orthogonal weight groups.

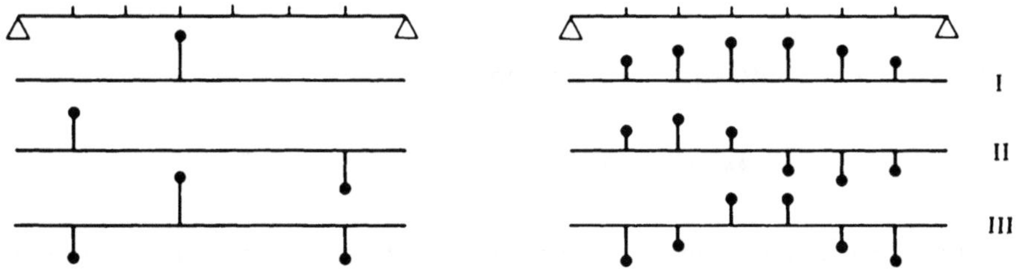

Fig. 4: Backward and forward orthogonal weight sets

The modal balancing being a step by step procedure, obviously no attempt will be made, to balance vibration modes which do not show up in the vibration response. Thus the balancing process is naturally terminated as soon as all relevant groups are installed, even if lets say S more groups are available. In such a case the plane forward influence coefficient approach yields an equation system, which is S times singular. But if suitable precautions are taken, when solving this singular equation system, even a forward orthogonal solution can be calculated.

3 Modal Balancing Without Testruns

As flexible rotors usually display a behaviour, which aqrees very well with the modal theory, it is in fact possible to exploit this theory more extensively than usually claimed. While the so called modal theory merely supplies some information on suitable weight groups, which allow a step by step balancing without much computational effort, it is even possible to balance a flexible rotor without previous testruns (6).

According to the modal theory the rotor can be modelled by a single degree of freedom system in the vicinity of a critical speed, which yields

$$m_i \ddot{w}_i + c_i \dot{w}_i + s_i w_i = \Omega^2 \hat{u}_i e^{j\Omega t} \tag{4}$$

w_i represents the modal deflection at the selected measuring points, m_i, c_i, s_i and $\hat{u}_i$ represent the generalized mass, damping, stiffness and unbalance for the i-th vibration mode, which is defined by the mode shape $\varphi_i(x)$.

Measured deflections $w(\Omega)$ however are not equal to one modal deflection $w_i(\Omega)$ alone but contain even components of other vibration modes. But if measurements are taken at different speeds in the vicinity of the critical speed ω_i only, we can assume that these contributions of other modes are constant, i.e. that they do not depend on Ω. This is expressed by equation (5)

$$\{ \hat{w}_i(\Omega) = \hat{w}(\Omega) - \hat{w}_0 \} e^{j\Omega t} \tag{5}$$

Substituting eq. (5) into eq. (4) we obtain

$$(-\Omega^2 m_i + j\Omega c_i + s_i)(\hat{w}(\Omega) - \hat{w}_0) = \Omega^2 \hat{u}_i \tag{6}$$

$\hat{w}(\Omega)$ being the time independent complex amplitude of the synchronous vibration measured at the selected measuring point and $\hat{w}_0$ the complex contribution from other mode shapes. This equation is valid for any rotor speed in the vicinity of ω_i. Thus we can measure at K different speeds Ω_k, $k = 1,2, \ldots K$ in order to obtain K equations

$$(-\Omega_k^2 m_i + j\Omega_k c_i + s_i)(\hat{w}_k - \hat{w}_0) = \Omega_k^2 \hat{u}_i \tag{7}$$

Calculations on the mode shapes and the generalized mass are quite reliable, while the damping and the exact critical speed cannot be calculated with sufficient accuracy to identify the generalized unbalance in eq. (7) . Therefore they have to be considered as unknown, as well as the generalized unbalance and the vibration amplitude $\hat{w}_o$, which contains the contributions from other modes. Thus equation (7) contains products of the type $s_i \cdot \hat{w}_o$ and $c_i \cdot \hat{w}_o$. But the problem can easily be linearized by introducing estimated values for s_i and c_i to begin with. After rearranging the equation system (7) we obtain:

$$
\begin{array}{c}
k = 1 \\ 2 \\ \vdots \\ k \\ \hline k = 1 \\ 2 \\ \vdots \\ k
\end{array}
\left[
\begin{array}{c|c|c|c|c|c}
-\Omega_k \hat{w}^{im} & \hat{w}_k^{re} & \Omega_k^2 m_i - s_{io} & -\Omega c_{io} & -\Omega_k^2 & 0 \\ \hline
\Omega_k \hat{w}_k^{re} & \hat{w}_k^{im} & -\Omega c_{io} & \Omega_k^2 m_i - s_{io} & 0 & -\Omega_k^2
\end{array}
\right]
\cdot
\left[
\begin{array}{c}
c_i \\ s_i \\ \hat{w}_o^{re} \\ \hat{w}_o^{im} \\ \hat{u}_i^{re} \\ \hat{u}_i^{im}
\end{array}
\right]
=
\left[
\begin{array}{c}
\Omega_k^2 m_i \hat{w}_k^{re} \\ \Omega_k^2 m_i \hat{w}_k^{im}
\end{array}
\right]
$$

Apparently K, the number of measurements, has to be $> = 3$. If more than three measurements are taken into account, the result can be computed from a least square procedure and will be more reliable.

Test results show, that it is possible to start with almost any values s_{io} and c_{io}, and that one or two iterations are generally sufficient to solve the problem.

The generalized unbalance being identified, balancing weights can be calculated for any suitable set of weights $\underline{t}$, which has to be scaled and rotated by an appropriate complex factor a_i, in order to match the generalized unbalance u_i according to the equation

$$\underline{\varphi}_i^t \, a_i \, \underline{t} + \hat{u}_i = 0 \tag{8}$$

which yields

$$a_i = \hat{u}_i \, / \, (\underline{\varphi}_i^t \, \underline{t}) \tag{9}$$

A word of caution is necessary when applying this method. It is assumed, that the phase angle measurements are correct. If sensors or amplyfiers introduce phase shifts these have to be accounted for appropriately.

The method has been tested both on experimental rotors and large ASEA generator rotors and yields remarkably good results.

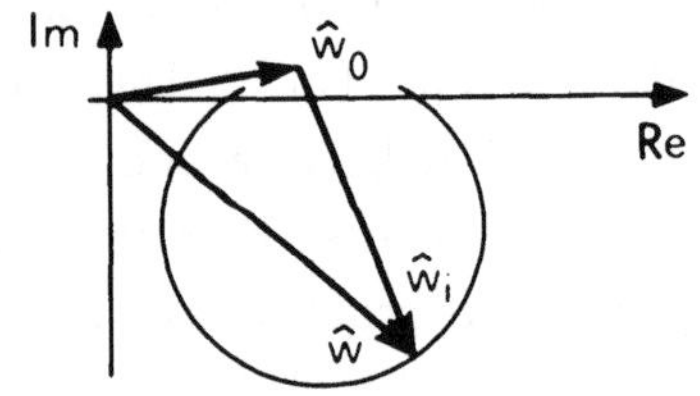

Fig. 5: Modal balancing without test runs

4 Influence Coefficient Techniques

The most crucial point in balancing of flexible rotors often lies in the assumed linearity and reproducibility of unbalance excited vibration in rotating machinery. This linearity often only exists in a statistical sense. This is especially so for large multispan shaft trains, but even applies to large turbo generators in a balancing pit.

This means that each run of the rotor supplies us with new statistically relevant information on the behaviour of the rotating system. Looking at the vibration response at comparable conditions (same speed, same operating conditions, same balancing weights), we might find, that the measured vibration amplitudes are spread out over a certain area according to figure 6

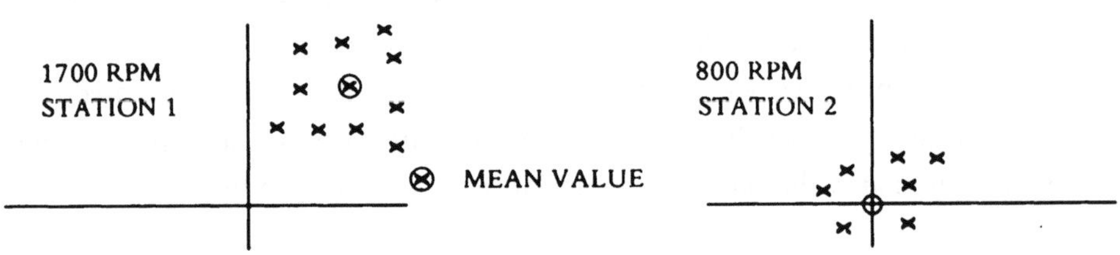

Fig. 6: Mean value and variation in vibration response

Apparently it would be reasonable to use the mean value of all observations rather than some individual value in balancing weight computations. But even in the more general case, (same speed, same operating conditions but different balancing weights) similar averaging methods can be employed.

4.1 Averaged influence coefficients

When determining influence coefficients it should always be kept in mind, that the initial unbalance distribution can change with time, due to setting effects, wear, corrosion, repair, overhaul and various other reasons. The influence of this changing initial unbalance on the influence coefficients is unwanted and most effectively eliminated by forming differences between consecutive trial runs only, as unknown changes in the initial unbalance are less likely, when the elapsed time interval between the considered trial runs is as small as possible.

Considering now the relationship between an arbitrary applied change in the balance weight distribution $\underline{\Delta u}_m$ and the corresponding change in the vibration response $\underline{\Delta w}_m$, we can write down the following equation

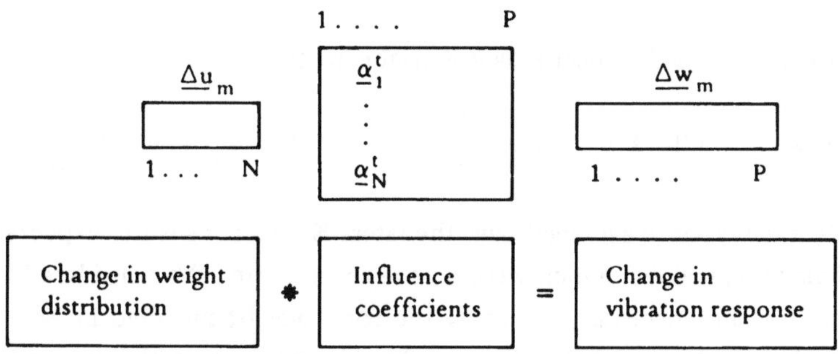

(10)

By establishing this equation for each observation available, we soon arrive at an overdetermined equation system which can be solved by the least square method.

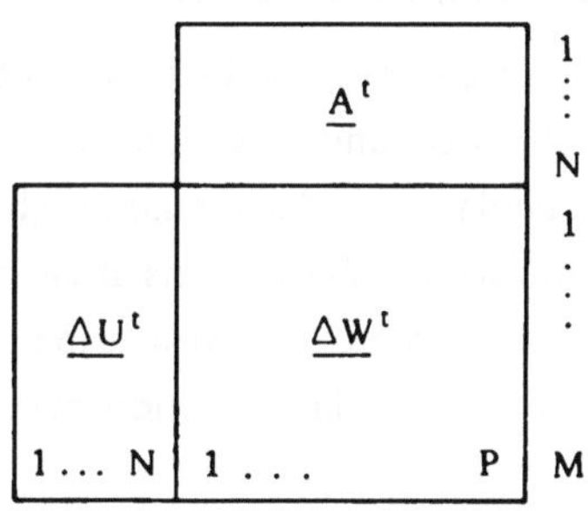

$$\underline{\Delta U}^t \ \underline{A}^t = \underline{\Delta W}^t \qquad (10)$$

As this information generally is at hand and becomes more and more comprehensive as time goes on, it is just a matter of efficient data organisation to obtain really reliable influence coefficient matrices. The equation system can readily be solved as soon as $M >= N$, the resulting averaged influence coefficients being

$$A^t = [(\Delta U)(\Delta U)^t]^{-1} \ \Delta U \, \Delta W^t \qquad (11)$$

4.2 Averaged trial runs

Likewise it is possible to average on the right hand side of the equation system for the balancing weights, in other words to identify the actual state of balance by averaging over the K latest representative runs available. This can even be done, if different balancing weights have been applied. As every measured response in the trial run $\underline{w}_i$ is composed of the effect of the initial unbalance $\underline{\alpha}_0$ and the influence of the attached balancing weights

$$\underline{w}_i = \underline{\alpha}_0 + \underline{A}\ u_i \tag{12}$$

we can calculate an averaged response over K trial runs to

$$\underline{w}_{av} = 1/K \sum_{i=1}^{K} w_i = 1/K \sum_{i=1}^{K} (\underline{\alpha}_0 + \underline{A}\ \underline{u}_i) = \underline{\alpha}_0 + \underline{A}\ 1/K \sum_{i=1}^{K} u_i \tag{13}$$

In other words, the response averaged over the latest K trial runs is the expected response of the averaged balancing weights which were on the rotor during the considered trial runs.

Of course it is not necessary to know the total weight installed in each plane. Often previously attached unknown balancing weights are present at the first trial run. These weights can be considered part of the initial unbalance distribution, so that the so called total weight is merely the total weight change relative to the first considered trial run.

4.3 The Evaluation of Singular Influence Coefficient Matrices

The influence coefficient matrix $\underline{A}$ can be composed of an arbitrary combination of columns $\underline{\alpha}_i$, which leaves it up to the distinction and intuition of the engineer, how many and which planes he wants to use for the calculation of balancing weights.

With respect to balancing of large multispan shaft trains it seems to be an accepted rule, to preferently use those planes, which are close to the measuring points displaying high vibration amplitudes. Thus the occurrence of singular equation systems is automatically avoided.

But as pointed out before, generally better balancing results are obtained, if all available balancing planes are used. This approach even tends to reveal the location of the unbalance and can be useful in field balancing of large shaft trains, because the calculated over all balancing weights might give more reliable suggestions as to which planes to use, than the above mentioned common sense approach.

A singular equation system can be detected during the solution process of the equation system by studying the relative pivot element size. Figure 7 shows the situation which is encountered at some step during the elimination process, if the equation system is singular (7).

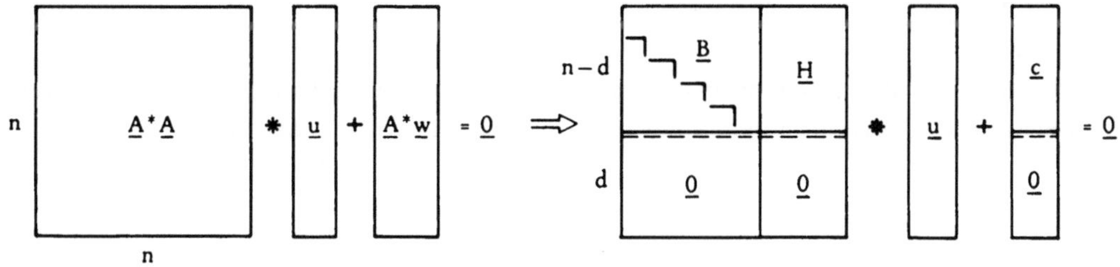

Fig. 7: Singular equation system

All elements under the broken line are comparatively small compared with the magnitude of the elements above it. Solving the upper part only yields a particular solution u_p and a set of d homogeneous solutions $\underline{H}$, d being the number of rows under the broken line, the deficiency of the matrix. The homogeneous solutions satisfy

$$\underline{B}\,\underline{H} = \underline{0} \quad \text{or} \quad \underline{H}^*\,\underline{B}^* = \underline{0} \tag{14}$$

so that the general solution is

$$\underline{u} = \underline{u}_p + \underline{H}\,\underline{a} \tag{15}$$

This equation calls for additional information to determine the arbitrary constants $\underline{a}$. Looking for a minimum solution satisfying

$$\underline{u}^*\,\underline{u} = \min \tag{16}$$

we can calculate by plane least squares method

$$\underline{a} = -(\underline{H}^*\,\underline{H})^{-1}\,\underline{H}^*\,\underline{u}_p \tag{17}$$

By substituting $\underline{a}$ from eq. (17) into eq. (15) we find

$$\underline{u}_{min} = (\underline{I} - \underline{H}^*(\underline{H}^*\,\underline{H})^{-1}\,\underline{H}^*)\underline{u}_p \tag{18}$$

Premultiplying this equation with $\underline{H}^*$ yields

$$\underline{H}^*\,\underline{u}_{min} = (\underline{H}^* - \underline{H}^*\underline{H}(\underline{H}^*\,\underline{H})^{-1}\,\underline{H}^*)\underline{u}_p = 0 \tag{19}$$

which means that the minimum weight distribution u_{min} is orthogonal to the homogeneous solutions H*.

This orthogonality relationship can be used directly to calculate a minimum weight set, which because of (14) has to be some linear combination of the matrix B*.

$$\underline{u} = \underline{B}^* \underline{x} \tag{20}$$

Substituting this into the original upper half equation (see figure 7) which reads

$$\underline{B}\,\underline{u} + \underline{c} = \underline{0} \tag{21}$$

we finally turn out with

$$\underline{u}_{min} = - \underline{B}^*(\underline{B}\,\underline{B}^*)^{-1}\,\underline{c} \tag{22}$$

Comparing this with the usual least square solution of an overdetermined equation system reveals an interesting duality between the solution of an overdetermined and an underdetermined equation system.

Thus we can summarize that a singular equation system $\underline{A}^*\underline{A}$ with a deficiency d allows to separate two groups of balancing weight sets, that are orthogonal to each other, the (n – d) columns of $\underline{B}^*$ containing those weight sets that have an important influence on the rotor behaviour in the investigated speed range and the d columns of H which contain homogeneous weight sets which only have a negligible influence in the investigated speed range. Hence these groups in $\underline{H}$ must be linear combinations of the higher modal weight groups, which are orthogonal to the dominating (n – d) mode shapes in the operating speed range. By composing the final balancing weight set as a linear combination of the first group only, we obtain a minimized balancing weight set, which is orthogonal to the homogeneous weight groups. Figure 8 gives an overview of the matrix relationships and their dimensions.

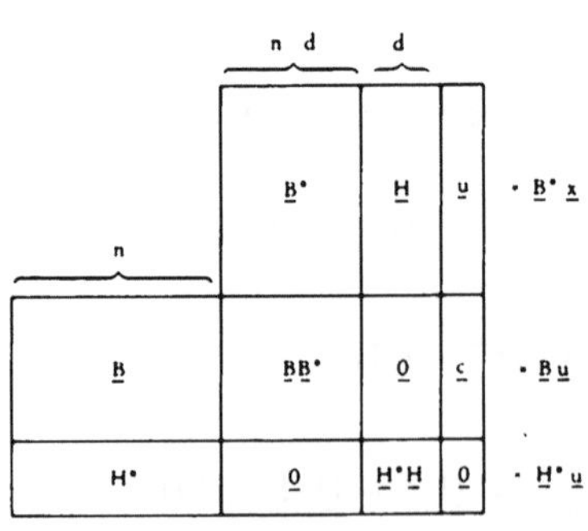

Fig. 8: Orthogonality relationship between weight set B* and H

However according to the modal theory this is not the appropriate orthogonality condition yet, which states, that the modal unbalance $u_i(x)$ is orthogonal to the mode shapes $\varphi_j(x)$ and not to the modal unbalances $u_j(x)$ and reads

$$\int_0^L u_i(x)\,\varphi_j(x)\,dx \; \begin{cases} = m_i & i = j \\ = 0 & i \neq j \end{cases} \tag{23}$$

the continuous i-th generalized unit unbalance being

$$u_i(x) = \mu(x)\,\varphi_i(x) \tag{24}$$

In analogy to the continuous case we obtain for the discrete case of N balancing planes the orthogonality

$$\underline{\mathring{u}}_i^{\,*}\,\underline{\varphi}_j \; \begin{cases} = m_i & i = j \\ = 0 & i \neq j \end{cases} \tag{25}$$

the discrete i-th generalized unit unbalance being

$$\begin{bmatrix} \\ \underline{\mathring{u}}_i \\ \\ \end{bmatrix} = \begin{bmatrix} \ddots & & \\ & M_{L,k} & \\ & & \ddots \end{bmatrix} \begin{bmatrix} 1 \\ \vdots \\ \underline{\varphi}_i \\ \vdots \\ N \end{bmatrix} \tag{26}$$

where $M_{L,k}$ is the lumped rotor mass associated with the balancing plane k and $\underline{\varphi}_i$ the deflection vector of the i-th mode shape φ_i at the N balancing planes. Thus we can even calculate the discretized mode shape $\underline{\varphi}_i$ if the associated generalized unit unbalance u_i is known.

Applying this to the homogeneous solution $\underline{H}$ of a singular influence coefficient matrix, which represent weight groups without influence on the rotor vibrations in the considered speed range, we can calculate the mode shape combinations associated with these groups $\underline{H}$ by premultiplying with the inverse of the lumped mass matrix

$$\underline{\phi} = \underline{M}_L^{-1}\,\underline{H} \tag{27}$$

This enables us to calculate low modal order weight groups which are orthogonal not to the groups $\underline{H}$ but to their associated mode shapes $\underline{\phi}$. This can be done by simply introducing the lumped mass matrix $\underline{M}_L$ into the orthogonality condition (14)

$$\underline{B}\,\underline{M}_L\,\underline{M}_L^{-1}\,\underline{H} = \underline{B}\,\underline{M}_L\,\underline{\Phi} = 0 \quad \text{or} \quad \underline{\Phi}^*\,\underline{M}_L\,\underline{B}^* = 0 \tag{28}$$

Composing the actual balancing weights $\underline{u}$ as a linear combination of the weight set $\underline{M}_L\ \underline{B}^*$ only, we can write similar to eq. (20)

$$\underline{u} = \underline{M}_L\,\underline{B}^*\,\underline{x} \tag{29}$$

Substituting this equation into eq. (21) we finally get

$$\underline{u}_{orth} = -\ \underline{M}_L\,\underline{B}^*\,(\underline{B}\,\underline{M}_L\,\underline{B}^*)^{-1}\,\underline{c} \tag{30}$$

This is a balancing weight combination which balances the identified low order modal unbalances and is orthogonal to all high order mode shapes identified by the homogeneous weight groups $\underline{H}$.

Thus we can summarize, that a singular equation system $\underline{A}^*\underline{A}$ with a deficiency d allows to separate two groups of balancing weight sets, which are mutually orthogonal to the set of deflections associated with the other weight set. According to figure 9 the (n – d) columns of $\underline{M}_L\underline{B}$ contain those weight sets which have an important influence on the rotor behaviour in the investigated speed range, and the d columns of $\underline{H}$ containing weight sets with negligible influence in this speed range. The corresponding shape sets $\underline{B}$ and $H^*\ M_L^{\ 1}$ contain (n – d) modes shapes with important contribution to the actual rotor deflection in the investigated speed range and d mode shapes with negligible contribution to this deflection. By composing the final weight set as a linear combination of the first weight group only we do not add unbalance to the d higher modal components, which could show up in the actual deflection shape, if the exciting unbalance is getting too large.

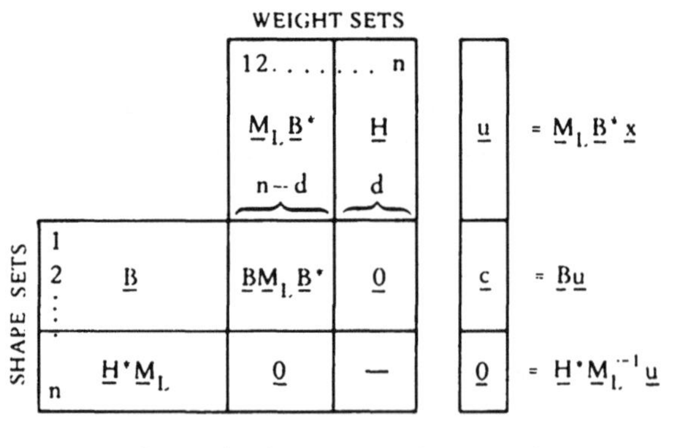

Fig. 9: Orthogonality between weight sets and shape sets

Off course we can even introduce an arbitrary weighing matrix G instead of the mass matrix $\underline{M}_L$, which makes it possible to minimize the balancing weight sets according to practical limitations, as there are restrictions in the magnitude of weight which can be installed. Balancing weights with small weighing factors will thus be minimized preferently.

This minimizing approach can also readily be applied to the total weight installed on the rotor. The orthogonal weight distribution $\underline{u}_{orth}$, which is equivalent to an arbitrary weight distribution $\underline{u}_o$ can be calculated to

$$\underline{u}_{orth} = \underline{M}_L \, \underline{B}^* \, (\underline{B} \, \underline{M}_L \, \underline{B}^*)^{-1} \, \underline{B} \, \underline{u}_0 \tag{31}$$

Such a weight redistribution can sometimes result in a considerable weight reduction and in a considerable reduction of the vibration level at operating speed.

The presented method of using all available balancing planes has been used in balancing large ASEA generators since 1978. This resulted in a drastic improvement of both production time and residual vibration level.

5 Rotors with Intial Bow in Single and Multispan Situations

If only the bearing forces are considered, the bent single span rotor will not behave differently from a straight rotor. Hence a bent single span rotor can be balanced, so that the dynamic bearing forces are entirely eliminated in the whole speed range. In this case the residual deflection consists of the initial bow only and will not change form at any speed.

The problem is slightly more involved if shaft deflection measurements are considered. In this case the measured shaft deflection, the total run out, contains a geometric run out and a dynamic deflection due to unbalance. Only the dynamic deflection should be compensated.

The geometric run out can easily be measured at low speed where the dynamic deflection is negligible. Then the dynamic deflection at high speeds can be identified by simply subtracting the geometric run out from the measured total run out.

When balancing a single span rotor with initial bent by assessing shaft deflections and using the modal method without test runs, we do not even need low speed run out measurements, because this method would automatically identify both the unbalance and constant run out term.

If the bearing conditions for a single span rotor are modified, the modal components of the residual unbalance change, due to changes in the mode shapes. But the total residual unbalance, that is the sum of all modal components is constant.

The situation is more involved when dealing with multispan rotors. Usually the bow will be different before and after mounting the rotor into its bearings due to the constraints on the rotor imposed by the bearing alignment. Let the initial bow before mounting be b(x). After mounting there will be a state of equilibrium between the internal forces of the constrained bow c(x) and the elastic forces of the deflected bearings.

Obviously the total elastic deflection $r_e(x, t)$ of the rotor due to the bearing constraints and the dynamic deflection will be

$$r_e(x, t) = r_c(x, t) + r_d(x, t) - r_b(x, t) \tag{32}$$

the total measurable deflection r(x, t) will be

$$r(x, t) = r_c(x, t) + r_d(x, t) \tag{33}$$

and the deflection of the center of gravity will be

$$r_m(x, t) = r_d(x, t) + r_c(x, t) + \varepsilon(x)\, e^{j\Omega t} \tag{34}$$

Thus the principle of virtual work, which yields the equation of motion reads

$$\int EI\, r''_c\, \delta r''\, dx + \int \mu\, \ddot{r}_m\, \delta r\, dx + \sum_k s_k\, r\, \delta r + \sum_k M_k\, \ddot{r}\, \delta r = 0 \tag{35}$$

Introducing equations (32) to (34) into (35) we find, that this equation contains the static equilibrium between the constrained shaft and the deflected bearings

$$\int EI\, r''_b\, \delta r''\, dx - \int EI\, r''_c\, \delta r''\, dx - \sum_k s_k\, r_c\, \delta r = 0 \tag{36}$$

Splitting off the static equilibrium (36) we obtain from eq. (35)

$$\begin{aligned} &\int EI\, r''_d\, \delta r''\, dx + \sum_k s_k\, r_d\, \delta r + \int \mu \ddot{r}_d\, \delta r\, dx + \sum_k M_k\, \ddot{r}_d\, \delta r = \\ &= \{\Omega^2 \int (\mu(x)\,(\varepsilon(x) + c(x))\, \delta r\, dx + \Sigma\, M_k\, c(x_k)\, \delta r(x_k)\}\, e^{j\Omega t} \end{aligned} \tag{37}$$

Equation (37) shows that not only the system behaviour, that is the mode shapes and critical speeds, are modified by the bearing conditions. Even the effective unbalance of the multispan rotor is influenced by the bearing conditions, if the rotor has an initial bow. Both the bearing mass and the bearing stiffness influence the new unbalance distribution.

An initial bow in a multispan rotor is usually due to coupling faults. It can introduce considerable unbalance, no matter how well the individual rotors are balanced before the final assembly.

REFERENCES

[1] Federn, K.: Grundlagen einer systematischen Schwingungsentstörung wellenelastischer Rotoren. VDI–Bericht 24 (1957) 9–27.

[2] Bishop, R.E.D.: The vibration of rotating shafts, J. Mech. Sc. 1,1 (1959) 66–77.

[3] Goodman, T.P.: A least squares method for computing balance corrections, ASME paper no. 63–WA–295 (1963).

[4] Rieger, N.F.: Computer program for balancing for flexible rotors, MTI Technical Report 67 TR 68 under Contract NAS3-10926.

[5] Tessarzik, Badgley, Anderson: Flexible rotor balancing by the exact point speed influence coefficient method, Vibration Conference Toronto (1971), Paper no. 71-Vibr-91.

[6] Gasch, R., Drechsler, J.: Modales Auswuchten elastischer Läufer ohne Testgewichtsetzungen, VDI–Bericht 320 (1978).

[7] Drechsler, J.: Computergestutztes Auswuchten elastischer Rotoren nach Eigenformtheorie und Einflusskoeffizientenverfahren, Ing Archiv 47 (1978).

PART IV

MEASUREMENT AND IDENTIFICATION

CHAPTER 4.1

VIBRATION MEASUREMENT AND MONITORING

V. Schlegel

1. Introduction

Successful vibration measurement and analysis require an intimate familiarity with types of measurement, transducer characteristics and application, plus the capabilities and limitations of the diagnostic instrumentation.

Since mechanical malfunctions have a tendency to disguise themselves with side effects and misleading disturbances, it is essential that the acquired data are reviewed in every reasonable manner prior to forming a conclusion.

The aim of this paper is to give a general review about methods and effects in measuring rotor vibrations, but not to present some interesting case histories.

2. Measurement devices and instrumentation

The major types of transducers used in turbomachinery analysis are displacement probes, velocity transducers and accelerometer. The proximity displacement probe can be contacting or non-contacting devices, whereas the velocity transducers and the accelerometer have to be contacting devices. The choice of proper transducers has to be done according to environmental conditions like:

- influence of oil, pressure, temperature and accessibility
- influence of size and diameter of shaft
- electrical properties of transducers (e.g. range of frequency and amplitude, accuracy, calibration procedures)
- number of transducers in one plane.

In the following the transducers are described in sequence and comprehensiveness according to their frequency in rotor vibration measurement.

2.1 Displacement transducers

<u>2.1.1 The eddy-current probe.</u> The probe (Fig. 1) radiates a radio frequency field in the immediate area ahead of the probe tip. As conductive material intersects the field, eddy currents are generated in the

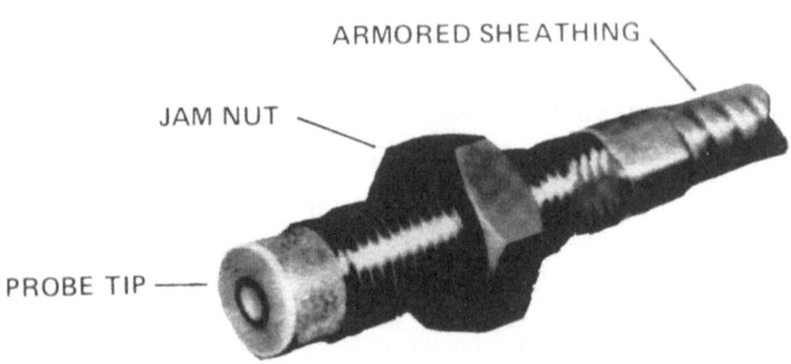

Fig. 1 Typical eddy-current probe

material, resulting in energy loss which is sensed by an oscillator circuit, which is modified to provide a linear elastical output signal. This probe can be used for static gap measurement (axial or radial position) or for dynamic motion (vibration). The essential part of the probe is a small coil of wire that is in the tip of a fiberglass body. The fiberglass body can be mounted in several different metal case designs to aid the installation. The range of the probe is a function of tip diameter, in general 5 mm or 8 mm. The larger the tip, the longer the linear measuring range. Usually the probes come with an integral coaxial cable and connector. Since the system operates according to inductive proximity principles, the probe does not contact the observed surface and is not affected by non-conducting materials such as air, oil, gas, plastics, etc. in the gap between the probe tip and the observed surface.

Due to the ability of the eccentricity position of the shaft in a journal bearing to change under varying conditions of machinery load, alignment, temperature of oil and shaft, it is important that the proximityprobe transducer system has a long linear range sufficient to allow for these eccentricity position changes to occur without having the shaft move outside of the linear range (Fig. 2).

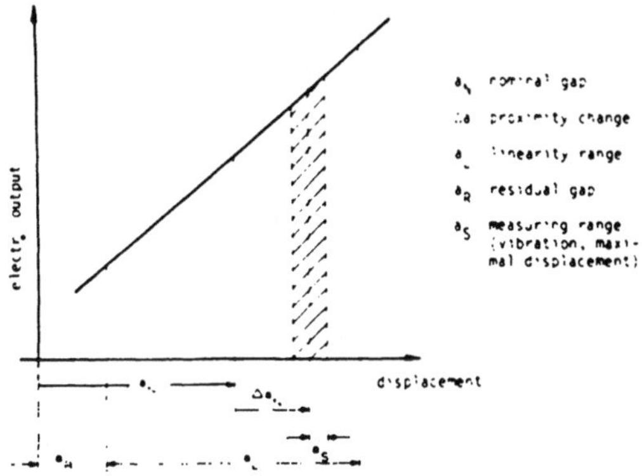

Fig. 2 Displacement curve of a rotor vibration measuring device with measuring ranges

A different aspect is necessary when measuring small vibration amplitudes, where not the linearity of the complete calibration curve is the decisive factor, but the differential linearity. The typical deviation of the calibration curve from the best possible straight line for a typical eddy-current probe is depicted in Fig. 3. Besides the properties of the eddy-current probes at fixed environmental conditions, there are deviations from the calibration curves caused by

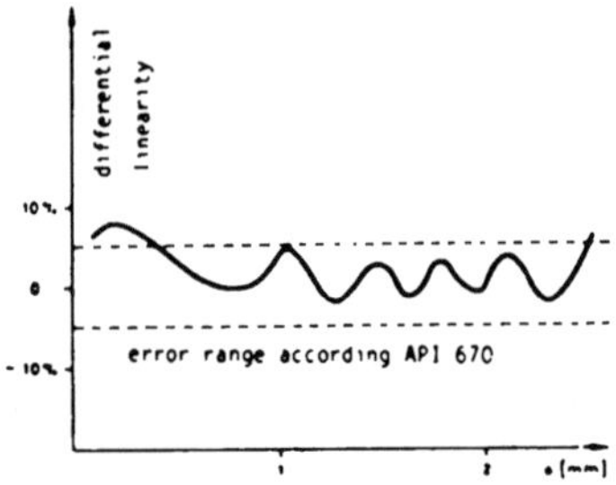

Fig. 3 Differential linearity of an eddy-current probe

- temperature influences on the ohmic and inductive resistance of the probe coil,
- thermal elongation of the probe body resulting in a displacement of the probe tip relative to the transducer fixation,
- influence of temperature and supply voltage on the oscillator/demodulator.

Each of these effects cause errors in the vibration measurement in the range of some percent.

Consideration must also be given to the shaft materials. The vibration signal is similar for most common shaft steel alloys. However, some materials have different surface areas which present varying values of resistivity, thus influencing the eddy currents in the surface. Depending on the frequency of the oscillator used the depth of the eddy currents amounts to 10 - 15 μm. But some shaft surface treatments (chrome plating, metalizing, etc.) are thin enough to allow the eddy currents to penetrate down to see two different materials.

The observed rotor surface must be free of all irregularities like scratches, rust and corrosion, out-of-roundness, chain marks etc. Such irregularities cause a change in probe gap which is not vibration, thus giving rise to a signal error. On the other hand it is not always advisable to grind the shaft surface of the rotor, because this can produce additional inhomogenities in conductivity.

2.1.2 The capacitive displacement probe. A plate shaped electrode in the transducer builds a condensor with the shaft, whose capacity depends on the distance between transducer and shaft. The capacity is measured by an oscillator/demodulator at high frequency. This measurement principle is applicable to all electrically conducting materials. The cable between transducer and oscillator must be short, similar to the eddy-current principle. This type of displacement probe is restricted to cases where the fluctuate permeability of oil within a turbomachine does not influence the gap between shaft and transducer.

2.2 Velocity transducers

The seismic transducers portion of the transducer system operates according to the inertial mass/moving case principle. The inertial mass is a copper-wire coil wound on a bushing and suspended by sensitive springs inside the transducer case. A permanent magnet is rigidly attached to the transducer case and physically located inside the coil. When the transducer is properly installed the vibration of the coil induces a current in the oil which proportional to the velocity of vibration, provided sufficient frequency of motion is present.

This type of transducer is mainly used for measuring bearing vibrations.

2.3 Acceleration transducers

The purpose of an acceleration transducer system (Fig. 4) is to measure structural motion in terms of acceleration. Its output is a voltage proportional to the accelerating along its sensitive axis of the surface on which it is mounted. The accelerometer uses a piezoelectric crystal situated between the accelerometer base and an inertial reference mass. When the crystal is strained (compression or tension force), a displaced electric charge is accumulated on the opposing major surface of the crystal, which has to be amplified. The crystal element acts as a precision spring to oppose the compression or tension force, and it supplies an electrical signal proportional to the applied force.

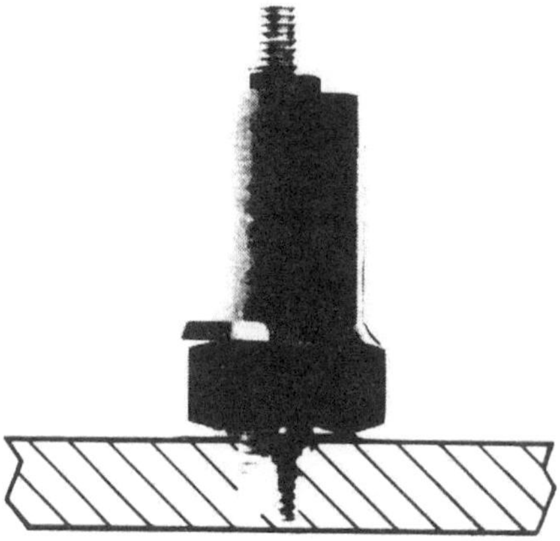

Fig. 4 Typical acceleration transducer

2.4. Phase reference

In most cases a transducer is installed on a machine observing a once-per-turn event (e.g. a gab, notch, etc.) on a rotating shaft providing a signal that occurs once per revolution of the shaft. This pulse gives a reference for vibration data. It is a reference mark and time signal for speed, phase angle, frequency measurement and all data acquisition.

In order to read phase angle accurately, instrumentation is required which filters the input signal synchronous to speed, because due to the irregularities of the shaft materials the vibration displacement signals contain very high harmonic components. Most of the commercially available phasemeters measure the zero crossings of the input signals by phase detecting, and without filtering they would produce an incorrect reading.

2.5. Vibration monitors

The monitor of a measurement system for rotor vibrations performs several functions:

- regulation and distribution of power to the transducers
- conditioning of the signal according to the type of the selected meter display
- conditioning vibration signals to proportional recorder compatible signals
- self-check and check for proper transducer functioning
- continuous alarm monitoring for excessive or out-of-limits vibration conditions.

3. Data presentation

3.1. Time domain

Vibration form can be separated into two categories: time base presentation and orbital presentation. Time base presentation is provided by displaying transducer inputs on the oscilloscope in the time base mode showing the complete waveform. If a motion, which is an exact ratio of rotative speed, is present a phase mark will appear to be locked onto the waveform of the vibration as displayed on an oscilloscope. If a non-synchronous waveform is present, this mark will appear to move over the waveform. This ability can be important in observation of whirl and whip phenomena as well as in many rotor rub situations, where the phase mark will be unstable and jittery.

The orbit presentation is provided by displaying the output from two separate probes at 90^0 angles to one another and generally 45^0 off the vertical axis in the upper part of the casing (Fig. 5).

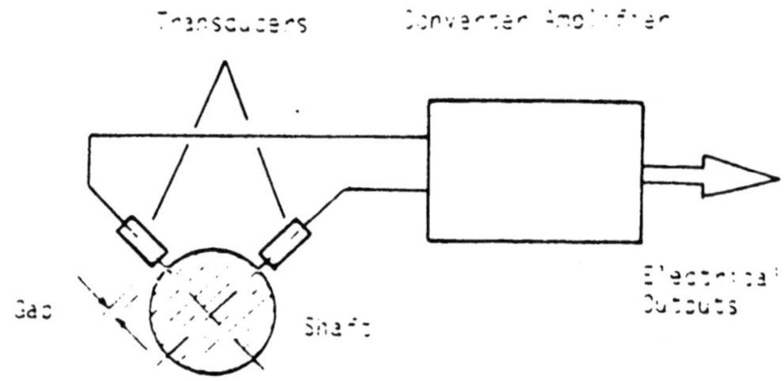

Fig. 5 Vibration measurement system for one measuring plane

In Fig. 6 the shaft orbit is depicted as an unfiltered motion. If a phase marker is used which provides a pulse with a decaying tail it is possible to identify the procession of the orbit if this pulse modulates the brightness of an oscilloscope.

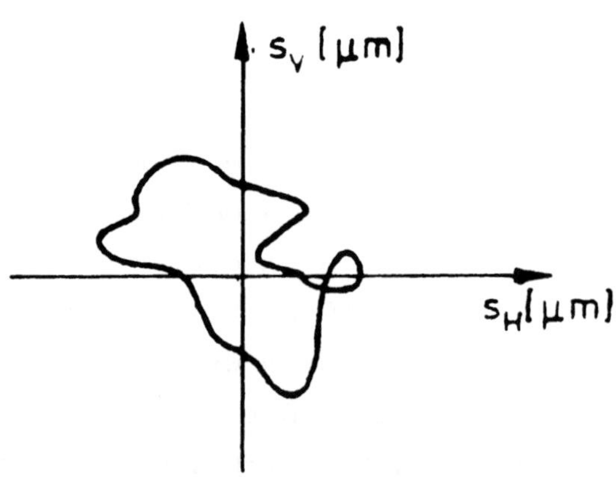

Fig. 6 Typical orbit

The deviation of the measured vibration signal from the shaft motion is due to the already mentioned mechanical and electrical irregularities of the shaft surface. Thus, only in the case of high vibration amplitudes the orbit seen on the oscilloscope resembles the elliptical or circular orbit theoretically predicted.

Where the 'fictitious' vibrations are present and amount to a major portion of the whole vibration signal, it is sometimes necessary to filter the high frequency components from the waveform before making phase angle measurement. This can be done by utilizing two matched tunable filters with both of them being exactly tuned to the synchronous speed of the rotor. Dual low pass or band pass filtering is a successful mean of observing orbital patterns with relatively high levels of fictitious vibration. However, it is necessary to proceed with caution since the phase angle is usually shifted by low or band pass filtering depending on the filter characteristics. But since phase measurement will usually not be made on the vibration orbits, this is no great disadvantage.

Since the filter is tuned to the rotational frequency of the rotor, the initial vector is eliminating that portion of shaft fictitious vibration which is coincident with the rotational frequency or to eliminate an initial bow in the shaft. All higher orders of rotor fixed non-vibrational components of the transducer signal are eliminated through the filter action of device. But care should be taken that no slow motion of the rotor-foundation system is present because according to the filter principle signals with very low frequencies are not sufficiently suppressed.

Once the initial vector has been compensated the remaining signal can be used to produce the true elliptical orbit.

The other equipment to smooth the vibration signal is the digital compensator. It provides the capability to digitally memorize a slow-roll waveform. Once a waveform is captured it will be subtracted from all future dynamic waveforms thus showing the whole vibration changes. Since no filtering or pre-conditioning of waveforms was done prior to the memory function, the digital compensator processes the displacement signal in its complex form. But devices of this kind demand that all the fictitious vibration signals must not change as a function of time, shaft speed, rotor axial position.

3.2. Frequency domain

Rotative machine response can be measured through the use of the Bode diagram, where rotative speed amplitude of a given measurement versus rotor speed is depicted along with the phase lag angle of that amplitude vector against rotational speed. This plot is most useful in showing the speed of various resonances of the machine. The presence of any initial bow or fictitious vibration, however, adds a constant vector to the vector of vibration which may severely alter the Bode plot.

Although sometimes the Bode diagram is an excellent way of portraying amplitude and phase, there is another method available for presenting the same information, the Nyquist plot.

On rotating machinery, the Nyquist (polar) plot is the amplitude and phase reading from a transducer showing the response of the machine to its residual or deliberate unbalance as a function of speed (Fig. 7). The polar plots are usually made of shaft relative motion but may also be shaft absolute displacement, casing velocity etc. They are much more descriptive and useful in displaying parameters of machine response than the Bode plots and therefore have been used in theoretical as well as in experimental studies. Their main advantage is the fact that the form of the polar is unchanged when an initial vector is encountered, while the amplitude and phase of the Bode diagram show major changes. In addition, the appearance of minor support resonances, which are barely discernable on the Bode diagram can be detected more easily.

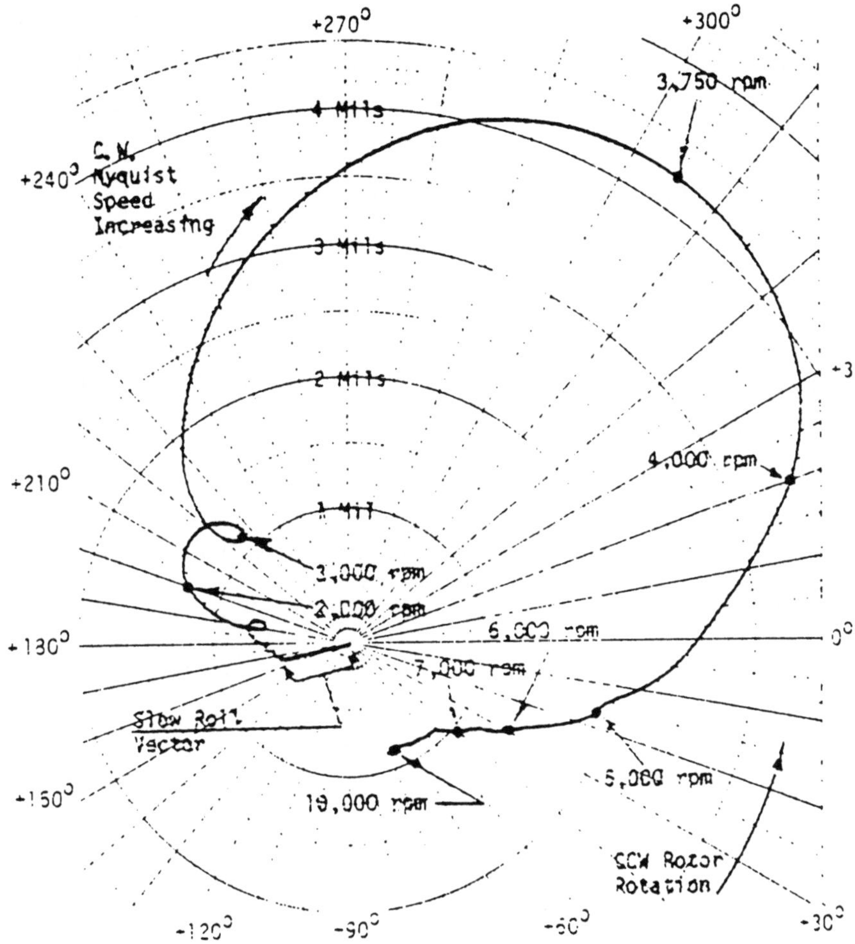

Fig. 7 Typical polar plot of rotor displacement during start-up

As a disadvantage the polar plot requires a manual identification of speed at various positions along the graph. This may be accomplished during the time the plot is generated or by comparing it to the standard Bode plot.

By simultaneously plotting two Nyquist diagrams of horizontal and vertical transducers, or the right and left transducer, which is more common, at one specific lateral location any variation in bearing damping and/or stiffness will show up as a deviation between the two generated plots. In the case of strongly nonisotropic stiffness and damping of the journal bearings the two plots will always differ.

3.3. Signature analysis

A very comprehensive way of presenting vibration data is known as 'signature analysis', where the vibration signals at all speeds and loads of the turbomachine are monitored and depicted in a graph containing the amplitudes of all harmonics of the vibration signal at any speed. Thus, the graph delivers a complete impression of the vibration levels of a machine (Fig. 8).

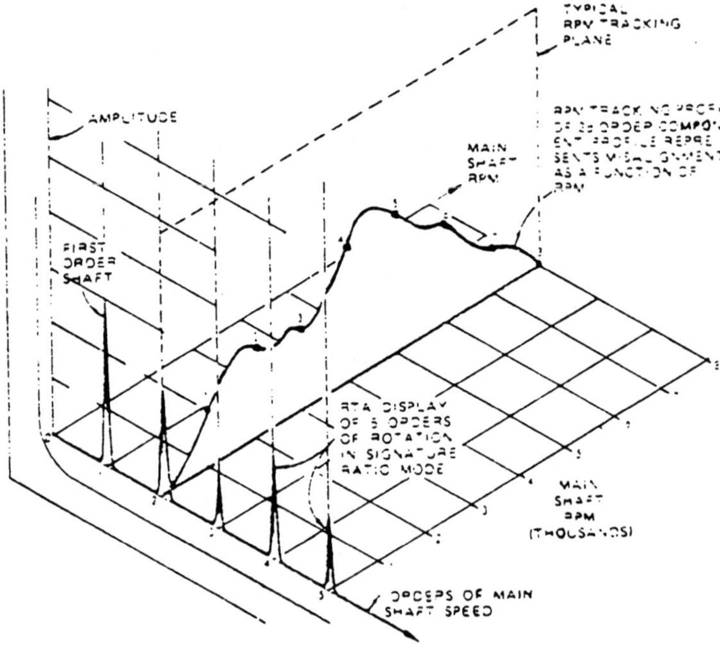

Fig. 8 Typical graph of spectrum vs. speed of a turbomachine

A second possibility of plotting the vibration spectrum is the cascade spectrum analysis. It is based on a series of spectrum plots from a single transducer depicting a change in one or more variables (e.g. vibration as a function of speed). The cascade plots provide a good deal of information regarding spectral content of the vibration signals, but they cannot be utilized to evaluate phase variations.

4. Interpretation and diagnosis of measuring results

The results of shaft vibration measurements can show a multitude of phenomena. Depending on the number and type of the instruments used in the measurement set-up certain characteristics of the shaft vibrations as well as their dependance on both, time and operating data can be determined.

The vibration monitoring which involves the extraction of fault signatures in most cases relies on the output produced by standard items of equipment for noise and structural analysis.

Data analysis serves two main purposes; first, it concerns itself with the extraction of fault signatures from background-wise and, secondly, it aims to predict which vibrations a defective machine is most likely to produce.

Most of the predictive work rests on some key assumptions regarding how the system will detune itself upon arrival of the fault.Table 1 [1] contains several examples for possible causes of shaft vibration. It can help to interpret the results of measurement. Without any claim to completeness, this table was compiled as an aid for decisions and references for the discussion and evaluation of simple problems. However, it should be kept in mind that all the symptoms are judged on a global basis without any connection to quantitative rotor parameters. This is left to the field of parameter identification.

5. Machinery surveillance and maintenance

A normal operating machine will generally have a stable amplitude reading of an acceptable low level. Any change in this amplitude reading indicates a change of the machine condition. But a simple investigation of amplitude and frequency alone does not provide sufficient information about machinery performance. Usually, a set of initial data is to be acquired with the machinery in a good state, so that any future change or deterioration in mechanical condition can be easily compared to the baseline information.

Most preventive maintenance programs are established with periodic inspection of machinery and permanent monitoring of machines. Generally, the parameters being monitored were amplitudes of vibration on the shaft velocity of bearings or acceleration of bearing house. Periodic checks of frequency of vibration are often included as a part of a preventive maintenance program.

A very difficult problem in vibration analysis is the establishment of severity limits. machines have different sensitivities and resistances to vibration forces. As an example a method of crack surveillance is depicted in Fig. 9 [2].

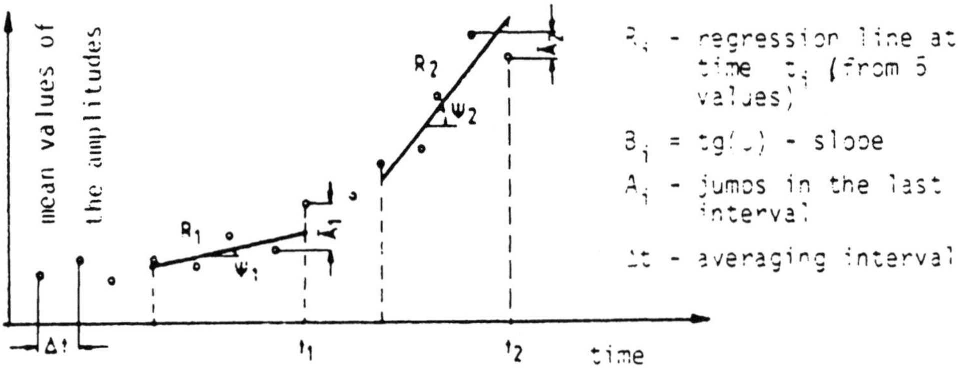

Fig. 9 Schematic representation of a crack identification for turbomachinery [2]

In order to interpret a certain range of changes in amplitudes during normal operation it is necessary to average the measured values of a given interval. So a regression line has to trip preset limits. This criterion is robust enough not to shut down a machine when a small jump in vibration amplitudes occurs and sensitive enough to identify a developing crack.

The measurement of eccentricity position can be an excellent indicator of bearing wear and unwanted conditions such as misalignment. Excentricity position is the steady state position of the shaft in the journal bearings. The measurement is accomplished by monitoring the slowly varying mean value of the vibration signal which should correspond to the mean distance of the centre of the shaft to the proximity probe. In practice this value is equated to the DC-output of the oscillator signal of the transducer system.

6. Guidelines and standards

For a better understanding of the behaviour of rotating machinery it is most desirable to standardize communication by means of a set of recommendations. So, in some countries guidelines and standards have been established.

ISO-standards exist on mechanical vibration of machines with operating speeds from 10 to 200 rev/s [3,4,5]. These standards deal with bearing vibrations measured by velocity transducers and state specifications in figures of vibration sensitivity.

In 1971 the first german guideline of the VDI appeared dealing with evaluation of vibrations of flexible shafts. In the meantime this guideline has been revised and has appeared as draft in 1979 [1]. It contains explanations of vibrations fundamentals and diagnosis samples to assist in operating turbomachinery in industry and power plants.

Besides guidelines for evaluation of vibration phenomena there is a german standard on vibration measurement instrumentation [6]. A similar paper has been issued by the API in the United States [7]. An ISO-standard on this subject is still to appear.

7. References

[1] VDI-guideline 2059 (draft 1979)

[2] Effects of transverse cracks on the vibration behaviour of turbine rotors (in German); VDI-Berichte 320 (1978)

[3] ISO-standard 2372

[4] ISO-standard 2954

[5] ISO-standard 3945

[6] DIN 45 670

[7] API-standard 670

[8] Bentley Nevada Corp., Application notes

[9] Spectral Dynamics Corp., Machinery analysis

[10] Schenk, Application notes

Cause of shaft vibrations	Time response of maximum displacement	Dependence of the maximum displacement on the operating conditions	Characteristic frequencies in the vibration spectrum	Shape of the kinetic shaft orbit	Graph of displacement versus time
Constant unbalance	Constant	Speed-dependent	Rotation frequency	Ellipse, in some cases also as circle or straight line	Sinusoidal
Thermal unbalance	Variable, depending on changes in the operating conditions (e.g.after start-up), often reaching a steady-state value	Dependent on power and power changes	Rotation frequency	Ellipse, in some cases also as circle or straight line	Sinusoidal
Unbalance due to corrosion, erosion or soiling	Slowly variable over long periods	Speed-dependent	Rotation frequency	Ellipse, in some cases also as circle or straight line	Sinusoidal
Unbalance due to rupture of rotor parts (e.g. rotor blades)	Sudden increase or decrease	Speed-dependent	Rotation frequency	Ellipse, in some cases also as circle or straight line	Sinusoidal

Table 1 Interpretation of the results of shaft vibration measurements [1]

Cause of shaft vibrations	Time response of maximum displacement	Dependence of the maximum displacement on the operating conditions	Characteristic frequencies in the vibration spectrum	Shape of the kinetic shaft orbit	Graph of displacement versus time
Self-excited vibrations, e.g. due to elastic hysteresis, blade tip excitation or bearing instability	As a rule, heavy fluctuations	Dependent on speed, power (in particular with blade tip excitation) bearing oil temperature (only in the case of bearing instability)	Frequency by magnitude often equal to the lowest lateral natural frequency of the shaft, taking into account the bearings	Irregular, rarely as a closed curve	Frequently similar to sinusoidal and fluctuating
Restraints (e.g. due to misalignment or coupling seizing) and loose bearing cups	Constant	Dependent on speed and power	Rotation frequency and its multiples (especially its double)	Frequently as a closed curve of varying shape (e.g. like an eight)	As a rule, generally periodic
Gearbox defect	Constant	Dependent on speed and power	Special gearbox frequencies with simultaneous rotation frequencies of input and output side	Irregular, usually not as a closed curve	Mostly periodic, rarely sinusoidal

Table 1 (continued)

Cause of shaft vibrations	Time response of maximum displacement	Dependence of the maximum displacement on the operating conditions	Characteristic frequencies in the vibration spectrum	Shape of the kinetic shaft orbit	Graph of displacement versus time
Electrical or magnetic faults on electric motors and generators	Constant or sometimes periodically beating	As a rule, dependent on power	Rotation frequency, mains frequency, double mains frequency, single and double slip frequency as modulation frequency	Frequently as elliptical curve	Sinusoidal or similar, sometimes amplitude-modulated

Table 1 (continued)

CHAPTER 4.2

IDENTIFICATION OF ROTOR PARAMETERS

V. Schlegel

1. Introduction

Turbomachines do not only possess the ability to convert fluid energy into rotational energy but also the undesirable attribute that their rotors vibrate. Therefore the mechanical behaviour of the rotor has to be analysed. Whereas the inertia distribution of the shaft can be determined from its geometry, the stiffness distribution cannot be obtained easily because of jumps in the diameter of the shaft. In modeling the system consisting of a rotor on its bearings, the task which presents most difficulty is the description of the dynamical properties of the oil film used in the journal bearings. In most cases a linear model, the simplest one, is chosen. The coefficients can either be obtained experimentally from very simplified test rigs with rigid rotors or by the solution of the linearised hydrodynamic theory for oil film bearings. A second uncertainty in calculating the vibrational behaviour of the rotor of a turbomachine is due to the difficulty of specifying the true distribution of the unbalance and the initial bow of a rotor.

Thus, it is not surprising that large differences between calculation and measurement are encountered when the rotor is run for the first time. A promising way of making calculation to approximate reality is the use of system identification.

2. Error definition

To enable one to compare the results of computation with those one can get from the system identification, the analytical model should have the same form as the experimental one.

Hence a linear partial differential equation is chosen as the mathematical description of the motion of a real rotor but with as yet unknown coefficient functions. Besides the bow of the rotor caused by its weight an initial bow will be taken into account. The assumed reference line from which the displacement is measured is, however, the static line of the unbent rotor, so that the motion $r(x,t)$ is defined as the displacement of the rotor from this line

$$C\ [r(x,t)] + \frac{\partial}{\partial t} D\ [r(x,t)] + \frac{\partial^2}{\partial t^2} M\ [r(x,t)] = F(x,t) \tag{1}$$

where

$$C = \frac{\partial^2}{\partial x^2}\left[A(x)\ \frac{\partial^2}{\partial x^2}\right] \tag{2}$$

and

$$F = f(x,t) + \sum_{k=1}^{n} F_k(t)\ \delta(x - x_k) \tag{3}$$

The stiffness operator is of fourth order with the coefficient function $A(x)$, which contains all the stiffness properties distributed in the axial and radial directions. The damping operator D stands for all those terms which describe viscous or gyroscopic forces. It is by no means a linear combination of the inertia and the stiffness operator and is not restricted to modal damping. The operator M includes the inertial effects of the mass distribution. The exciting forces F(x,t) of a rotor are composed of distributed unbalance forces and of pointwise-acting bearing forces.

The solution of this differential equation can only be examined by the actual motion of the real rotor points of measurement. Unfortunately the vibration of a rotor installed in the casing of a turbomachine can only be measured at very few, unevenly distributed locations, because of the restricted accessibility and the possibly high temperature within the machine. That is why it is impossible to find vibration modes by connecting the radial displacements from measurement point to measurement point by polygons or any method based thereon.

Thus, the vibrational behaviour of a rotor between measurement points has to be determined by suitably chosen interpolating functions (Fig. 1). For the purpose of parameter identification all complete, orthogonal functions sets can, in principle, be used.

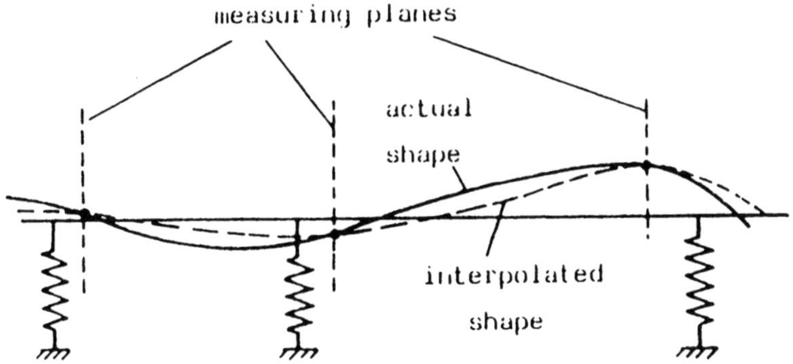

Fig. 1 Shape of vibration for a particular speed

However, it is appropriate and advisable to take the eigenfunctions of the calculated analytical model. The better the computed eigenfunctions approximate the motion of a real rotor, the smaller will be the difference between the real motion and the interpolation in the interval between two measurement locations. Depending on the expenditure spend on the calculation of the analytical model two kinds of approximating functions for the experimental model of a rotor are at our disposal. Either the real, decoupled eigenfunctions of the associated conservative system or the complex coupled eigenfunctions from a discretization with local shape functions can be chosen. But both of these can only serve as approximations, since the actual distribution of stiffness, damping and inertia differs from that used in the calculation.

The points of measurement can now be regarded as stations in a collocation, at which we demand that the chosen functions must fit the differential equation and the solution of the equation must agree with the measured displacement of the rotor, apart from statistical measuring errors. Let us now assume a solution in series form with shape functions $\varphi(x)$:

$$r_n(x,t) = \sum_{j=1}^{n} q_j(t)\ \varphi_j(x) \quad . \tag{4}$$

Here the weights q_j are time variable generalized coordinates. They contain the dynamical part of the solution and have to be determined by collocation in such a way, that the difference between the approximation and the solution will be as small as possible at the points of measurement.

Thus, we obtain:

$$\sum_{j=1}^{n} q_j\ C[\varphi_j(x_s)] + \sum_{j=1}^{n} \dot{q}_j\ D[\varphi_j(x_s)] + \sum_{j=1}^{n} \ddot{q}_j\ M[\varphi_j(x_s)] = f(x_s,t) \quad . \tag{5}$$

In matrix notation we obtain the ordinary differential equation of second order with generalized coordinates q_o:

$$M\ddot{q} + D\dot{q} + Cq = f(t) \tag{6}$$

The displacement of the rotor $r(x_s,t)$ is then given by the solution of this differential equation multiplied by the shape functions $\varphi(x)$:

$$r(x_s,t) = \varphi(x_s)^T \cdot q(t) \quad . \tag{7}$$

If we now express the error vector e by

$$\begin{bmatrix} e_1 \\ \cdot \\ \cdot \\ \cdot \\ e_m \end{bmatrix} = \begin{bmatrix} \hat{r}_1 \\ \cdot \\ \cdot \\ \cdot \\ \hat{r}_m \end{bmatrix} - \begin{bmatrix} \varphi_1(x_1) & \cdots & \varphi_n(x_1) \\ \cdot & & \cdot \\ \cdot & & \cdot \\ \cdot & & \cdot \\ \varphi_1(x_m) & \cdots & \varphi_n(x_m) \end{bmatrix} \begin{bmatrix} q_1(t) \\ \cdot \\ \cdot \\ \cdot \\ q_n(t) \end{bmatrix} \qquad m \geqq n \tag{8}$$

we can pose the identification problem the following way:

For given measurement values $\hat{r}(x_s,t)$ the matrices M, D and C are to be determined in such a way, that the error e with respect to all points of measurement will become minimal in the sense of a given criterion.

The identification problem has thus been converted by means of shape functions into a problem of parameter estimation, where the values of the shape functions are held constant during the estimation process. It is not necessary to know the exact values of the operators M, D and C beforehand because the error is defined after the discretization

3. Optimization method

Since the error vector depends on the parameters M, D and C in a non-linear way, we have a special case of the general optimization problem of finding the stationary value of a functional

$$J(x) = f\,(F(x)) \quad , \quad \forall\, x \in D \tag{9}$$

where

$$F : D \quad \mathbb{R}^n \rightarrow \mathbb{R}^m$$

and

$$f : G \quad \mathbb{R}^m \rightarrow \mathbb{R}^1$$

This functional consists of two successive mappings. The inner mapping F represents the model function and the other mapping the error law. To formulate an iteration algorithm for a functional of this kind with a nonlinear operator F, we need an approximation. If the operator is Frechét-differentiable, we can expand it in a Taylor series at the iteration point x_k and truncate the series after the linear term:

$$J(x) = f\,\{F x_k + F'(x_k)\,(x - x_k)\} \quad . \tag{10}$$

If the outer operator is twice differentiable we obtain the gradient of the functional at the iteration point c_k using the rule for successive operators

$$J'(x) = g_k(x) = f'\,\{F x_k + F'(x_k)\,(x - x_k)\}\; F'(x_k) \tag{11}$$

The symmetric Hessian matrix H can now be expressed by the operator F and its derivatives:

$$J''(x) = H_{g_k} = F'(x_k)^T\, H_f\,\{F x_k + F'(x_k)\,(x - x_k)\}\; F'(x_k) \ . \tag{12}$$

Hence we can write the Newton algorithm

$$x_{k+1} = x_k - [H_{g_k}\,(x_k)]^{-1} g^T(x_k) \tag{13}$$

of the functional under consideration

$$x_{k+1} = x_k - [F'(x_k)^T H_f(Fx_k)\, F'(x_k)]^{-1} F'(x_k)^T f'(Fx_k)^T \;. \tag{14}$$

In the case when the operator f consists of a quadratic form

$$f = \frac{1}{2} h^T h \quad , \tag{15}$$

the Hessian matrix of the operator f simplifies to the unit matrix. Thus, the algorithm reduces to the Gauß-Newton relation

$$x_{k+1} = x_k - [F'(x_k)^T F'(x_k)]^{-1} F'(x_k)^T F^T x_k \quad . \tag{16}$$

To adjust the unknown frequency function with matrices as arguments to measured data the operators in the iteration algorithm have to be replaced by their corresponding matrix expressions. The compact notation for derivatives and differentials of matrix functions with matrices or vectors as arguments, used by Vetter[1], proves to be extremely clear and useful for this purpose:

$$\begin{aligned} \mathcal{D}_{b_{k\ell}} A(B) &\doteq \frac{\partial a_{ij}}{\partial b_{k\ell}} \qquad k = 1,\dots,p \;\; ; \;\; \ell = 1,\dots,q \\ \mathcal{D}_B A(B) &\doteq {}_{b_{k\ell}}A \qquad k = 1,\dots,s \;\; ; \;\; \ell = 1,\dots,t \end{aligned} \tag{17}$$

Further, two linear mappings of matrices onto vectors will be employed in the following equations:

$$\text{rs}\; A = [a_{11} \dots a_{1n},\, a_{21} \dots a_{2n},\, \dots, a_{n1} \dots a_{nn}] \tag{18}$$

and

$$\text{cs}\; A = [a_{11} \dots a_{n1},\, a_{12} \dots a_{n2},\, \dots,\, a_{1n} \dots a_{nn}]^T \tag{19}$$

Both, the row operator rs as well as the column operator cs map any matrix onto a vector in an unique and reversible way. Here the expression 'matrix function A(B)' not only means a function, which is built from sums or products of matrices of the same order, but any mapping of vectors or matrices. Function and argument need not necessarily have the same dimension nor the same form.

We again expand the nonlinear matrix operator in a Taylor series and truncate the series after the linear term

$$A(B_k + dB) = A(B_k) + dA(B_k) +++ \quad . \tag{20}$$

With the notation just introduced we can express the complete matrix differential dA by the derivative and the Kronecker product at $d(rsB)$ and the unit matrix

$$dA(B_k) = \mathcal{D}_{rsB_k} A \, (d[rsB_k]^T \times I) \quad . \tag{21}$$

If we now express the linear approximation of the error E, where A contains the measured data, by

$$E = \hat{A} - A(B_k) - dA(B_k) \quad , \tag{22}$$

we can write the error functional J as the Euclidian norm

$$J = tr \ \{[L - dA(B_k)]^T [L - dA(B_k)]\} \quad . \tag{23}$$

The matrix L represents the difference between the measured data and the values of the function of the model at the iteration points. The trace operator can be replaced by the vector operators already mentioned

$$tr \ \{[\bullet] \, [\bullet]\} = rs[\bullet] \, cs[\bullet] \quad , \tag{24}$$

so that after a second substitution the functional J can be expressed by the column operator

$$J = \{cs^T [\ldots]cs[\ldots]\} \quad . \qquad (25)$$

Applying the column operator to the matrix differential, the Kronecker product vanishes. If we now make use of the linearity of the column operator, the functional can be written in its final form

$$J = [csL - \mathcal{D}_{rsB_k} \; csA(d[rsB_k]^T)]^T [\ldots] \quad . \qquad (26)$$

The form of this functional is equal to the term of Eq. 15, so we can write the Gauß-Newton algorithm as

$$rsB_{k+1} = (rsB_k)^T + \{\mathcal{D}^T_{rsB_k} \; csA \cdot \mathcal{D}_{rsB_k} \; csA\}^{-1} \mathcal{D}^T_{rsB_k} \; csA \cdot csL \quad . \qquad (27)$$

4. Frequency response functions

The vibrational behaviour of mechanical structures excited by sinusoidal forces can be described by functions of the following form:

$$r = N_{oo} \, (N_o + N_1\Omega + N_2\Omega^2)^{-1} \, (n_o + n_1\Omega + n_2\Omega^2) + n_{oo} \; . \qquad (28)$$

To fit such functions into the equations already derived we need the derivatives of some particular types of function. Applying the derivative operator and the column operator to inner products of matrices leads to expressions which contain Kronecker products.

Here and in the following equations the differentiation argument is dropped for clearness. Besides the derivatives of products we need the derivative of the inverse of an algebraic expression. Fortunately we can find a relation which circumvents the need to differentiate the inverse itself. The combination of both equations allows to evaluate the gradient and the Hessian matrix of the frequency response function:

$$\mathcal{D}cs(AB^{-1}C) = \{\mathcal{D}C\ C^T \times A)\}\{I \times cs(B^{-1})\} - \{(B^{-1}C)^T \times (AB^{-1})\}\mathcal{D}csB\ . \qquad (29)$$

In estimating the matrices in this function by the proposed iteration process the displacement vectors of the rotor vibration have to be measured at several frequencies Ω_j to avoid a singular coefficient matrix in the linear system of equations which yields the iteration step vector. Therefore, the errors at each frequency Ω_j have to be summed to give an overall error

$$J_{ges} = \sum_{j=1}^{\ell} \text{spur}\ (E_j^T E_j) \qquad (30)$$

Making use of the linearity of the column operator, we have a linear system of equations for calculating the iteration step vector

$$\{\sum_{j=1}^{\ell} \mathcal{D}^T csF_j \cdot \mathcal{D}csF_j\}\ d[rsB]^T = \sum_{j=1}^{\ell} \mathcal{D}^T csF_j \cdot csL_j \quad . \qquad (31)$$

The order of the coefficient matrix of this system is independent of the number of measurement frequencies and depends only on the number of elements in the function argument.

Since the variables in the frequency response function can assume complex values a linear, unique and reversible mapping RM

$$RM : D\ \mathbb{C}^{p \times q} \rightarrow \mathbb{R}^{2p \times 2q} \qquad (32)$$

is introduced to convert complex values into their real form

$$RM(B^k) = B^R = \begin{bmatrix} B^r & -B^i \\ B^i & B^r \end{bmatrix} \quad ; B^r, B^i \in \mathbb{R} \quad . \qquad (33)$$

Thus, all the equations can be extended from real matrix operators to complex matrices. This mapping is isomorphic from which follows that all computation rules including differentiation and inversion remain valid. This property is particularly important in evaluating the derivative of a complex function with respect to the real or imaginary part of an argument.

Since the exciting forces of a rotor are unknown we have to take some variable for reference. Generally the mass matrix is chosen, because this matrix can be calculated with the greatest accuracy. Denoting this reference by the subscript N we obtain the complex frequency response function of a rotor with unbalance and initial bow:

$$r = \Gamma (C_N + j D_N \Omega - 1 \Omega^2)^{-1} (b_N + u_N \Omega^2) + r_0 \quad . \tag{34}$$

The vector r_0 represents all those parts of the vibration with constant amplitudes which are caused by the measurement system and which have no statistical origin. Their source will be discussed later. The matrix Γ contains the constant shape functions. Using the mapping RM we can write the complex frequency response function in its real form. After having composed these matrices to hyper-matrices we perform the differentiation needed in the Gauß-Newton iteration.

5. Identification of eigenfunctions

When the iteration process has reached the minimum of the functional, we have found the desired mass, damping and stiffness matrix. We now solve the associated quadratic eigenvalue problem and obtain the eigenvalues of the identified system. The elements of the matrix X of the complex right eigenvectors of the discrete system can be regarded as weighting coefficients in the linear combination of the initially chosen shape functions to evaluate the eigenfunction of the tested rotor. There are three possibilities for this coordinate transformation:

- real shape functions and modal damping in the identified system

$$\psi(x) = \varphi(x) \cdot X \tag{35}$$

- real shape functions and arbitrary, viscous damping in the identified system

$$[\psi(x)\ \psi^*(x)] = \varphi(x) \quad [\tilde{X}\ \ \tilde{X}^*] \tag{36}$$

- complex shape functions and arbitrary, viscous damping in the identified system

$$[\psi(x)\ \psi^*(x)] = [\varphi(x)\ \varphi^*(x)] \begin{bmatrix} \tilde{X} & 0 \\ 0 & \tilde{X}^* \end{bmatrix} \quad . \tag{37}$$

6. Identification of unbalance distribution

While the eigenfunctions can be evaluated, at least approximately, the distribution of unbalance and initial bow remains unknown. Hence at the beginning of the iteration arbitrary vectors have to be chosen as vectors for the generalized unbalance and initial bow. The identification can only furnish the generalized vectors b_N and u_N which are referenced to the mass matrix, as already mentioned. The only use we can make of this is to confine the distribution of unbalance, if we demand that the distribution $e(x)$ should be expressed by a series of given functions $h(x)$

$$e(x) = \sum_{j=1}^{n} a_j \, h_j(x) \quad . \tag{38}$$

The required coefficients a_i are obtained from the solution of a linear system of equations:

$$K \cdot a = M_R \cdot u_N \tag{39}$$

where

$$k_{ij} = \int_0^{\ell} \varphi_i(x)\, h_j(x)\, dx \tag{40}$$

The functions are the shape functions and the matrix M_R is the mass matrix of the precalculation which serves as an approximation for the true mass matrix of the rotor.

7. Experimental investigations

In addition to testing an identification method by simulated data it is important to examine the behaviour of the algorithm by using measured data from an existing rotor. Thus, measurements were made on a test rotor which shows a greater resemblance to turbomachines than the usual small test rig (Fig. 2). This rotor is built of three separate shafts. It weighs 450 kg and has a length of 3 m. The displacement of the rotor was measured in four planes with two proximity probes in each plane positioned at 45 degrees to the vertical and at 90 degrees to each other.

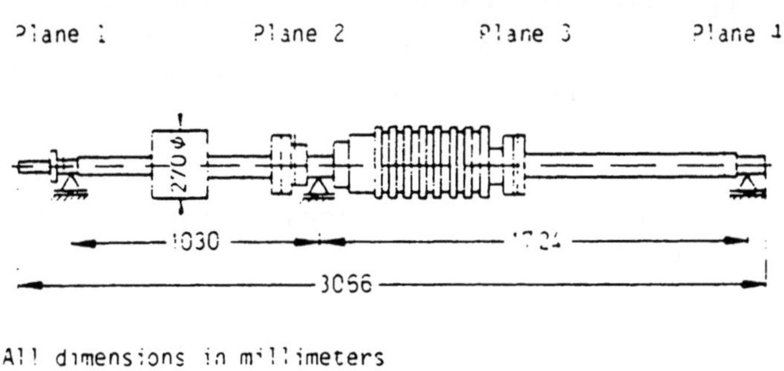

Fig. 2 Rotor of the model turbomachine

The electrical signals from the probes do not completely represent the real vibration of the rotor but contain additional parts which should be called 'fictitious vibrations' to avoid the misleading expression 'run put'. They can be produced by changes of the displacement signal which are not associated with vibrations, such as deviations of the cross-section of the shaft from circularity or by the physical character of the transducer. Among minor influences, the inhomogeneous conductivity of the shaft surface plays a major role when using proximity probes of the eddy current type.

The effects of these errors on the measured signals are depicted in Figure 3.

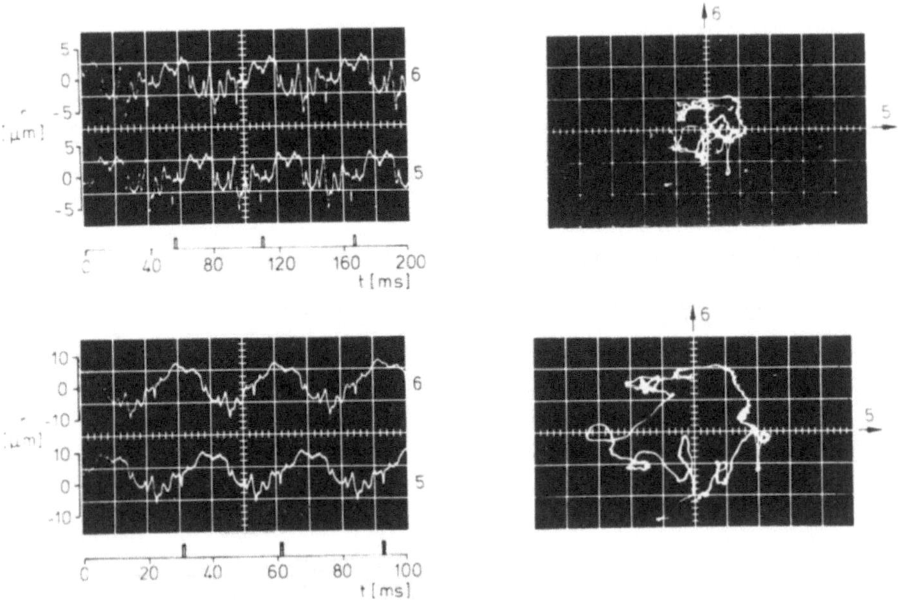

Fig. 3 Vibration signals at measuring plane 3 at two different speeds

Whereas the vibration amplitudes of the rotor increase considerably with increase of speed, the fictitious vibrations remain constant and can still be recognized. Further, one can notice that characteristic peaks in the signal maintain their phase relative to the reference pulse of the rotor, independently of rotor speed. On the right-hand side of Fig. 3 the signals from the two probes in the same plane are combined to give a closed orbit showing the shaft displacement during one revolution. The characteristic loops are caused by the fictitious vibration signals. Because in all cases the fictitious vibration remains constant in amplitude and phase it can be represented by complex and frequency invariant vectors. As for the identification process, we only need to take into consideration the first harmonic.

The most suitable procedure for extracting the first harmonic is to use the orthogonal correlation technique on the vibration signal and a rereference signal. This technique produces the real and imaginary part of the rotor vibration. Depicting these values in a Nyquist diagram we have the clearest presentation of the frequency response function, especially because the influence of an arbitrarily oriented initial vector can be seen more clearly in a Nyquist diagram than by separate plots of amplitude and phase.

When dealing with rotors with non-conservative and anisotropic journal bearings the Nyquist diagrams of the two vibration transducers in the same plane differ from each other. In order to show this difference the plots are depicted together. The phase angles of the two signals have been arranged in such a way that the angular position of the reference pulse coincides with the positive real axis.

This temparature of the oil in the journal bearings of a turbomachine influences the spring and damping constants of the bearing. The whole system 'rotor on journal bearings' cannot be looked upon as a linear system. But when we do so, the identification yields parameters belonging to that linear system which approximates the real system best in the sense of the quadratic mean . On the other hand the main advantage of this identification method is the ease with which nonlinear effects can be included in the error functional, as compared with the more popular modal method.

In Figure 4 the result of the identification together with the measured frequency responses is shown. It should be pointed out that the frequency responses of the two probes in this sample plane as well as the responses of all the other probes were simultaneously used to adjust the parameters of the model.

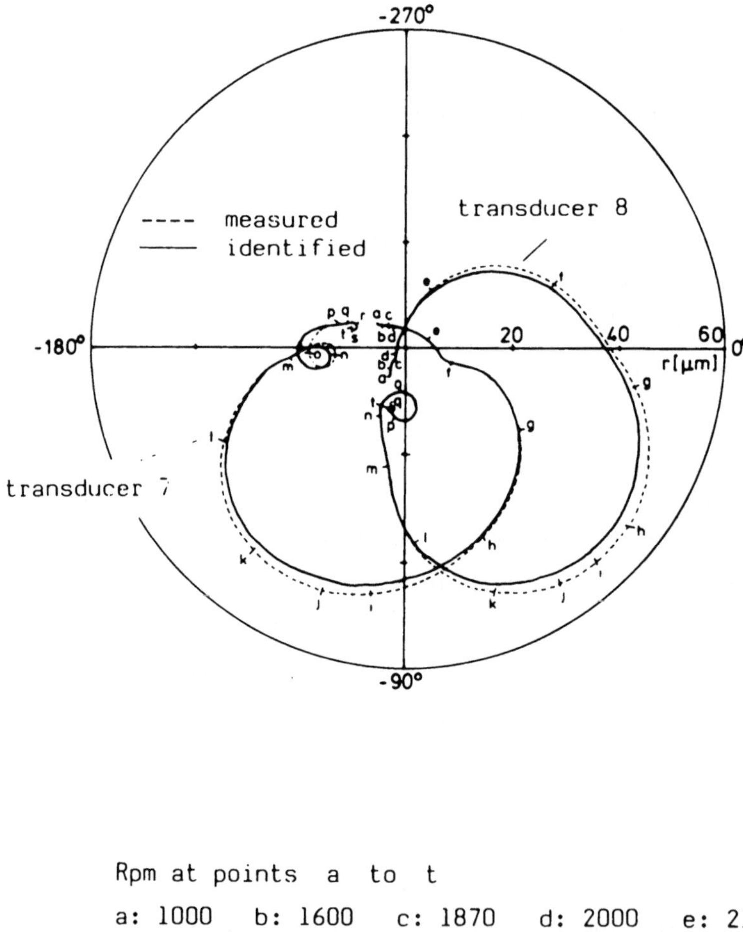

Rpm at points a to t

a: 1000	b: 1600	c: 1870	d: 2000	e: 2300
f: 2400	g: 2460	h: 2480	i: 2500	j: 2510
k: 2530	l: 2570	m: 2670	n: 2880	o: 2950
p: 3200	q: 3530	r: 3640	s: 3800	t: 4000

Fig. 4 Comparison of the measured and identified Nyquist plots of the measurement plane 4

8. Summary

The method presented in this paper requires no assumptions about the type of the damping matrix and is thus not restricted to modal damping. The system matrices are found by an iterative estimation process. The iteration step is evaluated by a modified Gauß-Newton algorithm, which is extended to matrix functions. By using the identified system matrices the corresponding quadratic eigenvalue problem can be solved to yield the eigenfunctions of the rotor as a linear combination of the chosen shape functions. Further the generalized vector of the unbalance can narrow the possible distribution of the unbalance.

Finally a particular error which specifically appears in rotor vibration measurement and which is subsumed under the expression 'fictitious vibration' can be included in the error function, thus separating this error from purely random errors.

9. References

1 Vetter, W.J.: Matrix Calculus and Taylor Expansion, SIAM Rev. 15 (1973) 2, 351-369.

CHAPTER 4.3

IDENTIFICATION OF MODAL PARAMETERS OF ROTORS

R. Nordmann

1 Introduction

The occurence of instability in rotating machinery may be caused by different effects: oil film in journal bearings, sealings, internal damping etc.
A machine designer wants to know, whether a machine will run stable during operation and what size the stability threshold speed will have. Furthermore he needs information about the parameters influencing the instability of a rotor.
Important informations about stability of a linear rotor-system are given by the complex eigenvalues, respectively the damping constants and natural frequencies. Together with the natural modes (eigenvectors) a valuation of the dynamic behaviour - free as well as forced vibrations - is possible. Eigenvalues and natural modes are called the "modal parameters" of a system. The modal parameters normally are calculated in the design stage of a rotating machine. Because of uncertain input data for the calculation the results have to be considered critically. Therefore

mechanical engineers also try to find out the modal parameters of built rotating machines or test rotors by measurements during operation. Such measurements indicate the real dynamic behaviour and possibly show in what way the previously employed models for calculation have to be modified.

Some years ago a combined experimental and analytical method was developed identifying modal parameters of nonrotating structures. This method is usually named "modal analysis". One of the assumptions in the method is the symmetry of matrices for the analytical model, which is not admissable in the case of rotating systems (nonsymmetry caused by journal bearings etc.). Therefore improvements are necessary for application of the method in rotating machinery. Such improvements are treated in this paper and an example is given investigating the eigenvalues of a Laval-shaft.

2 Modal Parameters of Rotors. The dynamic characteristics of a rotor in journal bearings - shown in Fig. 1 - can be described by the modal parameters: eigenvalues and natural modes.

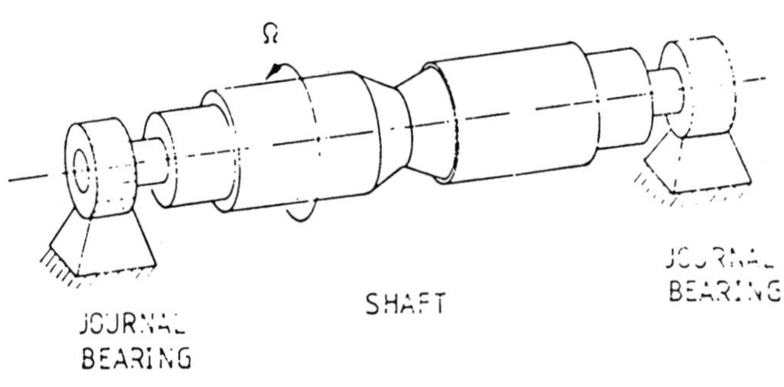

Fig. 1 Rotor in two journal bearings

They can be calculated by the homogeneous equations of the rotor

$$\underline{M}\,\ddot{\tilde{\underline{u}}} + \underline{C}\,\dot{\tilde{\underline{u}}} + \underline{K}\,\underline{u} = 0 \tag{1}$$

The solution of eq. (4.3-1) is of the form

$$\tilde{\underline{u}} = \underline{\phi}\, e^{\lambda t} \tag{2}$$

Substitution yields the quadratic eigenvalue problem

$$\{\lambda^2 \underline{M} + \lambda\, \underline{C} + \underline{K}\}\, \underline{\phi} = \underline{0} \tag{3}$$

with 2N eigenvalues λ_j and corresponding natural modes $\underline{\phi}_j$, if the order of the matrices is N. The eigenvalues as well as the eigenvectors mainly occur in conjugate complex pairs (real eigenvalues and eigenvectors are not considered),

$$\text{Eigenvalues:} \quad \lambda_j = \alpha_j + i\omega_j \;; \quad \bar{\lambda}_j = \alpha_j - i\omega_j \tag{4}$$

$$\text{Eigenvectors:} \quad \underline{\phi}_j = \underline{s}_j + i\underline{t}_j \;; \quad \bar{\underline{\phi}}_j = \underline{s}_j - i\underline{t}_j \tag{5}$$

The part of the solution, which belongs to such a conjugate complex pair can be written as

$$\tilde{\underline{u}}_j(t) = B_j\, e^{\alpha_j t}\{\underline{s}_j \sin(\omega_j t + \gamma_j) + \underline{t}_j \cos(\omega_j t + \gamma_j)\} \tag{6}$$

ω_j is the circular natural frequency of this part of the solution and α_j the damping constant. The damping constant determines, whether the solution $\tilde{\underline{u}}_j$ decreases ($\alpha_j < o$) or increases ($\alpha_j > o$).
The stiffness and damping coefficients of the oil film are functions of the running speed Ω. Therefore the modal parameters depend on the running speed, too.

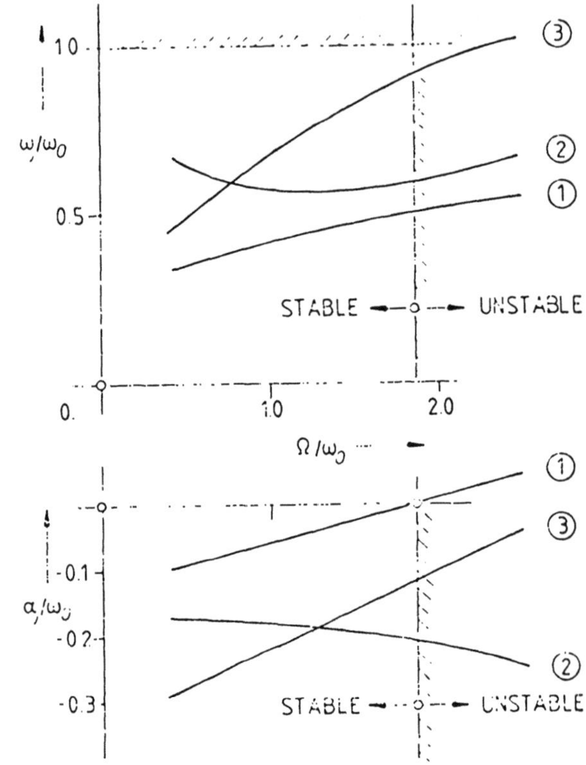

Fig. 2 Eigenvalues of the rotor (ω_o reference frequency)

In Fig. 2 the three lowest eigenvalues (circular natural frequencies ω_j and damping constants α_j) of the rotor (Fig. 1) are plotted versus the angular velocity Ω. The intersection between the damping constant α_1 and the abscissa establishes the threshold of instability.

The expression in parantheses { } of eq. (6) was defined as natural mode in chapter 1.[1], representing a time dependent curve in space. Fig. 3 shows the modal shapes corresponding to the three lowest natural frequencies. For any point of the shaft the plane motion is an elliptical orbit. These orbits are different in the individual natural modes. The shaded planes contain the major axis of the ellipses.
The following chapter deals with the problem, how to identify the described modal parameters.

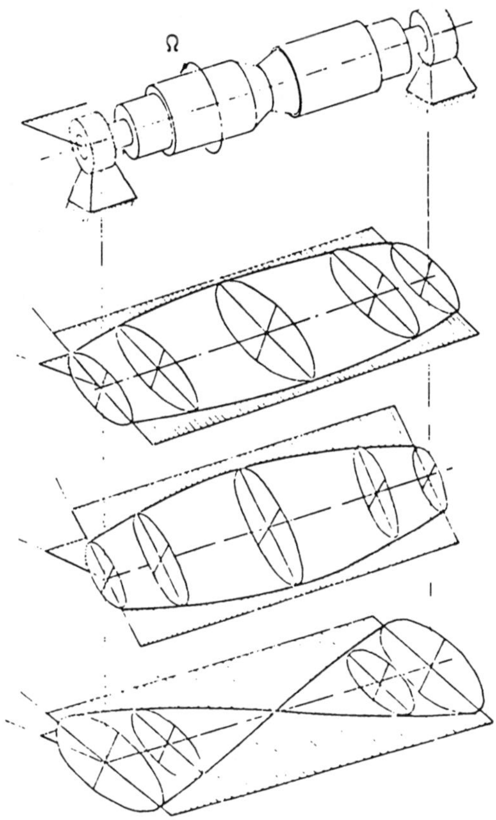

Fig. 3 Natural modes of the rotor in two journal bearings

3 Identification of Modal Parameters of Rotors. For some years past a combined experimental and analytical method identifying modal parameters of nonrotating elastic systems has been applied in aircraft-industry and recently in automotive industry, machine-tool industry and others. The aim of the method is to analyze a structure in its elementary modes and to determine their characteristics; hence the name "modal analysis". It is an identification procedure working in the frequency domain.

At first a number of measurement points with N measurement-coordinates are chosen for an actual structure. Test forces are applied systematically upon the structure at various points and the system response are measured.

After transformation (FFT) of the input and output signals into the frequency domain, frequency response functions can be determined.
It is also possible to represent analytical frequency response functions, which depend on the modal parameters of the system. The corresponding mathematical model normally assumes linearity, viscous damping and symmetric matrices etc.

The analytical functions are fitted to the measured functions by variation of the modal parameters. Results of this iterative fitting procedure are the modal parameters.
For application of the method three points are needed (Fig. 4)

- a mechanical and mathematical model with given structure (linearity, viscous damping, symmetric matrices etc.) but free parameters. Analytical frequency response functions belong to the mathematical model.
- a measuring device for the determination of frequency response functions. "Single point" excitation methods with a broadband-excitation are preferred today (Impact-excitation).
- a curve fitting procedure, which fits the analytical curves to the measured curves and calculates the modal parameters.

Important differences have to be considered in applying the method for rotating structures, for example

- nonsymmetry of matrices (change in the model-structure)
- speed dependence of modal parameters
- excitation during operation of the rotor.

In consideration of the above differences "modal analysis" is on principle also available for rotating structures.
For representation of the method we subdivide into the main parts

- Mechanical and mathematical model
  analytical frequency response

- Measurement of frequency response functions
- Determination of modal parameters by curve fitting.

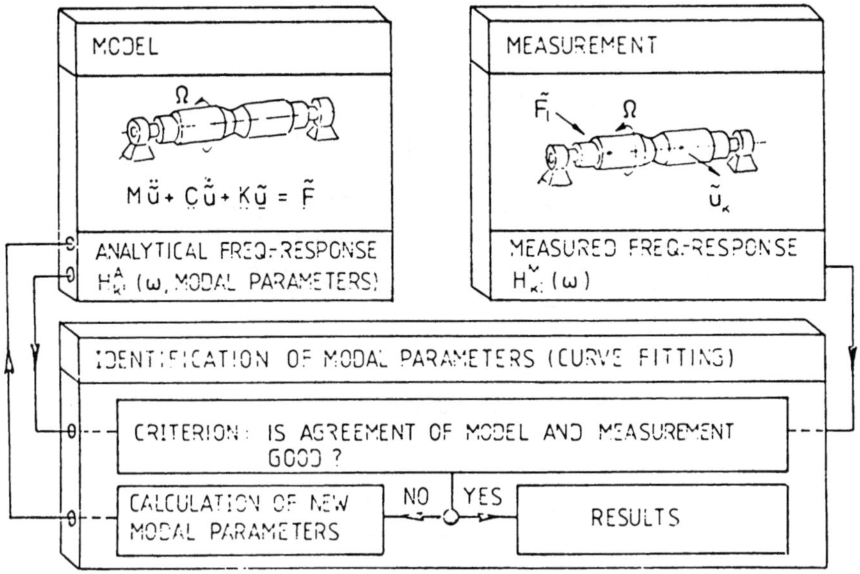

Fig. 4 Identification of modal parameters

Mechanical and mathematical model - analytical frequency response.

Equations of motion. We start with the mechanical model (Fig. 5), which can be represented by a linear system of N differential equations

$$\underline{M}\,\ddot{\tilde{\underline{u}}} + \underline{C}\,\dot{\tilde{\underline{u}}} + \underline{K}\,\tilde{\underline{u}} = \tilde{\underline{F}}(t) \qquad (7)$$

expressing the equilibrium of inertia-, damping- and stiffness forces as well as external forces. Damping and stiffness matrices are nonsymmetric and elements of them depend on the speed of the rotor.

Frequency response functions. If one excites the linear rotor in a certain point ℓ by means of a harmonic force (input signal), oscillating with a frequency ω (Fig. 6),

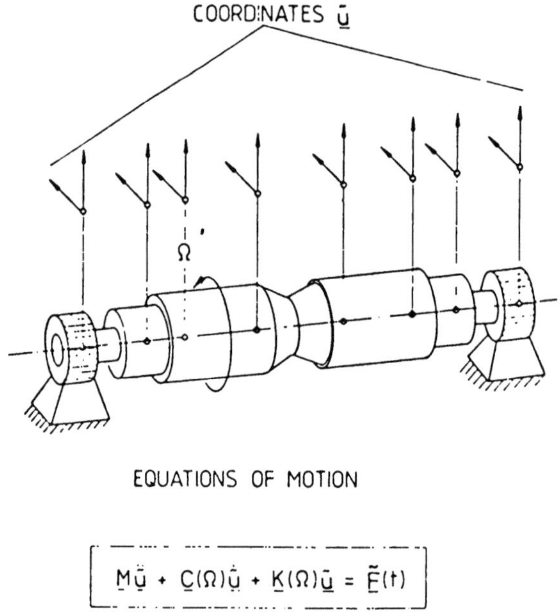

Fig. 5 Mechanical model

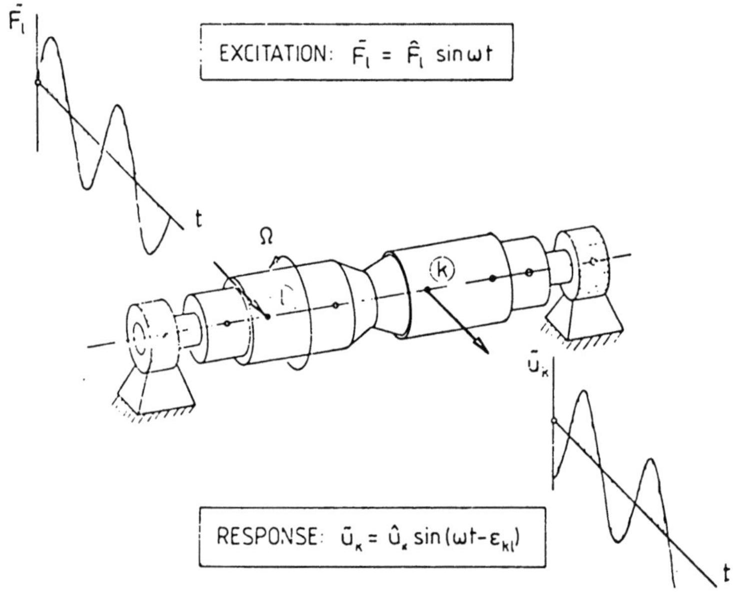

Fig. 6 Input and output signals at the rotor

$$\tilde{F}_\ell\ (t) = \hat{F}_\ell\ \sin\omega t = IM\ \{\hat{F}_\ell\ e^{i\omega t}\}\ , \tag{8}$$

further one measures the displacement (output signal) in another point k

$$\tilde{u}_k(t) = \hat{u}_k\ \sin(\omega t-\varepsilon_{k\ell}) = IM\ \{\hat{u}_k e^{i(\omega t-\varepsilon_{k\ell})}\}\ , \tag{9}$$

the response behaviour of the rotor can be characterized by the ratio of the amplitudes of the two signals $\hat{u}_k/\hat{F}_\ell$ and the phase $\varepsilon_{k\ell}$ between the two signals.

Both are frequency dependent functions and named

$\hat{u}_k/\hat{F}_\ell$ $\{\omega\}$ amplitude frequency characteristic

$\varepsilon_{k\ell}$ $\{\omega\}$ phase frequency characteristic

They are usually expressed in one complex frequency response function defining the complex ratio of output-signal to input-signal

$$H_{k\ell}\ \{\omega\} = \frac{\tilde{u}_k}{\tilde{F}_\ell} = \frac{\hat{u}_k e^{i(\omega t-\varepsilon_{k\ell})}}{\hat{F}_\ell e^{i\omega t}} = \frac{\hat{u}_k}{\hat{F}_\ell}\ e^{-i\varepsilon_{k\ell}} \tag{10}$$

$H_{k\ell}$ (ω) can be represented either by a polar diagram in the complex plane (Nyquist-plot) or as amplitude ratio and phase angle in function of the frequency (Bode-plot).

As mentioned above $H_{k\ell}$ (ω) is a function of the frequency ω, furthermore it depends on the system parameters (mass, stiffness, damping). On the other hand the frequency response is also representable by the modal parameters of the system. The next Fig. 7 shows an example of an amplitude frequency characteristic $H_{k\ell}$ in connection with the eigenvalues of the system. For constant running speed Ω of the rotor, the corresponding eigenvalues are found by the intersections between the curves $\omega_j(\Omega)$, $\alpha_j(\Omega)$ and the line Ω = constant.

If the frequency of excitation ω coincides with one of the natural frequencies ω_j, a peak is expected in the amplitude frequency characteristic.

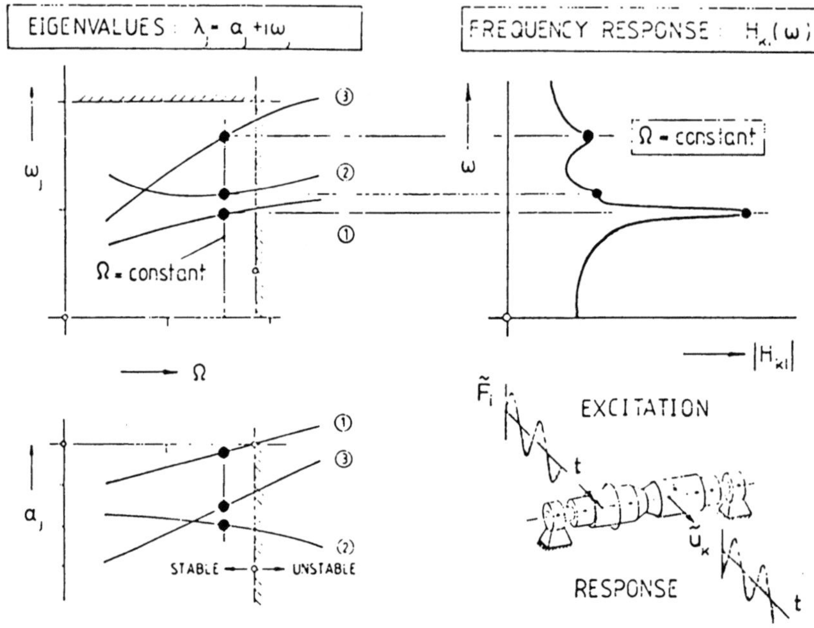

Fig. 7 Amplitude frequency characteristic and eigenvalues of a rotor

In the case of high damping constants α_j the corresponding peaks either will be very small or not occur at all. It is discernible that informations about modal parameters are contained in the frequency response. In the following chapter we shall express the frequency response by the modal parameters.

Expansion of the frequency response in terms of modal parameters.

It is well known that the response of a linear system can be represented in terms of the modal parameters (modal analysis). Expressing the frequency response, we start with the equations of motion (7) for a nonconservative rotor with N degrees of freedom

$$\underline{M}\,\ddot{\underline{\tilde{u}}} + \underline{C}\,\dot{\underline{\tilde{u}}} + \underline{K}\,\underline{\tilde{u}} = \underline{\tilde{F}}(t) \tag{11}$$

They can be converted to 2N first order differential equations with the dependent variables $\tilde{\underline{r}}$ constituting a 2N dimensional state vector (see chapter 1.1)

$$\underbrace{\begin{bmatrix} \underline{M} & 0 \\ 0 & -\underline{K} \end{bmatrix}}_{\underline{A}} \underbrace{\begin{bmatrix} \dot{\tilde{\underline{u}}} \\ \tilde{\underline{u}} \end{bmatrix}}_{\tilde{\underline{r}}} - \underbrace{\begin{bmatrix} 0 & \underline{M} \\ \underline{M} & \underline{C} \end{bmatrix}}_{\underline{B}} \underbrace{\begin{bmatrix} \ddot{\tilde{\underline{u}}} \\ \dot{\tilde{\underline{u}}} \end{bmatrix}}_{\dot{\tilde{\underline{r}}}} = \underbrace{\begin{bmatrix} 0 \\ -\tilde{\underline{F}} \end{bmatrix}}_{\tilde{\underline{f}}} \tag{12}$$

The system response can be expressed by an expansion in terms of the right eigenvectors $\underline{r}_j$ of the nonconservative system (see chapter 1.3)

$$\tilde{\underline{r}}(t) = \sum_{j=1}^{2N} \underline{r}_j \tilde{q}_j = \underline{R}\, \tilde{\underline{q}} \tag{13}$$

with the generalized coordinates $\tilde{q}_j$ and the modal matrix $\underline{R}$. Substituting in eq. (12) and premultiplying with the modal matrix of the left eigenvectors $\underline{L}^T$ we obtain decoupled equations of motion

$$\underline{L}^T\underline{A}\ \underline{R}\ \tilde{\underline{q}} - \underline{L}^T\underline{B}\ \underline{R}\ \dot{\tilde{\underline{q}}} = \underline{L}^T\tilde{\underline{f}}, \tag{14a}$$

respectively

$$\lambda_j \tilde{q}_j - \dot{\tilde{q}}_j = -\underline{\psi}_j^T \tilde{\underline{F}} \tag{14b}$$

$\underline{\psi}_j$ are the eigenvectors of the transposed eigenvalue problem.

In the case of harmonic excitation at point ℓ of the shaft - written in complex form

$$\tilde{\underline{F}}^T = \{0,0,\ldots.\hat{F}_\ell\ldots0,0\}\ e^{i\omega t} \tag{15}$$

the steady state solution of eq. (14b) is

$$\tilde{q}_j(t) = - \frac{\psi_{\ell j}\hat{F}_\ell}{\lambda_j - i\omega} e^{i\omega t} \tag{16}$$

and

$$\tilde{\underline{r}}(t) = \begin{bmatrix} \dot{\tilde{\underline{u}}} \\ \tilde{\underline{u}} \end{bmatrix} = \sum_{j=1}^{2N} \frac{\psi_{\ell j}\hat{F}_\ell}{i\omega - \lambda_j} \begin{bmatrix} \lambda_j \underline{\phi}_j \\ \underline{\phi}_j \end{bmatrix} e^{i\omega t} \tag{17}$$

Therefore with the response at point k

$$\tilde{u}_k(t) = \hat{u}_k e^{i\omega t} = \sum_{j=1}^{2N} \frac{\phi_{kj}\psi_{\ell j}}{i\omega - \lambda_j} \hat{F}_\ell e^{i\omega t} \tag{18}$$

the analytical frequency response $H_{k\ell}^A$ can be formulated in terms of the modal parameters

$$H_{k\ell}^A = \frac{\tilde{u}_k}{\tilde{F}_\ell} = \frac{u_k e^{i\omega t}}{\hat{F}_\ell e^{i\omega t}} = \sum_{j=1}^{2N} \frac{\phi_{kj}\psi_{\ell j}}{i\omega - \lambda_j} = \sum_{j=1}^{2N} \frac{a_{k\ell j}}{i\omega - \lambda_j} \tag{19}$$

The expression (19) consists of the eigenvalues λ_j and elements of left-eigenvectors $\psi_{\ell j}$ and right eigenvectors ϕ_{kj}.
For a rotor with N degrees of freedom NxN frequency response functions exist , assembled in the matrix $\underline{H}^A$ (Fig. 8).
It is important to note, that each row $\underline{Z}_k^T$ contains all of the left eigenvectors $\underline{\psi}_j$ and each column $\underline{S}_\ell$ contains all of the right eigen-vectors $\underline{\phi}_j$

$$\underline{Z}_k^T = \frac{\phi_{k1}}{i\omega - \lambda_1} \underline{\psi}_1^T + \frac{\phi_{k2}}{i\omega - \lambda_2} \underline{\psi}_2^T + \ldots\ldots \tag{20}$$

$$\underline{S}_\ell = \frac{\psi_{\ell 1}}{i\omega-\lambda_1}\,\underline{\phi}_1 + \frac{\psi_{\ell 2}}{i\omega-\lambda_2}\,\underline{\phi}_2 + \ldots\ldots \tag{21}$$

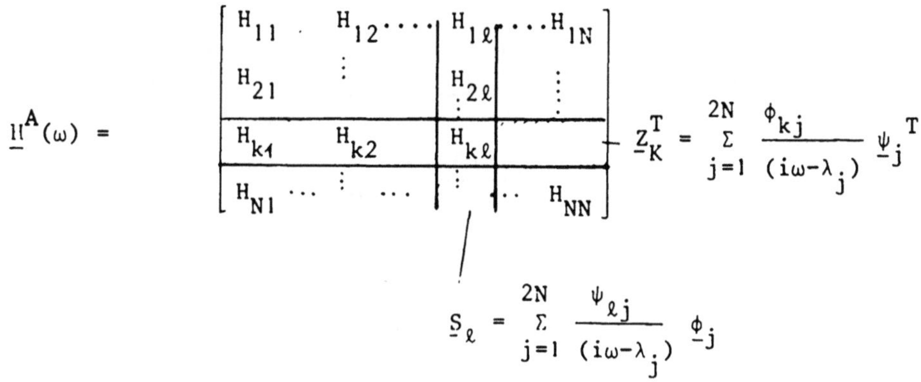

Fig. 8 Matrix of frequency response functions

One row $\underline{Z}_k^T$ and one column $\underline{S}_\ell$ of the frequency response matrix $\underline{H}^A(\omega)$ need to be measured in order to identify all of the modal parameters λ_j, $\underline{\phi}_j$, $\underline{\psi}_j$ of a rotor (Fig. 9).

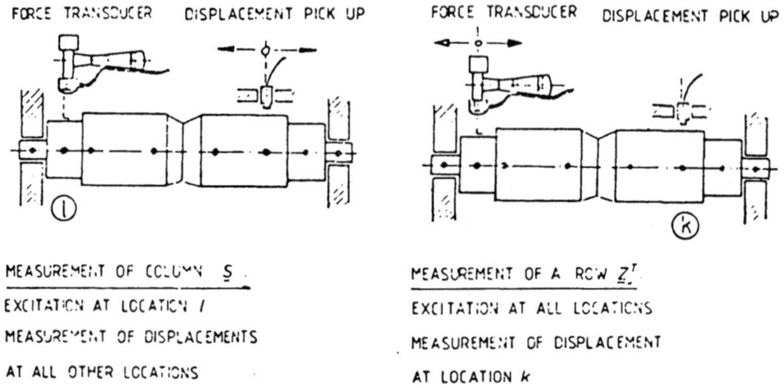

Fig. 9 Measurement of frequency response functions

It is sufficient to measure only one column, if eigenvalues and right eigenvectors will be determined. For determination of eigenvalues λ_j without natural modes, the whole information is contained in one frequency response $H_{k\ell}$ already. In the last case the structure has to be excited in one point and the response has to be measured in the same point or another point.
There are exceptions if the points of excitation or response are identical with node points of the natural modes.

Measurements of the frequency response functions. The frequency response functions can be determined by measurements, too.
In Fig. 10 the relationships between input and output of a linear system are represented in the time domain and in the frequency domain. In the time domain the unit impulse response is the connecting function, in the frequency domain the frequency response.
For the determination of the frequency response we take advantage of the

fact that the ratio of the Fourier transformed signals is equal to the frequency response. The signals $\tilde{F}_{\ell}(t)$ are measured in the time domain, transformed to the frequency domain by means of the Fast-Fourier-Transformation and the ratio is calculated. This procedure can be executed by efficient two channel Fourier analyzers.

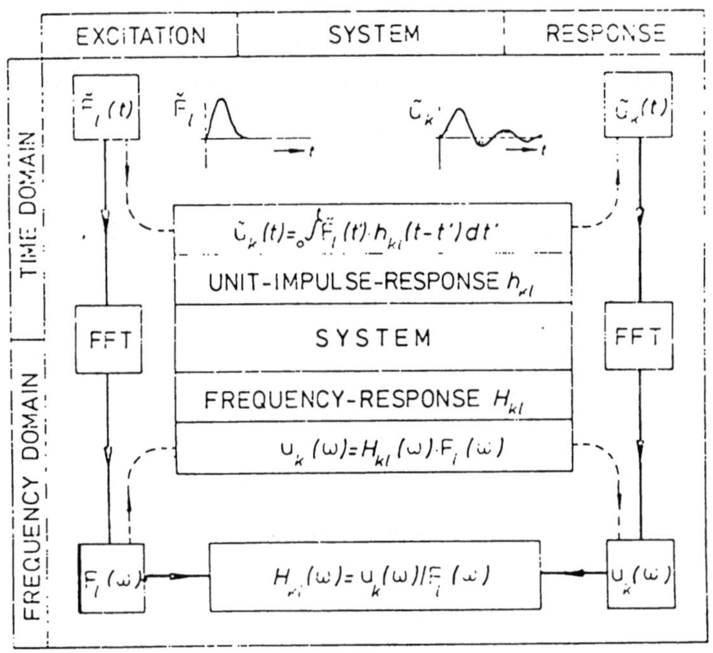

Fig. 10 Input-output relationships

One could excite the system with harmonic forces (Fig. 11). In this case the force signal as well as the response signal are sinusoidal signals and therefore the frequency spectrum of both signals as well as the frequency response have only one frequency line.

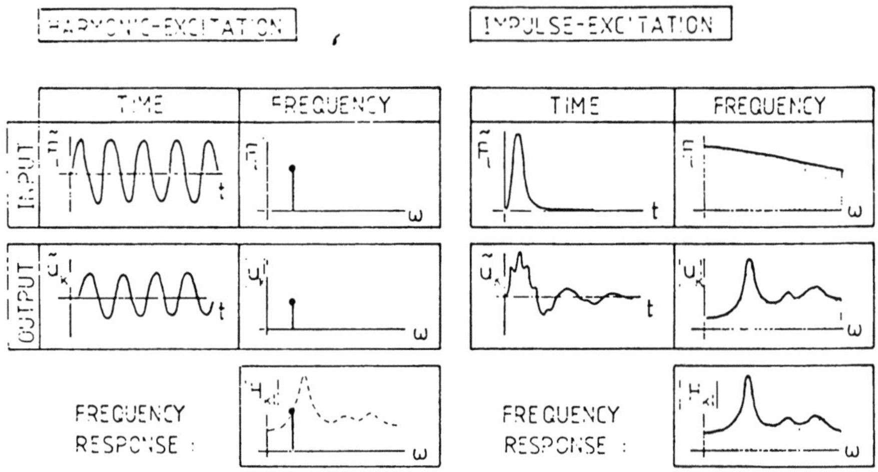

Fig. 11 Harmonic and impulse excitation

Applying forces with broadband characteristic in the frequency domain (impulse, random etc.) is more economical, because all frequencies of excitation are contained at once in a desired frequency range.
If the rotor is excited by an impulse, both the force and the system response are broadband functions in the frequency domain and the measurements can be carried out in a relative short time.
With a short impulse the energy is concentrated in higher frequencies and with a long impulse in lower frequencies. The pulse duration, the frequency content and the amplitude of force can be influenced by selection of a hammer mass, the flexibility of the impact cap and the impact velocity.

Fig. 12 shows in principle the measuring device. A special hammer excites the rotating shaft and the force is measured. The displacements of the shaft are picked up with inductive displacement pick-ups. Force and displacement signals are amplified and after analog-digital-conversion and Fast-Fourier-Transformation the frequency response functions can be calculated. For further treatment the functions are stored on a magnetic tape.

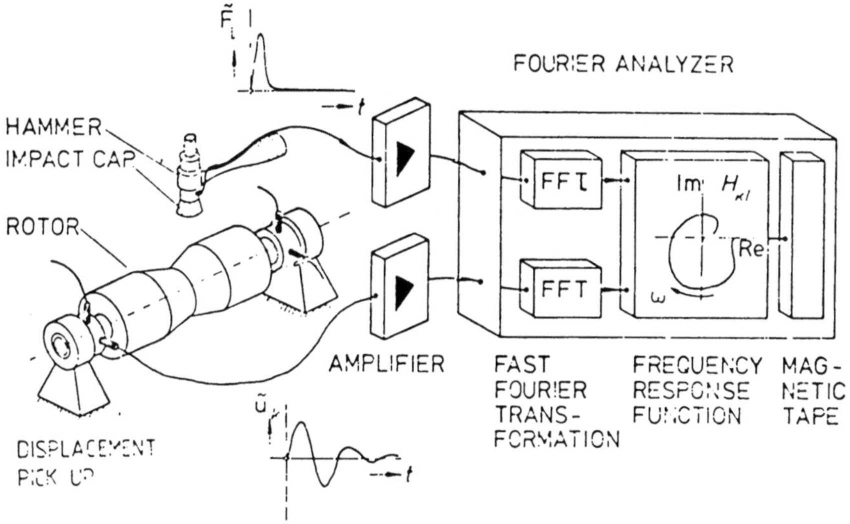

Fig. 12 Measuring device

Determination of modal parameters by curve fitting. For each rotor speed Ω the necessary frequency response functions $H_{k\ell}^{M}(\omega)$ are present from measurements (Fig. 13). The corresponding functions of the mathematical model are $H_{k\ell}^{A}(\omega, \lambda_j, a_{k\ell j})$.

The goal of the curve fitting procedure is finding such modal parameters λ_j, $a_{k\ell j}$ - respectively $\underline{\phi}_j$, $\underline{\psi}_j$ - to get best agreement between the functions $H_{k\ell}^{M}$ and $H_{k\ell}^{A}$.
A possible criterion is the minimization of the scalar function

$$E = \sum_p \{H_{k\ell}^{M}(\omega_p) - H_{k\ell}^{A}(\omega_p, \lambda_j, a_{k\ell j})\}^2 \qquad (22)$$

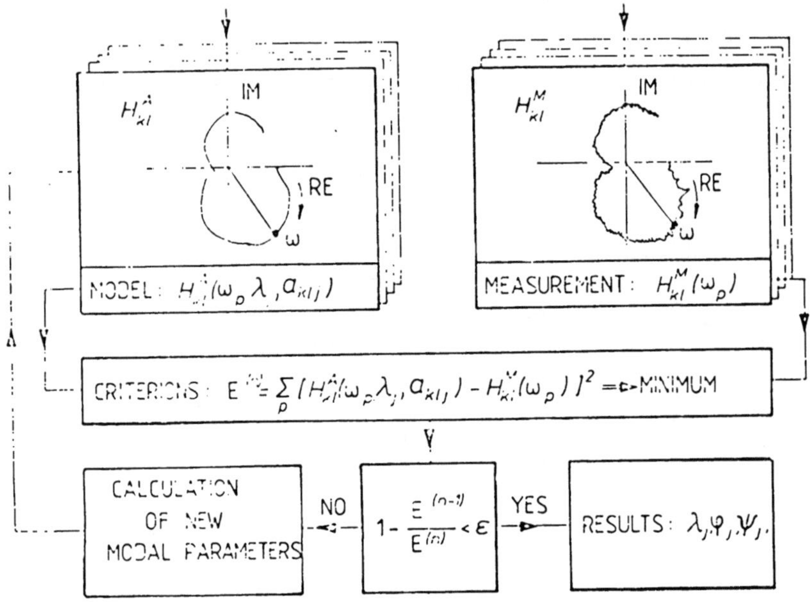

Fig. 13 Curve fitting

This can be interpreted as a least squares criterion for the error. The condition of minimization of the function E leads to equations for the unknowns λ_j and $a_{k\ell j}$. These equations are solved by an iterative procedure (linearization). After each step n the variation of the scalar function E is controlled. If the relative variation of E is less than a number ε, the last values of λ_j, $a_{k\ell j}$ - respectively $\underline{\phi}_j$, $\underline{\psi}_j$ are taken as modal parameters for this speed.
At the beginning of the procedure a starting vector of the unknowns must be chosen.

Critical remarks. Tne above described method finding all modal parameters of a nonconservative rotor is applicable, if all measurement points are accessible - for example in test rotors. In this case eigenvalues λ_j, as well as natural modes (eigenvectors) $\underline{\phi}_j$, $\underline{\psi}_j$ and possibly the system parameters (mass, damping, stiffness) can be determined.

In practise of rotating machinery there are only few points at the rotor, which allow to excite the system and to measure the response during operation. The determination of all modal parameters, the natural modes included, is not possible in this case.
If only the eigenvalues are needed (stability, resonances), one point for excitation and one point for measuring the response are sufficient to determine at least one frequency response and the corresponding eigenvalues.

Difficulties may arise in large turbomachinery, where a hand hammer normally will be to small. Other exciting mechanisms have been developed e.g. magnetic hammers, pneumatic hammers, snapping a loaded strap etc. Other difficulties may occur in the case of eigenvalues with high damping constants (overdamping), which may be equivalent with missing peaks in the frequency response.

4 Example - Eigenvalues of a Test Rotor. For testing the method, measurements were carried out at an excisting test rig: Laval-shaft in two cylindrical journal bearings. This test rotor had only displacement pick ups near the bearings. Therefore the measurement of one column of the frequency response matrix $\underline{H}^{A}$ and the following determination of the natural modes $\underline{\phi}_j$ was not possible. So this investigation was limited to the identification of the eigenvalues and the stability threshold speed of the rotor. The results were compared with theoretical results.

Rotor test rig. Fig. 14 shows the rotor test rig. It consists of a cylindrical shaft (diameter 50 mm, length 1000 mm) with a disk (mass 53 kg, diameter 300 mm, width 100 mm) at the center of the shaft.
A d.c.-electric motor with speed control drives the shaft, which is running in two cylindrical journal bearings with a length-diameter ratio B/D = o.8.
The motor and the movable bearing pedestals are mounted on a concrete foundation with elastic springs. The foundation mass is 6000 kg.
The hammer for pulse excitation has a mass of 1,2 kg. Each bearing has

two displacement pick ups for measuring the displacements in horizontal and vertical directions.

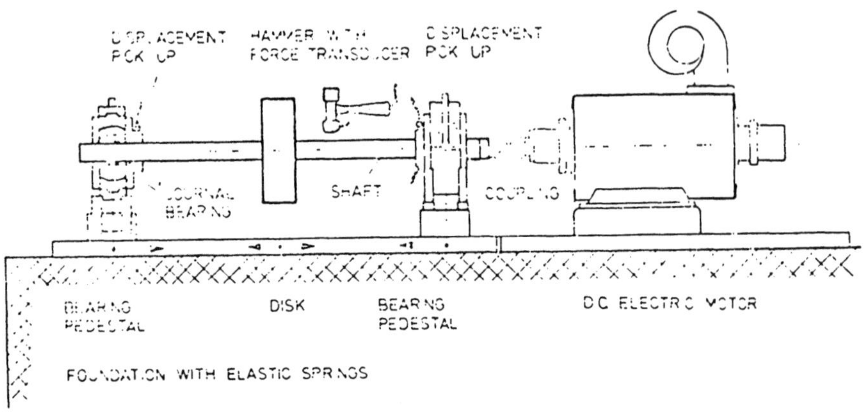

Fig. 14 Rotor test rig

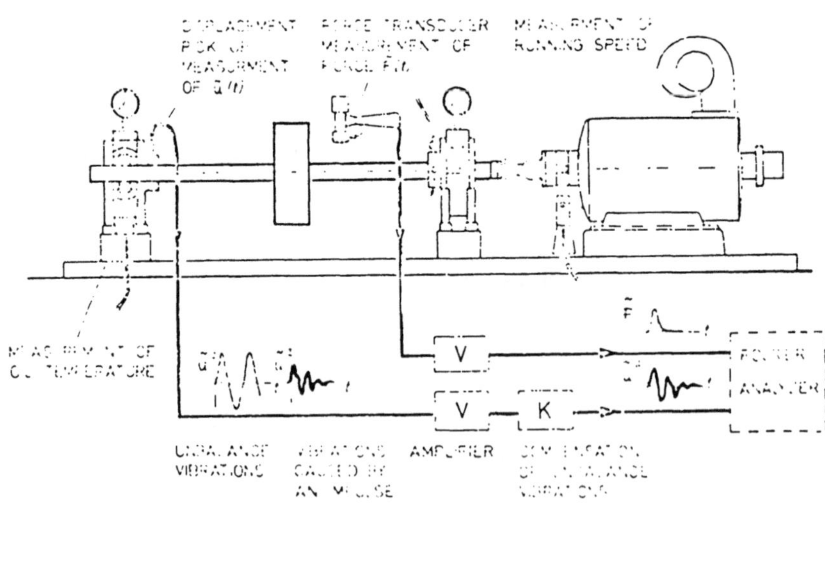

Fig. 15 Measured quantities

Measured quantities. The most important measured quantities are the exciting force and the displacements of the shaft (Fig. 15). The last ones contain besides the impulse response a residual synchronous response caused by unbalance. This part of the signal, which disturbs the signal processing, is eliminated by an electronic circuit. Further measured quantities are the rotor speed and the oil temperature.

Some results. Fig. 16 shows the natural frequencies ω_j and the damping constants α_j versus the running speed Ω. Without the oil film (Ω = o) the natural frequency is 42.5 Hz. Two speed dependent eigenvalues were measured for the rotating shaft in the considered frequency range. The flexibility of the oil film reduces the natural frequencies but the oil film stiffness is relatively high compared with the shaft. Contrary to the damping constant α_2 the damping constant α_1 changes strongly with the running speed Ω. The intersection between the damping constant α_1 and the abscissa establishes the threshold of instability. Instability occurs at a frequency of the rotating shaft 74 Hz with a natural frequency of 32 Hz.
Besides the measured eigenvalues also calculated eigenvalues are plotted in Fig. 16. A finite element program was used for calculation. All important effects were taken into account.
The comparison of measured and calculated values shows a good agreement. Further results are shown in Fig. 17 with other distances between bearings. There is again a good agreement between measurement and calculation.

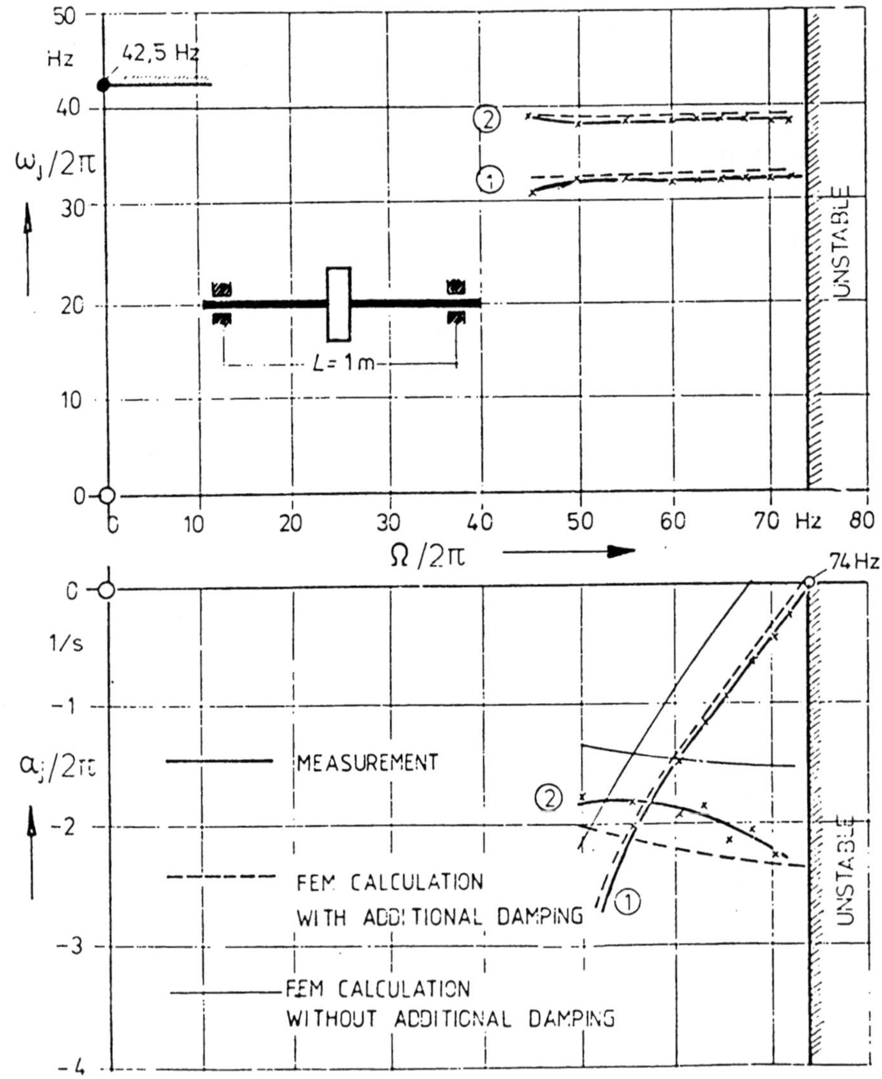

Fig. 16 Eigenvalues of the test rotor

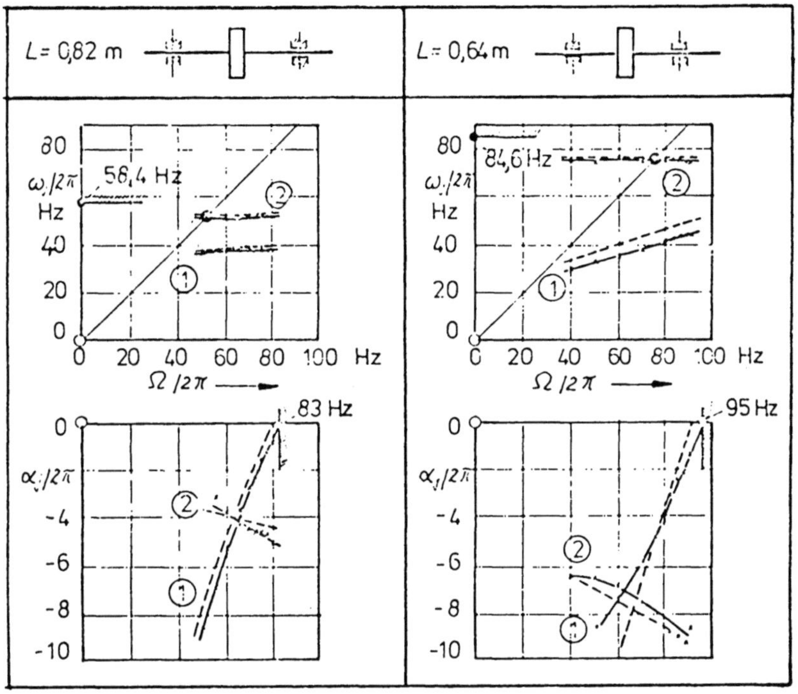

Fig. 17 Eigenvalues of the test rotor for different distances between bearings

CHAPTER 4.4

IDENTIFICATION OF STIFFNESS AND DAMPING COEFFICIENTS OF JOURNAL BEARINGS BY MEANS OF THE IMPACT METHOD

R. Nordmann

1 Introduction

It is well known from the preceding chapters, that vibrations of rotors with journal bearings are influenced by the dynamic behaviour of the oil film. Therefore in rotor dynamics investigations oil film characteristics have to be considered. In linear theory the dynamic behaviour of the journal bearings can be described by four stiffness and four damping coefficients. These coefficients can be found either by calculations or by experimental methods.

A review about calculating and experimental methods for the determination of oil film coefficients is given by Lund [1]. For calculation there exist several numerical methods to solve the Reynolds equations. Concerning experimental methods generally input signals (forces) and output signals (displacements) of the dynamic system are measured and the unknown parameters of the system are calculated by means of input-output-relationships.

Most test rigs have a single test bearing, which floats on a very stiff rotating shaft [2,3]. The test bearing is excited by harmonic forces and the displacements of the test rig are measured. The determination of bearing coefficients is possible when there are at least two independent sets of measurements, which can be obtained by two independent sets of forces.

The here presented method is different in some essential points. It was first of all approved at a small existing test rig. A rigid rotor, running in journal bearings, is excited by a hammer (pulse testing). Input signals (forces) and output signals (displacements of the rotor) are transformed into the frequency domain and the complex frequency response functions are calculated. Analytical frequency response functions, which depend on the bearing coefficients, are fitted to the measured functions. Stiffness and damping coefficients are results of this iterative fitting process.

2 Determination of Stiffness and Damping Coefficients. The new method was approved at a rotor, which is very stiff compared with the stiffness of the bearings. For this real system a linear mechanical system can be modeled. The equations of motion for this mechanical system represent the mathematical model. We suppose that the structure of the mathematical model is known, whereas the parameters - the bearing coefficients - are to be determined. At first the frequency response functions of the real system are measured. Then in an iterative procedure the frequency response functions of the mathematical model are fitted to the measured functions by variation of the bearing coefficients.
For representation of the method we subdivide in the main parts

- Mechanical and mathematical model
- Measurements of the frequency response functions
- Curve fitting (Parameterdetermination).

Mechanical and mathematical model. The mechanical model consists of a symmetrical rigid rotor with mass m, running with angular velocity Ω in two equal journal bearings. The dynamic behaviour of the bearings is characterized by the four stiffness coefficients k_{xx}, k_{xy}, k_{yx}, k_{yy} and the four damping coefficients c_{xx}, c_{xy}, c_{yx}, c_{yy}.

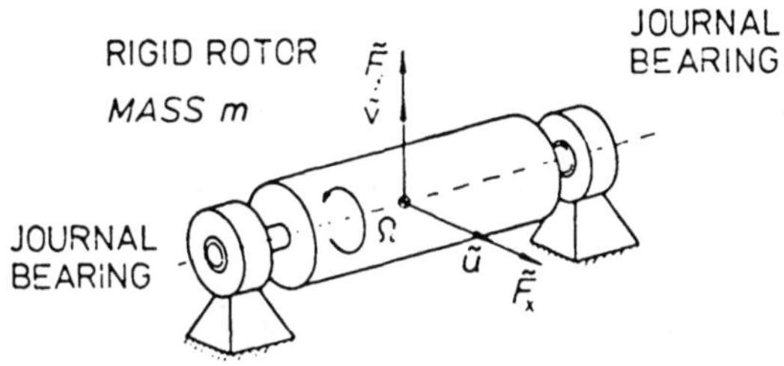

Fig. 1 Mechanical model of a rigid rotor in journal bearings

The motions of the rigid rotor can be described by the displacements $\tilde{u}(t)$ and $\tilde{v}(t)$ of the centre of gravity. Because the exciting forces $\tilde{F}_x(t)$ and $\tilde{F}_y(t)$ are applied only in the centre of the rotor (Fig. 1), the system responds only with translatory motions; rotations about the x-axis and y-axis are omitted. The displacements of the journals are equal to the displacements of the center of gravity.
The equations of motion for the mechanical model describe the equilibrium of the forces of inertia, the oil film forces and exciting forces

$$m\ddot{\tilde{u}} + 2\,(c_{xx}\dot{\tilde{u}} + c_{xy}\dot{\tilde{v}} + k_{xx}\tilde{u} + k_{xy}\tilde{v}) = \tilde{F}_x(t) \tag{1}$$

$$m\ddot{\tilde{v}} + 2\,(c_{yx}\dot{\tilde{u}} + c_{yy}\dot{\tilde{v}} + k_{yx}\tilde{u} + k_{yy}\tilde{v}) = \tilde{F}_y(t)$$

With the following definitions

$$\omega_o^2 = g/\Delta r, \quad \tau = \omega_o t, \quad w = \Omega/\omega_o \tag{2}$$

$$F_{stat} = mg/2, \; H_o = \Delta r/2F_{stat} \; , \; d(\;)/d\tau = (\;)'$$

the equations of motion can be written with the non-dimensional bearing coefficients

$$\begin{aligned} \tilde{u}'' + \frac{\beta_{xx}}{w}\tilde{u}' + \frac{\beta_{xy}}{w}\tilde{v}' + \gamma_{xx}\tilde{u} + \gamma_{xy}\tilde{v} &= H_o\tilde{F}_x \\ \tilde{v}'' + \frac{\beta_{yx}}{w}\tilde{u}' + \frac{\beta_{yy}}{w}\tilde{v}' + \gamma_{yx}\tilde{u} + \gamma_{yy}\tilde{v} &= H_o\tilde{F}_y \end{aligned} \tag{3}$$

The nondimensional bearing coefficients are defined as

$$\gamma_{xx} = k_{xx}\frac{\Delta r}{F_{stat}} \; ; \quad \beta_{xx} = c_{xx}\frac{\Delta r \;\; \Omega}{F_{stat}} \tag{4}$$

For a given bearing these nondimensional coefficients are only dependent on the Sommerfeld number So, respectively on the static equilibrium position of the journal center. We can represent the coefficients in the nondimensional matrices

$$\underline{\gamma} = \begin{bmatrix} \gamma_{xx} & \gamma_{xy} \\ \gamma_{yx} & \gamma_{yy} \end{bmatrix} ; \quad \underline{\beta} = \begin{bmatrix} \beta_{xx} & \beta_{xy} \\ \beta_{xy} & \beta_{yy} \end{bmatrix} \tag{5}$$

If we introduce a Sommerfeld number So_o with angular velocity ω_o, we are able to describe the system by the parameters So_o and w.

Frequency response of a rigid rotor in journal bearings. From the equations of motion the unknowns $\tilde{u},\tilde{v}$ can be computed for different load cases. Applying harmonic forces successively in the two directions, four frequency response functions of the model can be determined. If the exciting force is F_y (with the nondimensional frequency $\eta = \omega/\omega_o$)

$$\tilde{F}_y(\tau) = \hat{F}_y \sin \omega t = \hat{F}_y \sin \frac{\omega}{\omega_o} \omega_o t = \hat{F}_y \sin \eta\tau = \mathrm{IM}\,\{\hat{F}_y e^{i\eta\tau}\} \tag{6}$$

the following statement with the complex amplitude $\hat{u}^*$

$$\tilde{u}(\tau) = \mathrm{IM}\,\{\hat{u}^* e^{i\eta\tau}\} = \mathrm{IM}\,\{\hat{u}\, e^{i\varepsilon_{xy}} e^{i\eta\tau}\} \tag{7}$$

leads to the complex frequency response

$$H_{xy} = \frac{\hat{u}^* e^{i\eta\tau}}{\hat{F}_y e^{i\eta\tau}} = \frac{\hat{u}}{\hat{F}_y} e^{i\varepsilon_{xy}} = V_{xy}\, e^{i\varepsilon_{xy}} \tag{8}$$

which is the response in x-direction caused by a force in y-direction. We obtain four frequency response functions

$$\begin{aligned}
H_{xx} &= V_{xx}\, e^{i\varepsilon_{xx}} = (\gamma_{yy} - \eta^2 + \frac{\beta_{yy}}{w}\eta i)\; H_o/N(\eta) \\
H_{xy} &= V_{xy}\, e^{i\varepsilon_{xy}} = -\,(\gamma_{xy} + \frac{\beta_{xy}}{w}\eta i) \cdot H_o/N(\eta) \\
H_{yx} &= V_{yx}\, e^{i\varepsilon_{yx}} = -\,(\gamma_{yx} + \frac{\beta_{yx}}{w}\eta i) \cdot H_o/N(\eta) \\
H_{yy} &= V_{yy}\, e^{i\varepsilon_{yy}} = (\gamma_{xx} - \eta^2 + \frac{\beta_{xx}}{w}\eta i)\; H_o/N(\eta)
\end{aligned} \tag{9}$$

with the denominator

$$
\begin{aligned}
N(\eta) = {} & (\gamma_{xx} - \eta^2 + \frac{\beta_{xx}}{w}\eta i)\,(\gamma_{yy} - \eta^2 + \frac{\beta_{yy}}{w}\eta i) \\
& - (\gamma_{xy} + \frac{\beta_{xy}}{w}\eta i)\,(\gamma_{yx} + \frac{\beta_{yx}}{w}\eta i)
\end{aligned}
\tag{10}
$$

The four functions contain the ratios of the amplitudes V_{xx}, V_{xy}, V_{yx}, V_{xy} - the amplitude frequency characteristics - and the phase frequency characteristics ε_{xx}, ε_{xy}, ε_{yx}, ε_{yy}.

In the following the frequency response functions will be represented as nondimensional functions, for example

$$
H_{xy} = \frac{H_{xy}}{H_o} = \frac{V_{xy}}{H_o}\, e^{i\varepsilon_{xy}} = \bar{V}_{xy} e^{i\varepsilon_{xy}} \tag{11}
$$

Fig. 2 shows as an example the four amplitude characteristics $\bar{V}_{xx}$, $\bar{V}_{xy}$, $\bar{V}_{yx}$, $\bar{V}_{yy}$ for a rotor with cylindrical bearings B/D = 0.8 and constant parameters So_o and w.

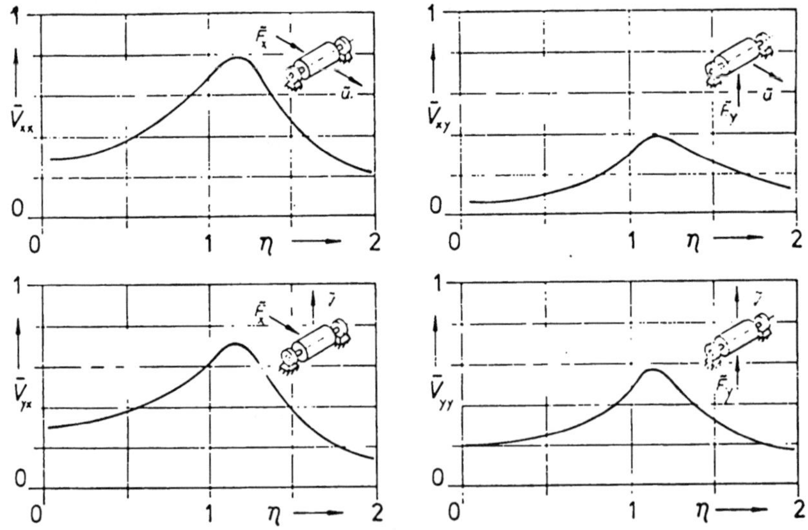

Fig. 2 Amplitude frequency characteristics of a rigid rotor in journal bearings

Measurements of the frequency response functions. The frequency response functions can be determined by measurements. Similar to the calculation procedure, one could excite the system with harmonic forces successively. Working with impulse forces instead of harmonic forces is more economical, because an impulse force contains all exciter frequencies at once in a determined frequency range.
For the determination of the frequency from measured input and output signals we take advantage of the fact that the quotient of the Fourier transformed signals is equal to the frequency response. The signals $\tilde{F}_x(t)$ $\tilde{F}_y(t)$, $\tilde{u}(t)$, $v(t)$ are measured in the time domain, transformed to the frequency domain by means of the Fast Fourier Transformation and the respective quotient is calculated.
Fig.3 shows in principle the measuring equipments. A special hammer excites the rotating shaft and the acceleration of the hammer is measured. Under certain assumptions the acceleration is a measure of the impact force.

With a short pulse the energy can be concentrated in higher frequencies and with a long pulse in lower frequencies. The pulse duration, the frequency content and the amplitude of force can be influenced by selection of the hammer mass, the flexibility of the impact cap and the impact velocity.

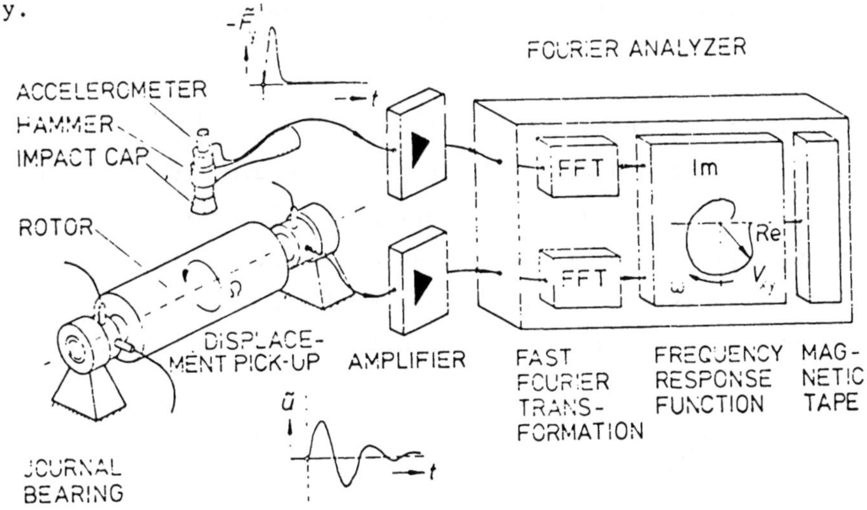

Fig. 3 Equipment for frequency response measurements

The pulse duration increases with hammer mass and flexibility of the impact cap. Correspondingly the frequency content decreases with hammer mass and the flexibility of the cap. The force amplitude increases with the hammer mass and the impact velocity, but decreases with the flexibility of the cap.

With plastic material at the contact surface the friction force between hammer and shaft is powerfully reduced.
The displacements of the shaft are measured with inductive displacement pick ups near the bearing location. Force and displacement signals are amplified and after analog-digital-converting and Fast Fourier Transformation the frequency response functions can be calculated. For further treatment the functions are stored on a magnetic tape.

Curve fitting. For determination of the parameter matrices $\underline{\gamma}$ and $\underline{\beta}$ one does not need to work with the complete frequency response functions (amplitude frequency and phase frequency characteristics), because all of the unknown parameters γ_{xx}, γ_{xy}, γ_{yx}, γ_{yy} and β_{xx}, β_{xy}, β_{yx}, β_{yy} are already contained in the amplitude frequency characteristics $\bar{V}_{xx}$, $\bar{V}_{xy}$, $\bar{V}_{yx}$, $\bar{V}_{yy}$. Working only with these four functions the numerical effort in the curve fitting procedure is reduced. It could be shown by a numerical simulation that this simplification gives good results for the unknown parameters.

For each rotor speed the four functions $\bar{V}_{xx}^{M}$, $\bar{V}_{xy}^{M}$, $\bar{V}_{yx}^{M}$, $\bar{V}_{yy}^{M}$, are known from measurements (Fig. 4). The corresponding functions of the mathematical model depend on the bearing coefficients. The goal of the curve fitting procedure is finding such elements of the matrices $\underline{\gamma}$ and $\underline{\beta}$ to get best agreement between calculated and measured functions. A possible criterion is the minimization of the scalar function

$$Z = \sum_{i=x,y} \sum_{k=x,y} \sum_{p} \{\bar{V}_{ik}(\eta_p,\underline{\gamma},\underline{\beta}) - \bar{V}_{ik}^{M}(\eta_p)\}^2 \tag{12}$$

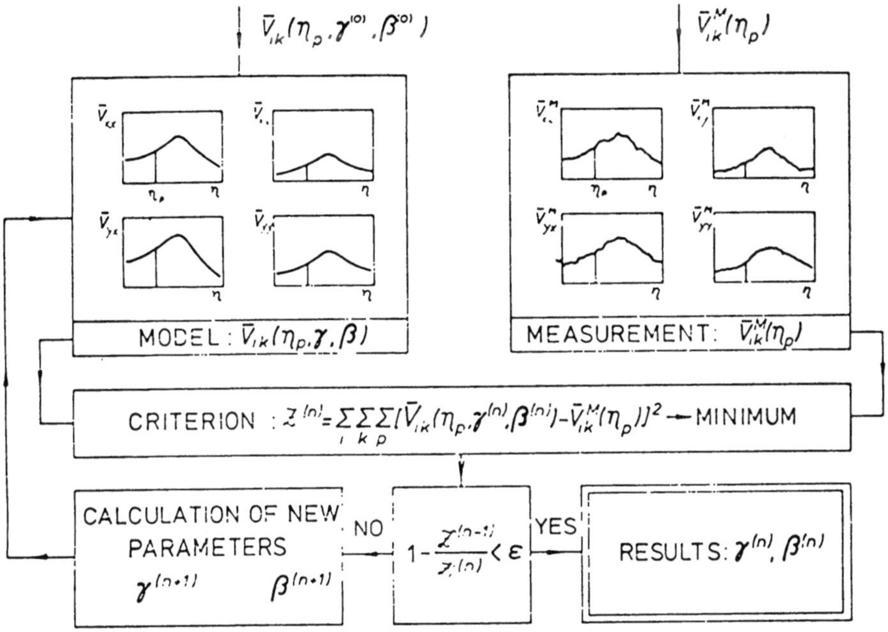

Fig. 4 Determination of bearing coefficients

This can be interpreted as a least squares criterion for the error. The condition of minimization of the function Z leads to nonlinear equations for the unknowns $\underline{\gamma}$ and $\underline{\beta}$. These equations are solved by an iterative procedure (linearization). After each step n the variation of the scalar function Z is controlled (Fig. 4). If the n-th step the relative variation of F is less than a number ε, the last values of $\underline{\gamma}$ and $\underline{\beta}$ are taken as stiffness and damping coefficients for this case.

At the beginning of the procedure a starting vector of the unknowns $\underline{\gamma}^{(o)}$ and $\underline{\beta}^{(o)}$ must be chosen. It can be found from the measurement curves.

3 Measurements at a Rotor Test Rig

Rotor test rig. Fig. 5 shows the rotor test rig. The rotor in his middle range (length 475 mm) has a diameter of 150 mm, the diameter of the journals is 50 mm. A d-c electric motor with speed control drives the shaft, which is running in two cylindrical bearings with B/D = 0.8.

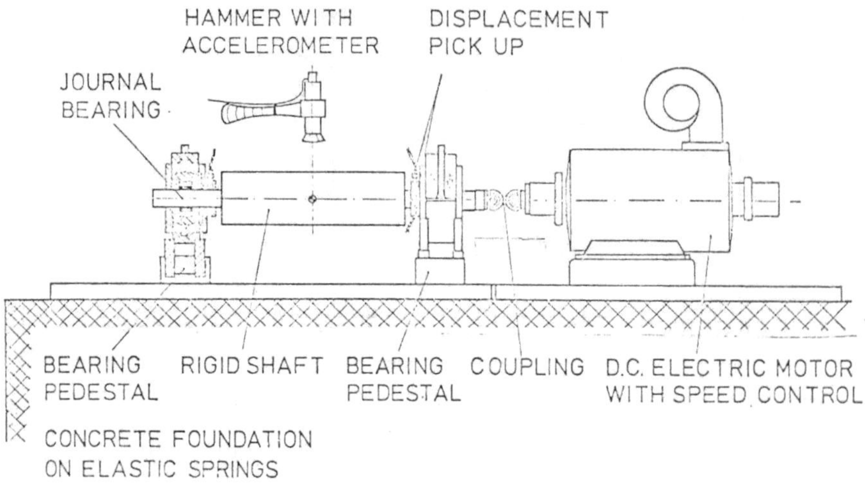

Fig. 5 Rotor test rig

The radial clearance of the bearings is 210 μm. The employed oil has a lubricant viscosity of $9{,}5.10^{-2}$ Ns/m^2 (20°C). The average bearing load is 1,75 bar. The oil temperature can be regulated in a range 15°-60°C. The motor and the bearing pedestals are mounted on a concrete foundation with elastic springs. The foundation mass is 6000 kg.

The hammer for pulse excitation has a mass of 1,2 kg. Each bearing has two inductive displacement pick ups for measuring the displacements in horizontal and vertical directions.

Measured quantities. The most important measured quantities are the exciting force and the displacements of the shaft. The last ones contain besides the pulse response a residual synchronous response caused by unbalance. This part of the signal, which disturbes the signal processing, is eliminated by an electronic circuit. Further measured quantities are the rotor speed, the outlet temperature of oil and the pressure of the oil at the bearing inlet.

Some results of bearing coefficients. For the described rotor in cylindrical bearings measurements were carried out in a speed range from 1500 to 5400 rev/min (maximum of motor speed). For each constant speed the four functions V_{xx}^M, V_{xy}^M, V_{yx}^M, V_{yy}^M and the other measured quantities as speed, oil temperature etc. were determined.
The values of γ_{xx}, γ_{xy}, γ_{yx}, γ_{yy} and β_{xx}, β_{xy}, β_{yx}, β_{yy} were fitted and the corresponding Sommerfeld number was calculated. In Fig. 6 and 7 the nondimensional damping and stiffness coefficients are represented versus the Sommerfeld number So.

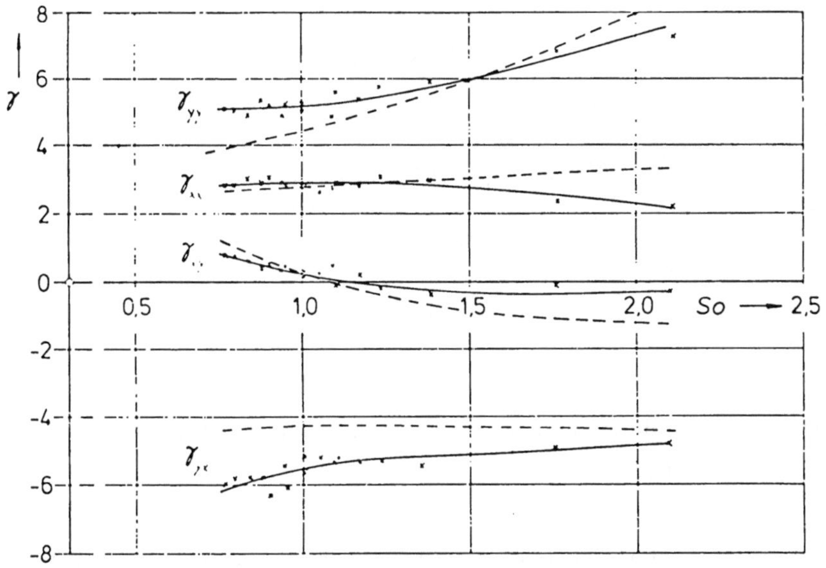

Fig. 6 Nondimensional stiffness coefficients of a cylindrical bearing B/D = 0.8

Because of the speed limitation and in default of variation possibilities for other parameters, Sommerfeld numbers less than 0.75 could not be realized at this rotor test rig.
Figures 6 and 7 show that the scatter is relatively small for the stiffness coefficients and greater for the damping coefficients, especially for β_{yx} and β_{yy}. Besides the measured bearing coefficients earlier results from Glienicke for the same type of bearing are represented with

dashed lines. The agreement is relatively good, only some of the values have greater deviations.

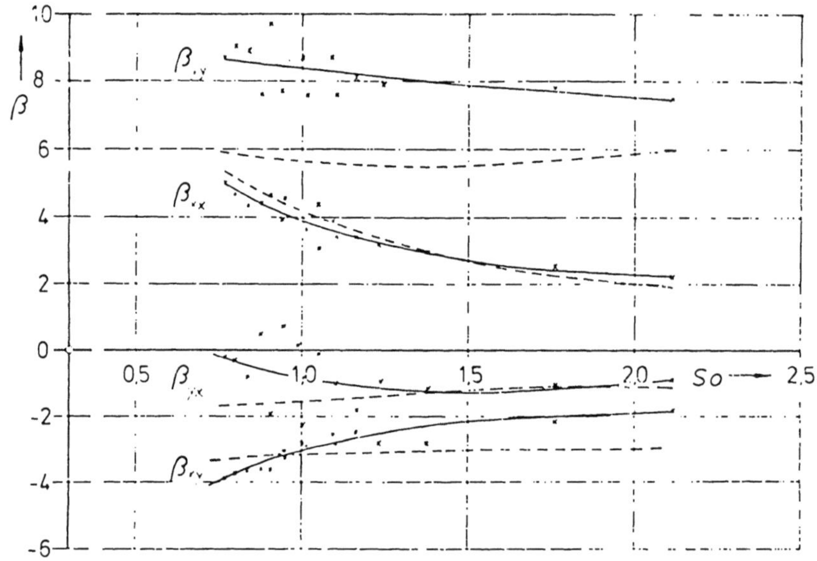

Fig. 7 Nondimensional damping coefficients of a cylindrical bearing B/D = 0.8

4 Control of the results. A possible testing of the results can be realized by driving the real system and the mechanical model with same system parameters by the same input and by studying the different behaviour of the two systems.
In our case the amplitude frequency characteristics and the essential eigenvalue of the two systems are compared.

Comparison of the amplitude frequency characteristics. For given system parameters So_0 and w the four functions $\bar{V}_{xx}$, $\bar{V}_{xy}$, $\bar{V}_{yx}$, $\bar{V}_{yy}$ of the mechanical model are calculated, employing the measured stiffness and damping coefficients. Then at the rotor test rig the four functions are measured for the same parameters So_o and w. In Fig. 8 we compare the measured and calculated functions for the parameters So_o = 1,65 and w = 1,75. There is a good agreement for all four functions. Certainly we have to notice that such functions are compared, which were used to determine the bearing coefficients in a fitting process with the criterion to get best agreement between calculation and measurement.

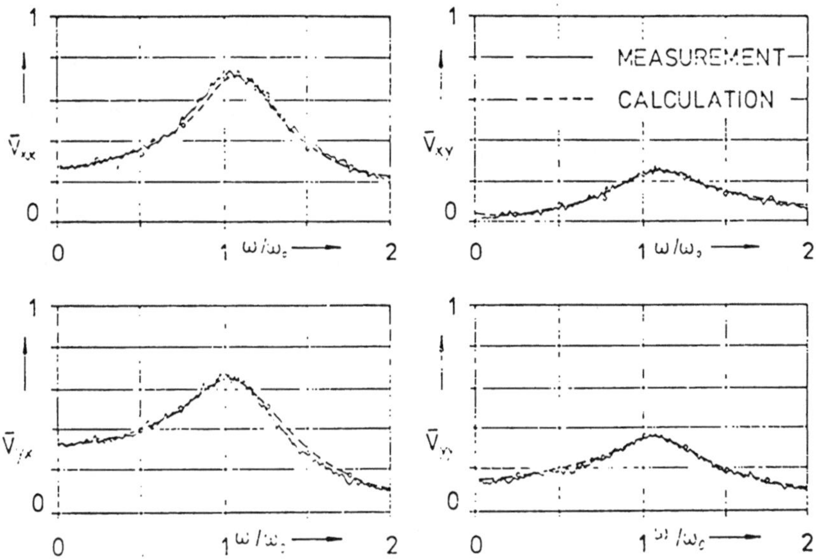

Fig. 8 Comparison of amplitude frequency characteristics (So_o = 1,65, w = 1,75)

Comparison of measured and calculated eigenvalues. If we replace the dimensionless frequency ηi in the denominator $N(\eta)$ (eq. 10) by the complex number λ, $N(\lambda)$ is the equation for determination of the system eigenvalues. The eigenvalues are either real or complex. If they are complex they appear in conjugate complex pairs: $\lambda_j = \alpha_j \pm i\omega_j$ with the decay constant α_j and the circular natural frequency ω_j.

The rigid rotor in journal bearings has four eigenvalues, but two of them have only little practical meaning. The corresponding eigensolution decays very rapidly. The remaining eigenvalues generally are conjugate complex, we call them $\lambda_1 = \alpha_1 + i\omega_1$ and $\bar{\lambda}_1 = \alpha_1 - i\omega_1$. Their corresponding eigensolution is a decreasing or increasing oscillatory motion, which is observable.

For a given system parameter $So_o = 1{,}65$ the eigenvalues λ_1, $\bar{\lambda}_1$ first of all are calculated for different rotor speeds, employing again the measured bearing coefficients. The experimental determination of the eigenvalues λ_1, $\bar{\lambda}_1$ can be realized in the time domain by measuring the decay rate and the natural frequency directly from the natural vibrations. On the other hand calculation of the eigenvalues from measured frequency response curves in the frequency domain is possible.

Figures 9 and 10 show the measured and calculated (dashed line) non-dimensional decay constant α_1/ω_o and the non-dimensional natural frequency ω_1/ω_o versus the rotor speed Ω/ω_o. Again there is a good agreement between measured and calculated values.

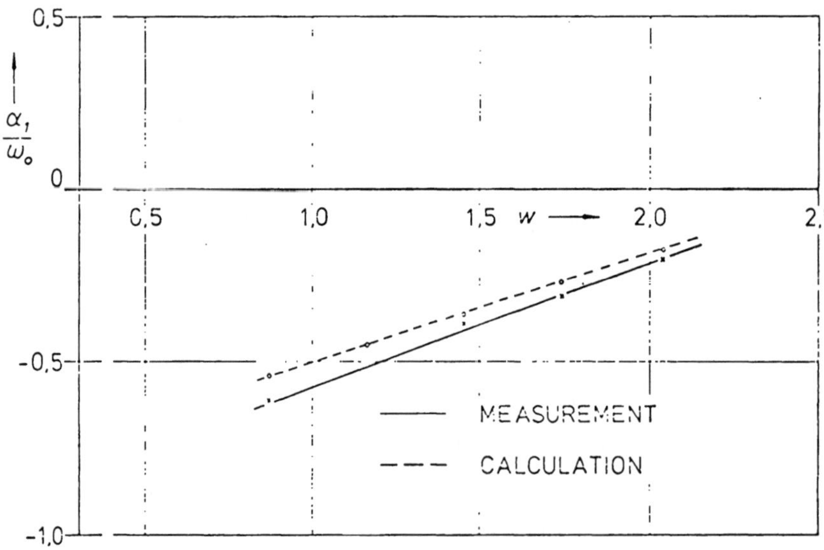

Fig. 9 Nondimensional decay constant of a rigid rotor in cylindrical bearings B/D = 0,8; $So_o = 1{,}65$

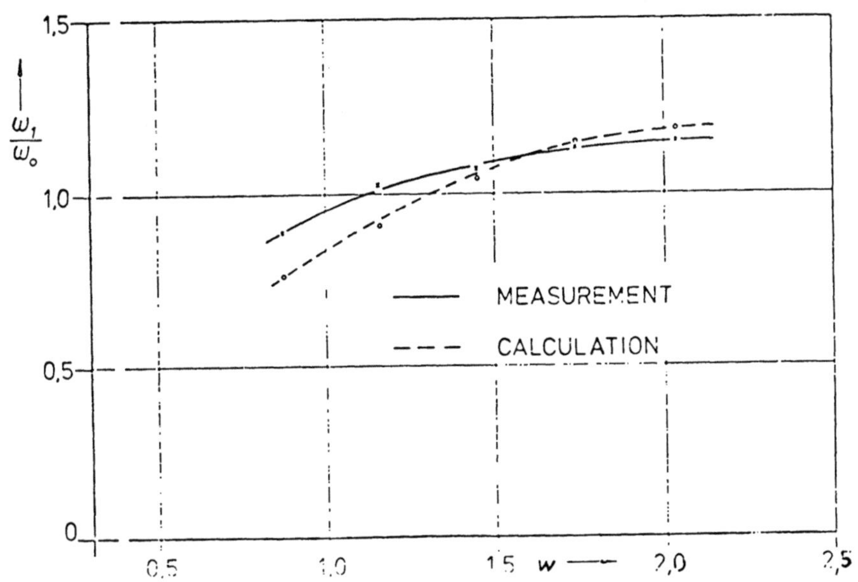

Fig. 10 Nondimensional natural frequency of a rigid rotor in cylindrical bearings B/D = 0.8; So_o = 1.65

References:

[1] Lund, J.W.: Evaluation of stiffness and damping coefficients for fluid film bearings. Shock and vibration digest (1978).

[2] Glienicke, J.: Feder- und Dämpfungskonstanten von Gleitlagern für Turbomaschinen und deren Einfluß auf das Schwingungsverhalten eines einfachen Rotors. Dissertation TH Karlsruhe, (1966).

[3] Mitchell, J.R.; Holmes, R. and von Ballegooyen H.: Experimental determination of a bearing oil film stiffness. Instn. Mech. Engr. Proc. 18/, Pt. 3k, (1965-1966) pp 90-96.

CHAPTER 4.5

EXPERIMENTAL DETERMINATION OF BEARING STATIC PROPERTIES

Z.A. Parszewski

As it was explained in chapter 2.1 the bearing static characteristics is given by the relations

$$S = S(c); \qquad c = c(\alpha) \tag{1}$$

giving Sommerfield's number S as function of the eccentricity ratio c and corresponding locus of equilibrium positions of the journal centre in the bearing.

Following gives the results of computation as well as experimental results obtained on a laboratory research stand (Fig. 1) used for experimental measurement of static and dynamic characteristics of bearings ([6], [7], [8]chapter 3.1 and [4] chapter 2.4). The bearing casing 1 together with the sleeves (Fig. 2) is suspended on a rigidly supported rotating shaft 2. Static load P_o is applied by means of the springs and screws and the static displacement components

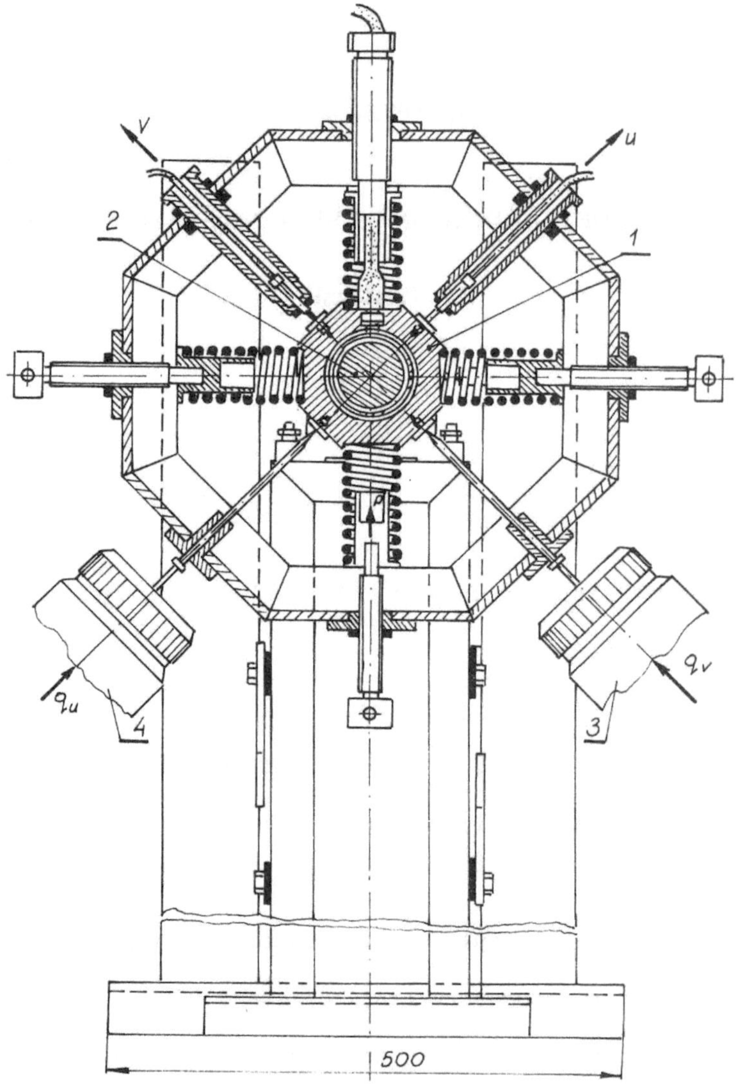

Fig. 1

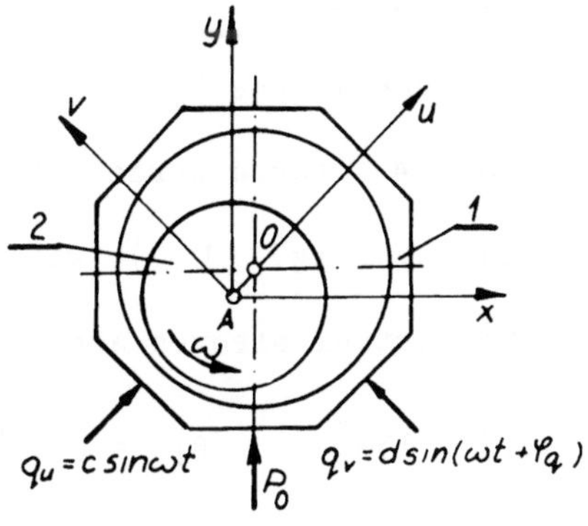

Fig. 2

u_o, v_o are measured at various shaft speeds.

The displacement components u, v are measured by two electrodynamic transducers, orientated at 45° to the load direction (Figs. 1 and 2). These displacements give the following components in the x and y directions

$$x_o = \frac{\sqrt{2}}{2}(u_o - v_o) \; ; \qquad y_o = \frac{\sqrt{2}}{2}(u_o + v_o) \tag{2}$$

The eccentricity ratio c and the angle α_o describing the journal centre position in equilibrium are hence

$$\alpha_o = 2\pi - \delta_o - \mathrm{arctg}\,\frac{y_o}{x_o} \; ; \qquad c = \frac{\sqrt{x_o^2 + y_o^2}}{r} \tag{3}$$

Nondimensional load capacity S of a bearing, defined as nondimensional load (P_o) corresponding to the equilibrium position of the journal at a given speed is

$$S = \frac{P_o(\frac{r}{R})^2}{2RL\eta} \tag{4}$$

The described measurements give hence the locus of journal centre equilibrium positions in the bearing and corresponding nondimensional load capacity (as function of eccentricity ratio)

$$c = c(\alpha_o) \; ; \qquad S = S(c) \tag{5}$$

These relations represent the static characteristics of the bearing.

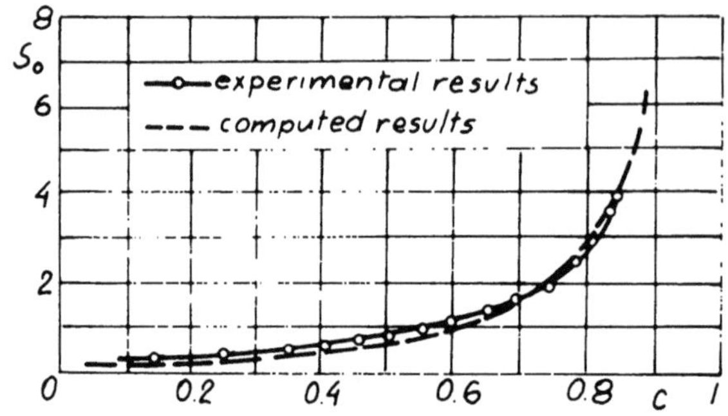

Fig. 3

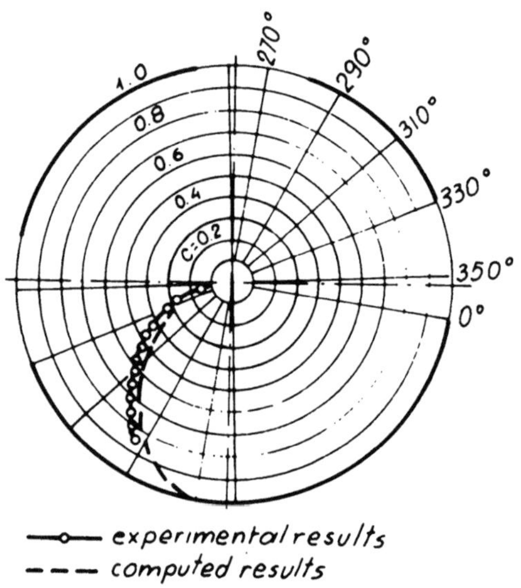

Fig. 4

This characteristics [7] is given in the Figs. 3 and 4 for a three lobe bearing (Fig. 1), with the partial sleeves arcs of

$\theta_{11} = 150^\circ$ $\theta_{12} = 65.5^\circ$

$\theta_{13} = 47^\circ$ and

$$\frac{L}{2R} = \frac{L}{D} = 0.83$$

Figures 3 and 4 give also the comparison of experimental (continuous lines) and computed (broken lines) results.

Corresponding curves [8], for bearings with eccentric sleeves (Fig. 5) are given in Figs. 6 and 7 .

The parameter $f = f_1(f_1 + r_1)$ characterises the eccentricity of the sleeves (Fig. 5).

Bearings as described are used for example, in the generators of Dolmel-Wroclaw production.

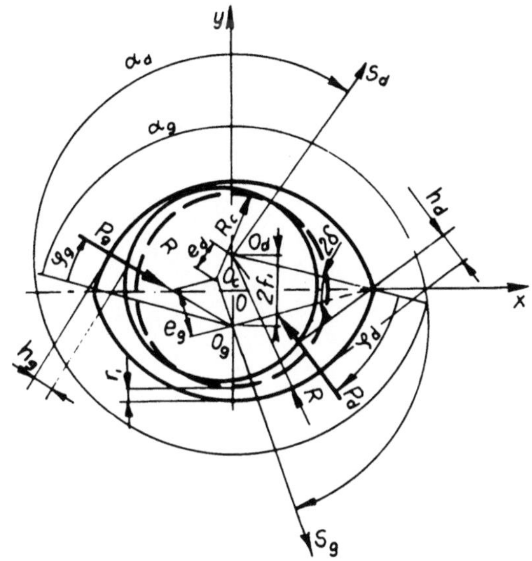

Fig. 5

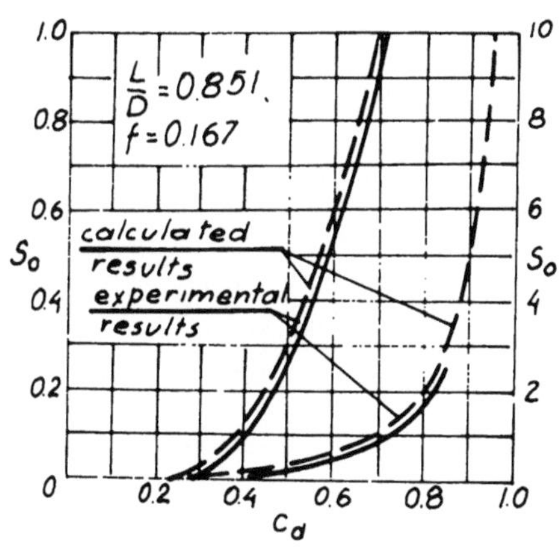

Fig. 6

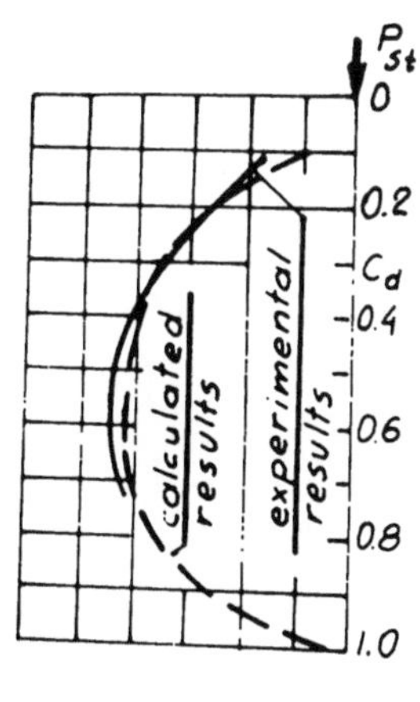

Fig. 7

CHAPTER 4.6

EXPERIMENTAL DETERMINATION OF BEARING DYNAMIC PROPERTIES

Z.A. Parszewski

The bearing dynamic characteristics (chapter 2.1) is given by the stiffness and damping matrices

$$[K] = \begin{bmatrix} K_{xx} & K_{xy} \\ K_{yx} & K_{yy} \end{bmatrix} \tag{1}$$

$$[C] = \begin{bmatrix} C_{xx} & C_{xy} \\ C_{yx} & C_{yy} \end{bmatrix} \tag{2}$$

The elements of both the matrices are functions of the steady state (equilibrium) parameter (hence of eccentricity ratio).

These coefficients (elements) can be obtained experimentally in an indirect way on the experimental stand (Fig. 1) described in the preceeding chapter ([6],[7], [8] chapter 3.1 and [4] chapter 2.4.).

Vibrations are excited at the bearing casing around each equilibrium position (eccentricity ratio c at a given load P_o and speed ω) by use of the electromagnetic exciters 3 and 4 (Fig. 1).

The exciting forces q_u and q_v and corresponding dynamic displacements u and v of the form

$$u = a \sin (\omega t + \theta) \; ; \quad v = b \sin (\omega t + \theta) \tag{3}$$

are measured for vibrations excited around consecutive journal centre equilibrium positions, given by the static characteristics $c = c(\alpha_o)$ of the bearing.

Elements of the stiffness and damping matrices are found from the set of equations of vibrations of the bearing casing

$$m\ddot{x} + \overline{K}_{xx}x + \overline{K}_{xy}y + \overline{C}_{xx}\dot{x} + \overline{C}_{xy}\dot{y} = q_x \tag{4}$$

$$m\ddot{y} + \overline{K}_{yx}x + \overline{K}_{yy}y + \overline{C}_{yx}\dot{x} + \overline{C}_{yy}\dot{y} = q_y$$

written for two different sets of measured data, corresponding to two different sets of exciting forces q.

m - is the mass of the bearing (the casing together with the sleeves) and

$$q_x = \frac{\sqrt{2}}{2}(q_u - q_v) \;; \qquad q_y = \frac{\sqrt{2}}{2}(q_u + q_v)$$

$$x = \frac{\sqrt{2}}{2}(u - v) \;; \qquad y = \frac{\sqrt{2}}{2}(u + v) \tag{5}$$

Nondimensional stiffness and damping coefficients of the bearing oil film

$$K = \frac{Sr}{P_o}\overline{K} \;; \qquad C = \frac{S\omega}{P_o}\overline{C} \tag{6}$$

are given ([7] chapter 2.1) in Fig. 1 and Fig. 2 for the three lobe bearing (Fig. 1 Ch. 2.1) with the partial sleeves arcs

$\theta_{11} = 150^o$; $\theta_{12} = 65.5^o$; $\theta_{13} = 47^o$ and $\frac{L}{2R} = \frac{L}{D} = 0.83$.

These figures give also the comparison of experimental (continuous lines) and computed (broken lines) results.

Corresponding curves ([8] chapter 2.1) for bearings with eccentric sleeves (Fig. 5 Ch. 4.5) are given in Figs. 3 and 4 .

The parameter $f = f_1/(f_1 + r_1)$ characterises the eccentricity of the sleeves (Fig. 5 Ch. 4.5).

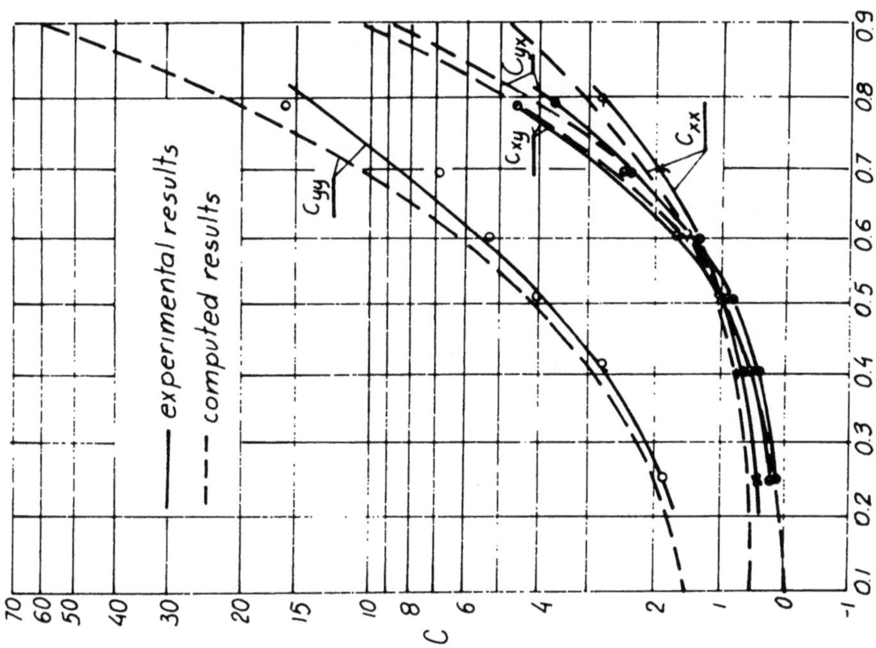

Fig. 2

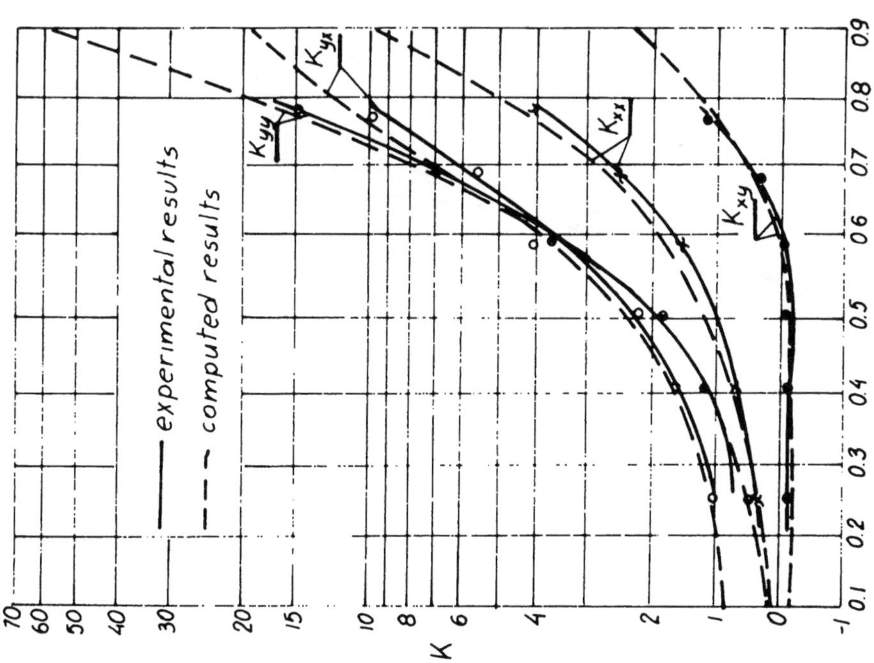

Fig. 1

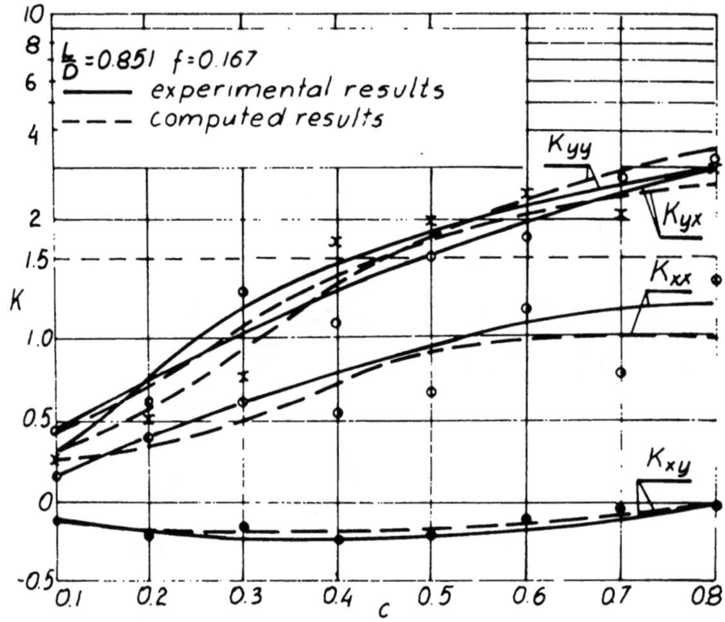

L/D = 0.851 f = 0.167
experimental results
computed results
K_yy
K_yx
K_xx
K_xy
K
c

Fig. 3

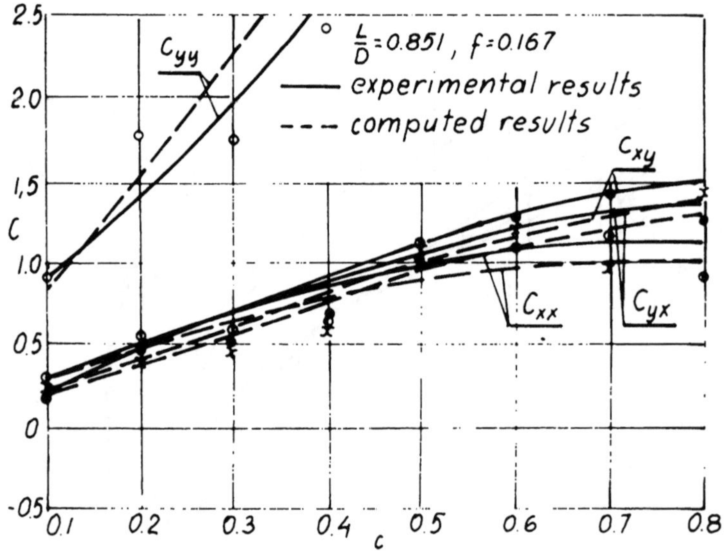

L/D = 0.851, f = 0.167
experimental results
computed results
C_yy
C_xy
C_xx
C_yx
C
c

Fig. 4

PART V

MISCELLANEOUS TOPICS

CHAPTER 5.1

THE VIBRATIONAL BEHAVIOUR OF A ROTATING SHAFT CONTAINING A TRANSVERSE CRACK

B. Grabowski

Part 1

Crack Models and the Vibration of a Rotor with Single Disk on Massless Shaft Containing a Transverse Crack

1. Introduction

Frequently cracks in turbine rotors were found. Nevertheless, until now it is not really known, how cracks can be recognized early enough so that large consecutive damage can be prevented.

The latest greater damage in Germany occured in the nuclear power plant Würgassen. For one year the plant had to be put out of operation because two new rotors had to be produced. Both the LP-rotors have had a crack in the middle of the shaft. At running speed there was noticed a higher level of vibration amplitudes and during rundown the resonance

*) Since 1981 companion of the "Ingenieurbüro für Maschinendynamik, MAHRENHOLTZ + PARTNER", 3000 Hannover 1, Freundallee 23, RF Germany

amplitudes were very large, but these vibrations have not been connected with a crack. The crack has been discovered accidentally. This makes clear, how less known is the result of a crack on the vibrational behaviour of a shaft. Certainly there has to be take into account, that in practice cracks are comparatively rare. On the other hand in 1922 Stodola [1] has already on principle shown the effect of a crack and these investigations were continued by other authors [2,3,4].

As far as known the first measurement results for a rotating shaft with a crack were published by Mayes and Davies [17] in September 1980. Like in Würgassen an abnormal increase in shaft vibration was noticed at many other cracked rotors. But in some cases the warning signs were too small and some rotor broke, in some cases turbine plants virtually exploded and fragments of the shafts flew away up to some hundred meters.

Henry and Okah-Avae [7] also present cases, in which deep cracks have been found without any influence in vibration amplitudes.

The cause for shaft vibrations due to a crack is the asymmetric cross section at the crack position in connexion with the self-weight of the shaft.

Well known is the effect of an asymmetric cross section of shafts of bioplar generators on the vibrational behaviour. Kellenberger [2] probably was the first who investigated such a system in 1958 a paper of later date is published in January 1980 [16] by Inagaki et.al.

2. Crack Models

2.1 Gaping Crack. First of all the most simple assupmtion is an always gaping crack so that the stiffness of the shaft in a body fixed coordinate system does not change. On principle this is the same as with a shaft of a bipolar generator.

Fig. 1 shows the static deflection of a shaft with gaping crack due to the self-weight. Because of its dimension size the shaft can be assumed as a simple model for a large turbine rotor. Refered to a mean static deflection in the vertical direction the shaft lifts and goes down two times per revolution according to the effective moment of inertia resp. the stiffness. At each angle of rotation the shaft has a horizontal deflection except at vertical or horizontal directions of the

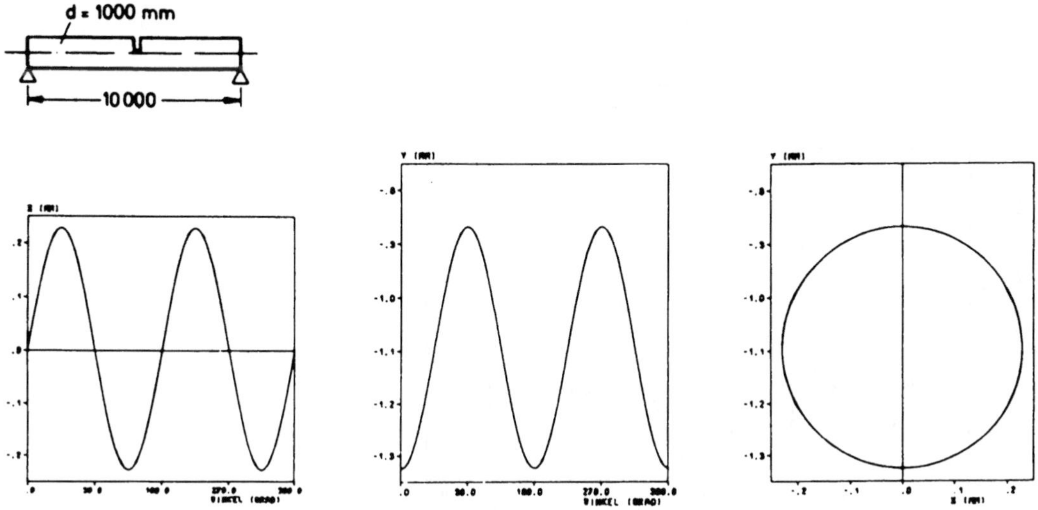

Fig. 1 Self-weight deflection at shaft center for a gaping crack of 50 percent depth

main axes and this two times per revolution too. This causes a superposition of the mean static deflection with a circular motion. The shaft moves on the circular orbit twice per revolution. However, this is valid only for slow rotations, at faster motions the effect of inertia of the mass of the shaft has to be regarded.

2.2 Breathing Crack. A real crack will show another behaviour. Depending on the self-weight of the shaft the crack area opens and closes [Fig. 2]. For a location above the horizontal diameter, the crack is subjected to only compression and the entire cross-section is supporting. At further rotation a part of the crack area opens.

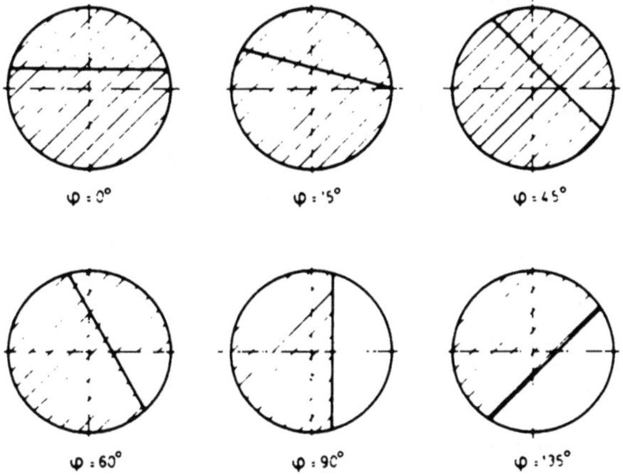

Fig. 2 Crack model cross-sections showing rotation angle-dependent stressed regions (shaded areas). After Grabowski [15].

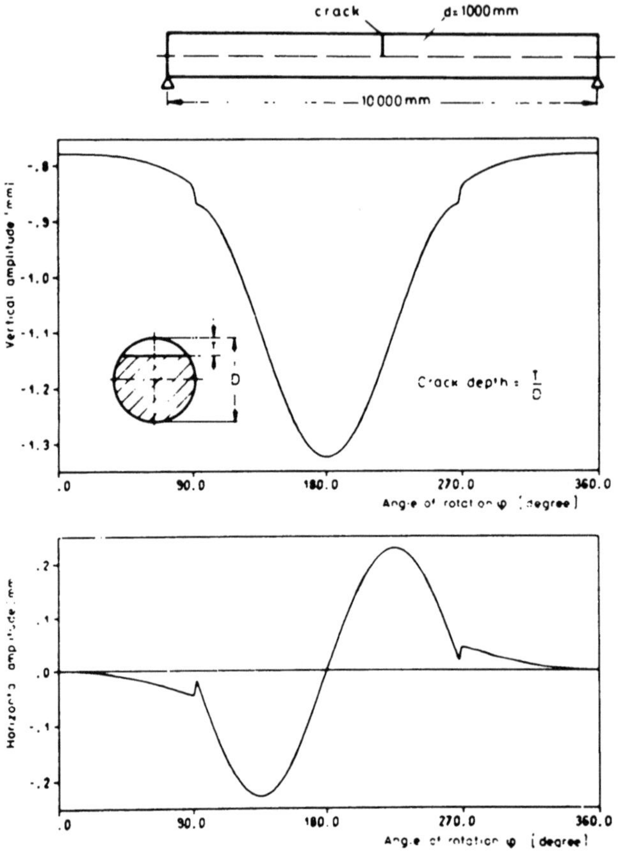

Fig. 3 Self-weight deflection at shaft center for a "breathing" crack of 50 percent depth. After Grabowski [15].

In Fig. 3 the crack is closed at the angle of φ = 0°. When the shaft is rotating the crack area opens only slowly. At the angle of φ = 90° it is assumed that the crack is suddenly gaping completely. That is the cause for the point of discontinuity at φ = 90° and φ = 270°. But this assumption has praktically no influence on one vibrational behaviour calculated later on. In this model a breathing crack between φ = 90° and φ = 270° shows the same behaviour as the always gaping crack. During one rotation the shaft moves on an onion-shaped orbit, only once per revolution.

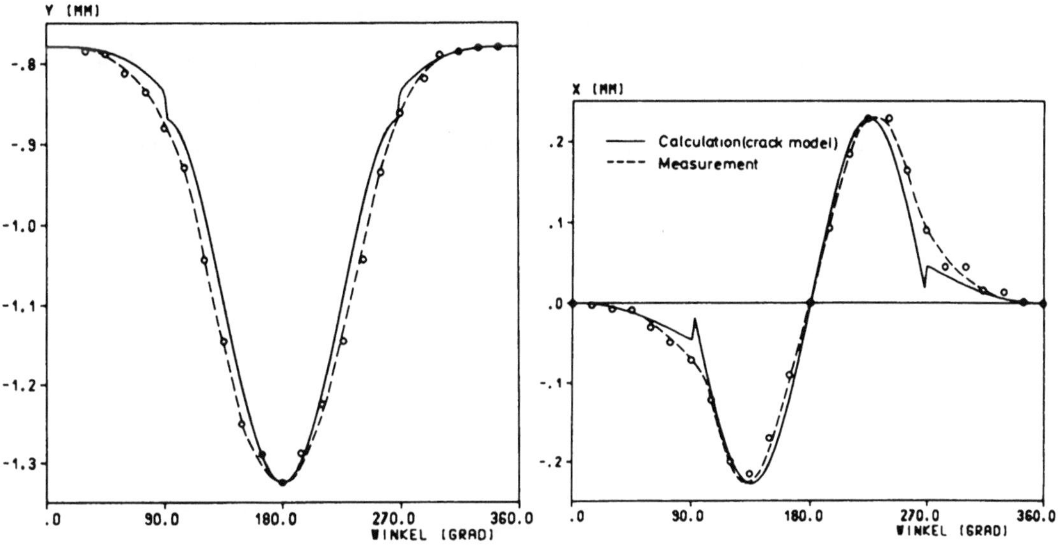

Fig. 4 Self-weight deflection at shaft center for a breathing crack of 50 percent depth

In Fig. 4 the results of this crack model are compared with experimental results of Ziebarth et.al. [10]. The level of the amplitudes is adapted, only the shape of the curves shall be compared. There is a good agreement.

2.3 Change of Stiffness due to a Crack. In the longitudinal direction of the shaft the extension of a crack is very small and of a gaping crack too. But the actual reduction of the stiffness at the crack position does not jump, the change is continuous along the neighbouring range. For steps in shafts the 45°-approximation is used with good results up to now. In order to simplify the mathematical model the wedge-shaped cut-out is replaced by a square cut with L = T [Fig. 5].

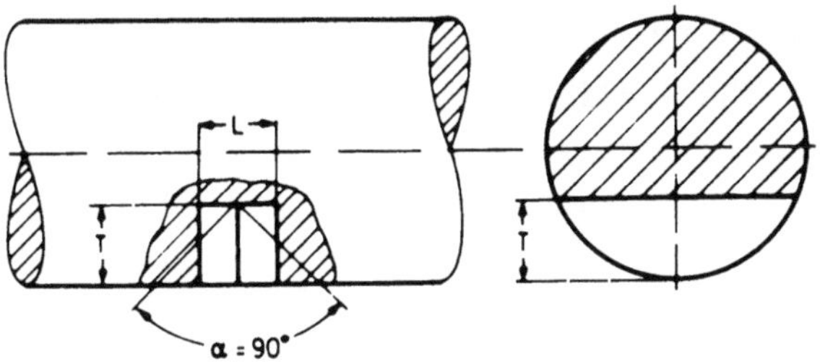

Fig. 5 Simulation of the decreased-section of the open crack. After Grabowski [15]

In addition to the development of the theoretical crack model the stiffness of a cracked shaft has been investigated. At a shaft of 46 mm diameter and 300 mm length [Fig 6] the crack was simulated by a thin milled transverse cut of 0,5 mm width. This shaft has been loaded in such a kind, that in the neighourhood of the crack a constant bending moment was produced.

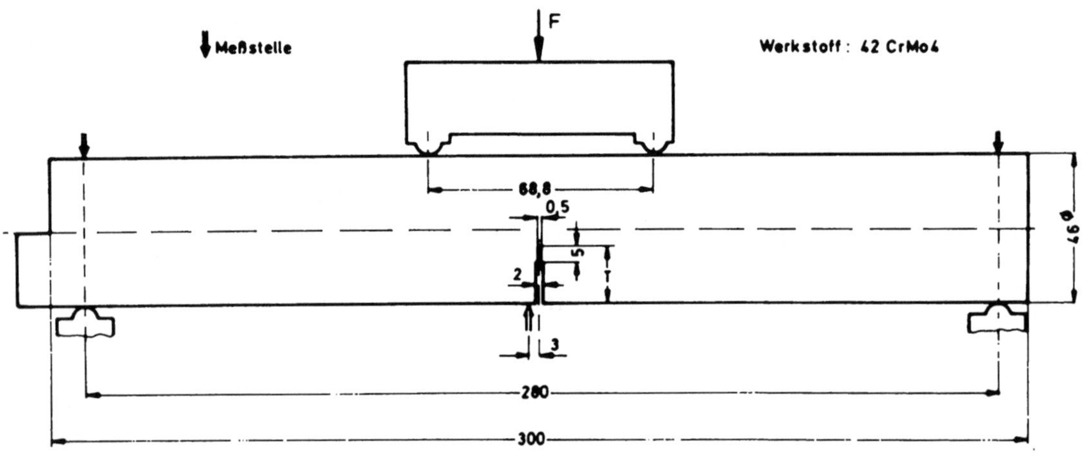

Fig. 6 Shaft with milled cut

Parallel to these measurements the stiffness has been calculated. At the crack position a square cut with L = T was assumed, but only for crack depths until 50 percent. For deeper cracks L has to be reduced analogically.

In Fig. 7 the results of measurement and calculation are compared. The total compliance of the shaft depending on the crack depth is plotted. The difference between measurement and calculation amounts less than 10 percent, except the crack depths of 20 and 70 percent. In consideration of the simple crack model even with these differences the agreement is remarkable good.

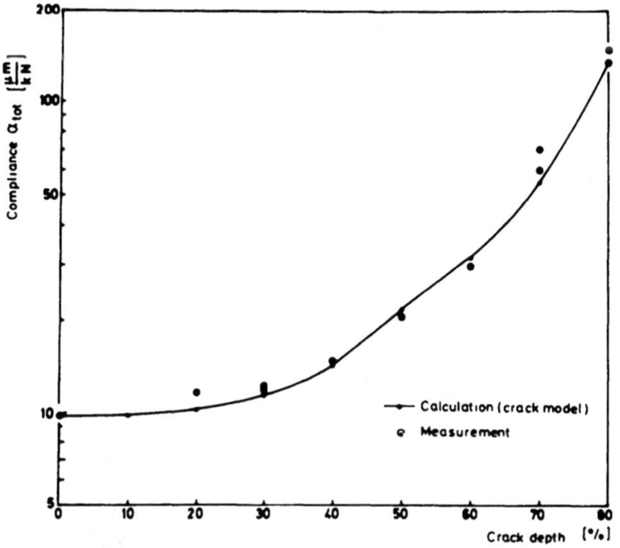

Fig. 7 Compliance α_{tot} of the shaft depending on crack depth

3. Vibration of a Rotor with Single Disk on Massless Shaft of Unequal Stiffness (Gaping Crack)

3.1 Equations of Motion in Space-fixed System of Coordinates. The differental equations of a rotor with single disk on massless shaft with symmetric cross-section in rigid bearings are very well known. With an asymmetric shaft the force F_S changes depending on the angle of rotation Ωt [Fig. 8]. Because of this just the stiffness matrix changes. The coefficients are well known as the moments of inertia in rotated coordinate system. It is essential, that in addition to a mean constant stiffness a stiffness difference appears, which changes two times per revolution and excites twice per revolution vibrations.

In publications often a simplified formulation is used with a mean stiffness k_m. Than the factor μ describes the normalized difference between the stiffness k_ξ and k_η and the mean stiffness.

However, the equations cannot be solved in this form with time dependent periodical coefficients.

$$\begin{bmatrix} F_{Sx} \\ F_{Sx} \end{bmatrix} = -\frac{48E}{\ell^3} \begin{bmatrix} \frac{1}{2}(I_\eta+I_\xi)+\frac{1}{2}(I_\eta-I_\xi)\cos 2\Omega t & \frac{1}{2}(I_\eta-I_\xi)\sin 2\Omega t \\ \frac{1}{2}(I_\eta-I_\xi)\sin 2\Omega t & \frac{1}{2}(I_\eta+I_\xi)-\frac{1}{2}(I_\eta-I_\xi)\cos 2\Omega t \end{bmatrix} \begin{bmatrix} x \\ y \end{bmatrix}$$

$$\begin{bmatrix} F_{Sx} \\ F_{Sy} \end{bmatrix} = -k_m \begin{bmatrix} 1+\mu\cos 2\Omega t & \mu\sin 2\Omega t \\ \mu\sin 2\Omega t & 1-\mu\cos 2\Omega t \end{bmatrix} \begin{bmatrix} x \\ y \end{bmatrix}$$

with

$$k_\xi = \frac{48EI_\eta}{\ell^3}, \quad k_\eta = \frac{48EI_\xi}{\ell^3}$$

$$k_m = \frac{k_\xi + k_\eta}{2}$$

$$\mu = \frac{k_\xi - k_\eta}{2c_m}$$

$$> k_\xi = k_m(1+\mu), \quad k_\eta = k_m(1-\mu)$$

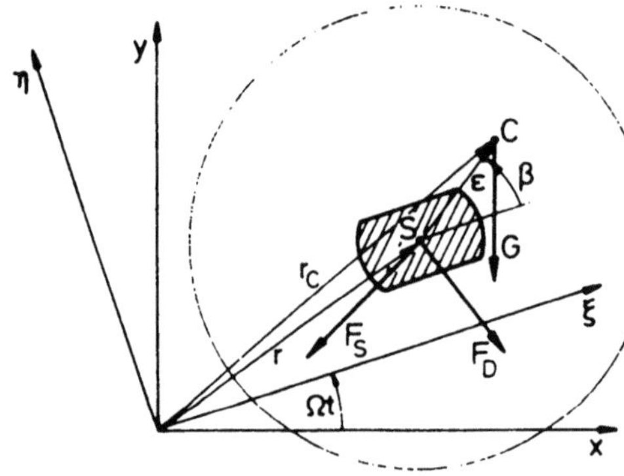

$$\begin{bmatrix} m & 0 \\ 0 & m \end{bmatrix}\begin{bmatrix} \ddot{x} \\ \ddot{y} \end{bmatrix} + \begin{bmatrix} d & 0 \\ 0 & d \end{bmatrix}\begin{bmatrix} \dot{x} \\ \dot{y} \end{bmatrix} + c_m \begin{bmatrix} 1+\mu\cos 2\Omega t & \mu\sin 2\Omega t \\ \mu\sin 2\Omega t & 1-\mu\cos\Omega t \end{bmatrix}\begin{bmatrix} x \\ y \end{bmatrix} = \begin{bmatrix} 0 \\ -G \end{bmatrix} + m\varepsilon\Omega^2 \begin{bmatrix} \cos(\Omega t+\beta) \\ \sin(\Omega t+\beta) \end{bmatrix}$$

Fig. 8 Rotor with single disk on massless asymmetric shaft (space fixed system of coordinates)

3.2 Equations of Motion in Rotating System of Coordinates. If the equations of motion are formulated in rotating coordinates ξ and η the stiffness will be time invariant. And if we use the main axes as coordinates the deviation moment is not needed [Fig. 9]. For that the velocity and acceleration must be formulated in rotating coordinates [Fig. 10]. The equations of motion in Fig. 11 now contains exclusively constant coefficients. The equations are easy to solve and the stability of the system can be investigated too.

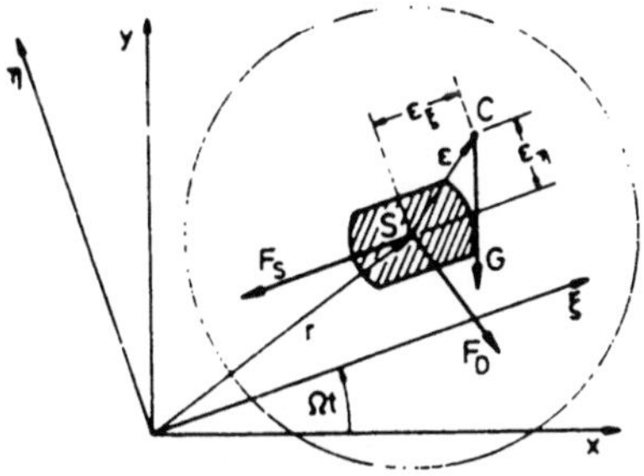

$$-\vec{F} = m \cdot \ddot{\vec{r}}_C$$

$$\begin{bmatrix} F_{S\xi} \\ F_{S\eta} \end{bmatrix} = -\frac{48E}{l^3} \begin{bmatrix} I_\eta & 0 \\ 0 & I_\xi \end{bmatrix} \begin{bmatrix} \xi \\ \eta \end{bmatrix} = -\begin{bmatrix} c_\xi & 0 \\ 0 & c_\eta \end{bmatrix} \begin{bmatrix} \xi \\ \eta \end{bmatrix}$$

$$\begin{bmatrix} G_\xi \\ G_\eta \end{bmatrix} = -G \begin{bmatrix} \sin \Omega t \\ \cos \Omega t \end{bmatrix}$$

Fig. 9 Rotor with single disk on massless asymmetric shaft (rotating system of coordinates)

Rotating coordinates: $\vec{\rho},\ \vec{\rho}_C$

$$\vec{r} = \vec{\rho}$$

$$\dot{\vec{r}} = \vec{\Omega} \times \vec{\rho} + \dot{\vec{\rho}}$$

$$\ddot{\vec{r}} = \vec{\Omega} \times (\vec{\Omega} \times \vec{\rho}) + 2\vec{\Omega} \times \dot{\vec{\rho}} + \ddot{\vec{\rho}}$$

$$r_C = \begin{bmatrix} \xi_C \\ \eta_C \end{bmatrix}$$

$$\dot{r}_C = \begin{bmatrix} 0 & -\Omega \\ \Omega & 0 \end{bmatrix} \begin{bmatrix} \xi_C \\ \eta_C \end{bmatrix} + \begin{bmatrix} \dot{\xi}_C \\ \dot{\eta}_C \end{bmatrix}$$

$$\ddot{r}_C = -\Omega^2 \begin{bmatrix} \xi_C \\ \eta_C \end{bmatrix} + 2 \begin{bmatrix} 0 & -\Omega \\ \Omega & 0 \end{bmatrix} \begin{bmatrix} \dot{\xi}_C \\ \dot{\eta}_C \end{bmatrix} + \begin{bmatrix} \ddot{\xi}_C \\ \ddot{\eta}_C \end{bmatrix}$$

Fig. 10 Rotor with single disk on massless shaft (rotating system of coordinates)

$$\begin{bmatrix} F_{D\xi} \\ F_{D\eta} \end{bmatrix} = -d \left(\begin{bmatrix} 0 & -\Omega \\ \Omega & 0 \end{bmatrix} \begin{bmatrix} \xi \\ \eta \end{bmatrix} + \begin{bmatrix} \dot{\xi} \\ \dot{\eta} \end{bmatrix} \right)$$

$$-\vec{F} = m \cdot \vec{a}_C$$

$$\begin{bmatrix} \xi_C \\ \eta_C \end{bmatrix} = \begin{bmatrix} \xi + \varepsilon_\xi \\ \eta + \varepsilon_\eta \end{bmatrix}$$

$$\begin{bmatrix} \dot{\xi}_C \\ \dot{\eta}_C \end{bmatrix} = \begin{bmatrix} \dot{\xi} \\ \dot{\eta} \end{bmatrix}, \quad \begin{bmatrix} \ddot{\xi}_C \\ \ddot{\eta}_C \end{bmatrix} = \begin{bmatrix} \ddot{\xi} \\ \ddot{\eta} \end{bmatrix}$$

$$\begin{bmatrix} m & 0 \\ 0 & m \end{bmatrix} \begin{bmatrix} \ddot{\xi} \\ \ddot{\eta} \end{bmatrix} + \begin{bmatrix} d & -2\Omega m \\ 2\Omega m & d \end{bmatrix} \begin{bmatrix} \dot{\xi} \\ \dot{\eta} \end{bmatrix} + \begin{bmatrix} k_\xi - m\Omega^2 & -d\Omega \\ d\Omega & k_\eta - m\Omega^2 \end{bmatrix} \begin{bmatrix} \xi \\ \eta \end{bmatrix} = -m\,g \begin{bmatrix} \sin \Omega t \\ \cos \Omega t \end{bmatrix} + m\,\Omega^2 \begin{bmatrix} \varepsilon_\xi \\ \varepsilon_\eta \end{bmatrix}$$

Fig. 11 Rotor with single disk on massless asymmetric shaft (rotating system of coordinates)

The stability condition becomes

$$\left(1 - \frac{\Omega^2}{\omega^2}\right)^2 - \mu^2 + 4\,D^2\,\frac{\Omega^2}{\omega^2} > 0 \qquad (3.2\text{-}1)$$

with

$$\omega^2 = \frac{k}{m}\,. \qquad (3.2\text{-}2)$$

3.3 Numerical Results. In Fig. 12 the stability map is plotted. For a constant crack depth the unstable range is located between the two eigenfrequencies $\omega_\eta^2 = \frac{k_\eta}{m}$ and $\omega_\xi^2 = \frac{k_\xi}{m}$. In addition of the normalized stiffness difference the corresponding crack depth of a bilateral gaping crack is marked. Already at a crack depth of approx. 25 percent a viscous damping ratio of D = 0.1 is needed, so that the system does not become unstable.

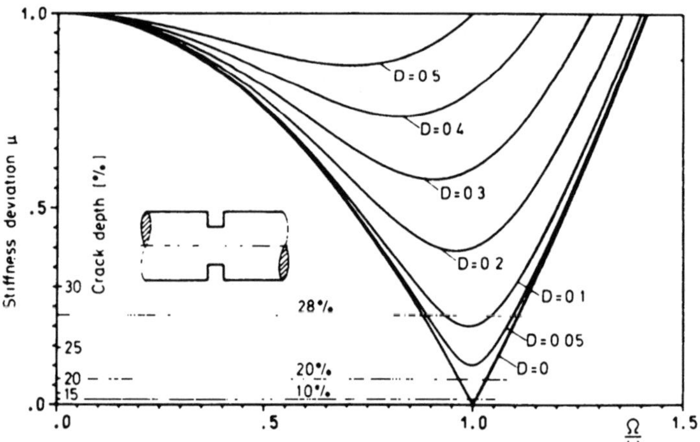

Fig. 12
Stability map for the massless shaft with a single disk containing a gaping crack, viscous damping; rigid support

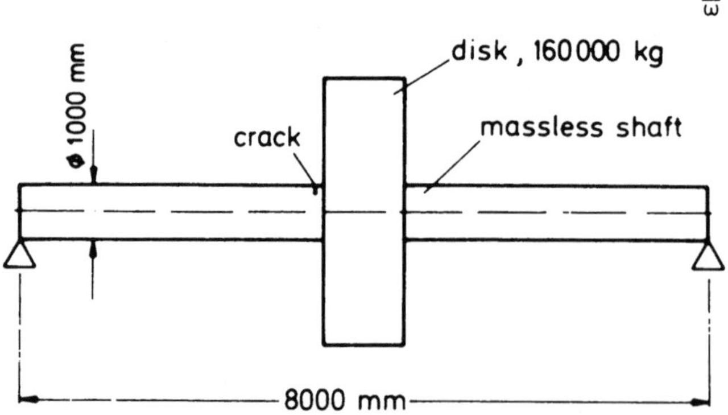

Eccentricity : 10 μm
Critical speed (uncracked): 735 min^{-1}

Fig. 13 Rotor model

To calculate the amplitude of the forced vibration due to unbalance, only the stiffness matrix of the differential equation is nessessary. This matrix completely contains the stability condition too.

To calculate some theoretical results, the simple model for the rotor of a LP-turbine was choosen [Fig. 13]. The Figures 14, 15 and 16 show the vibration amplitudes as a result of unbalance and crack depending on the speed. At a crack depth of 10 percent the vibration is still stable [Fig. 14]. As for a shaft without crack the maximum of the once per revolution vibration amplitude (Ω) amounts to: $R \approx \frac{\varepsilon}{2D}$.

The amplitude does not change depending on crack depth until the system becomes unstable. Furthermore there do not arise two resonance points in the once per revolution vibration as could be assumed as a result of the both different stiffnesses. Just if the system becomes unstable the stationary solution delivers two resonance speeds.

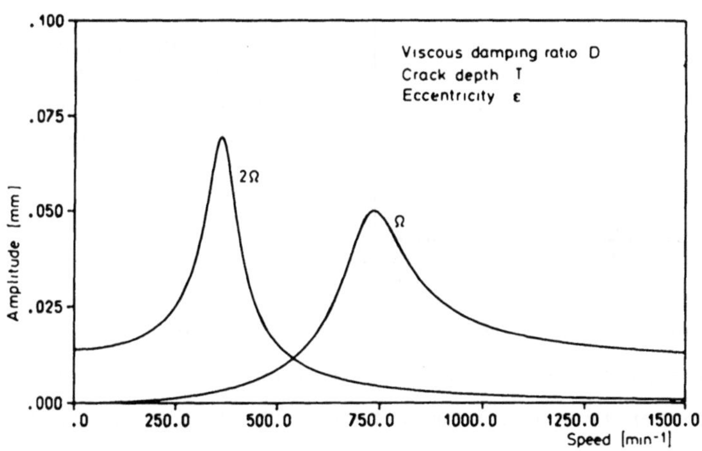

Fig. 14 Vibration amplitudes of a simple rotor with gaping crack D = 0.1, T = 10 %, ε = 10 μm

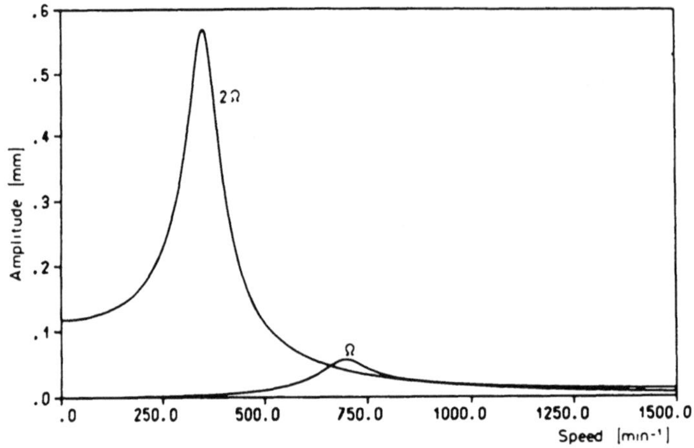

Fig. 15
Vibration amplitudes of a simple rotor with gaping crack D = 0.1, T = 20 %, ε = 10 μm

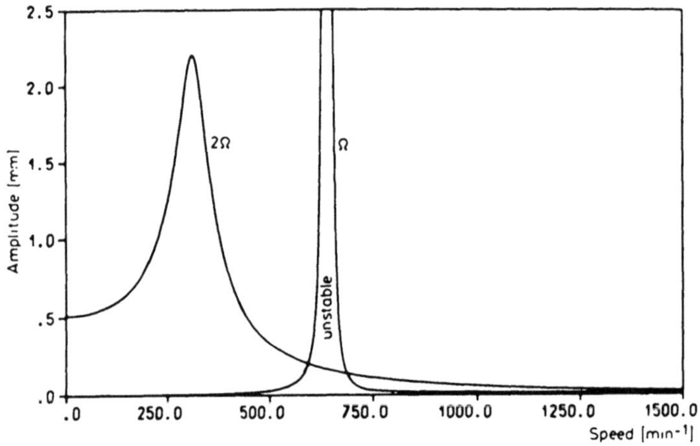

Fig. 16
Vibration amplitudes of a simple rotor with gaping crack D = 0.1, T = 28 %, ε = 10 μm

Although the crack depth with 10 percent of the diameter is small, the twice per revolution amplitudes are greater than the amplitudes due to unbalance. As the once per revolution amplitudes, the twice per revolution amplitudes are directly inversely proportional to the damping ratio. And also the amplitude grows with crack depth. The resonance speed amounts to one half the critical speed of the unbalance vibrations.

For a crack depth of 20 percent [Fig. 15] the vibration amplitudes due to unbalance do not change, however, the twice per revolution resonance amplitudes are approx. eight times greater than for the crack depth of 10 percent. At a crack depth of 28 percent an unstable speed range arises [Fig. 16] between the two eigenfrequencies, which belong to the greatest and the smallest stiffnes of the shaft.

Fig. 17 shows four calculated orbits from Tondl [3]. Transmitted to the rotor in Fig. 14 the orbits a, b, c and d are applicable approx. at speeds of 250, 350, 600 an 700 rpm.

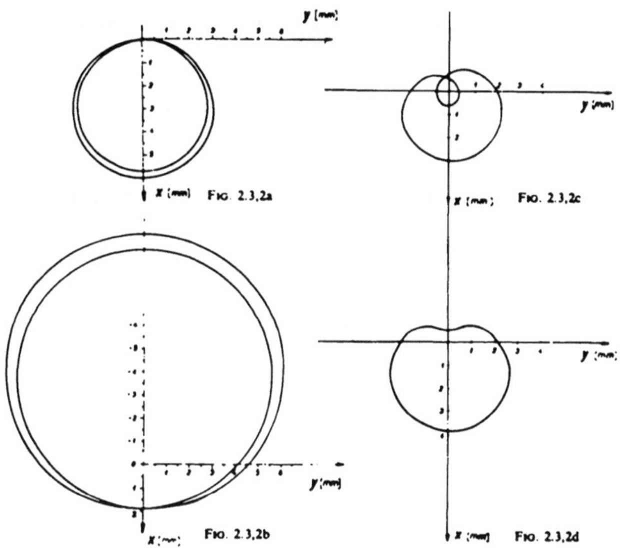

Fig. 17
Vibration orbits.
After Tondl [3]

<u>4. Some Comments on the Model for a Gaping Crack.</u> In the crack model the moments of inertia are proportional to the static deflection as a result of the crack: In space fixed coordinates the moments of inertia change depending on angle of rotation. The Fourier analysis of this function for one rotation shows the harmonical components. These components correspond to the factor μ in the equations of motion. With a gaping crack only the second harmonic excists [Fig. 18]. With a breathing crack [Fig. 19] the first harmonic component dominates. It shows a progressive increase as a function of the crack depth. The twice per revolution part has a definite maximum at approximately 35 percent crack depth and it is altogether essential smaller than with the gaping crack. The higher harmonics are small. At the other axis the difference is not important, here the twice per revolution part even dominates for small crack depths. It is similar for the deviation moment.

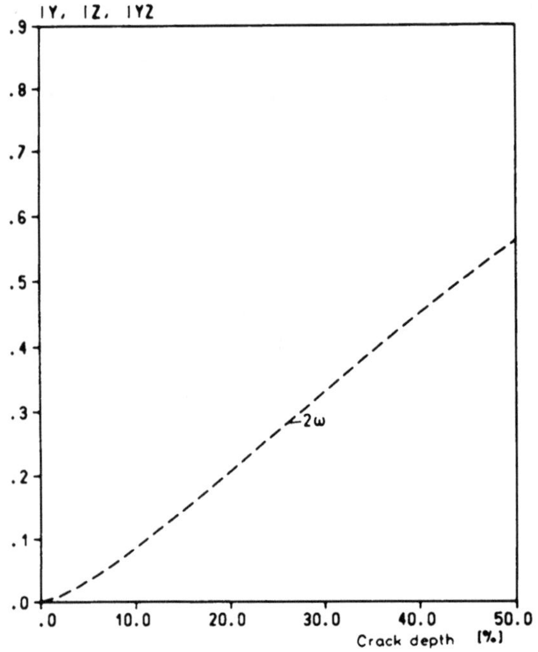

Fig. 18 Harmonic parts of normalized second moments of area, gaping crack

For a breathing crack it is supposed in this model, that the crack area opens and closes only due to self-weight bending as the shaft rotates. Some other authors assume that the opening and closing of the crack in addition to the deflection due to the self-weight of the rotor also depends on the vibration amplitudes. The mean static deflection of a shaft which can be assumed as a model for a large LP-turbine amounts to 1 mm [Fig. 3]. At running speed vibration amplitudes of approx. 50 μm can be assumed.

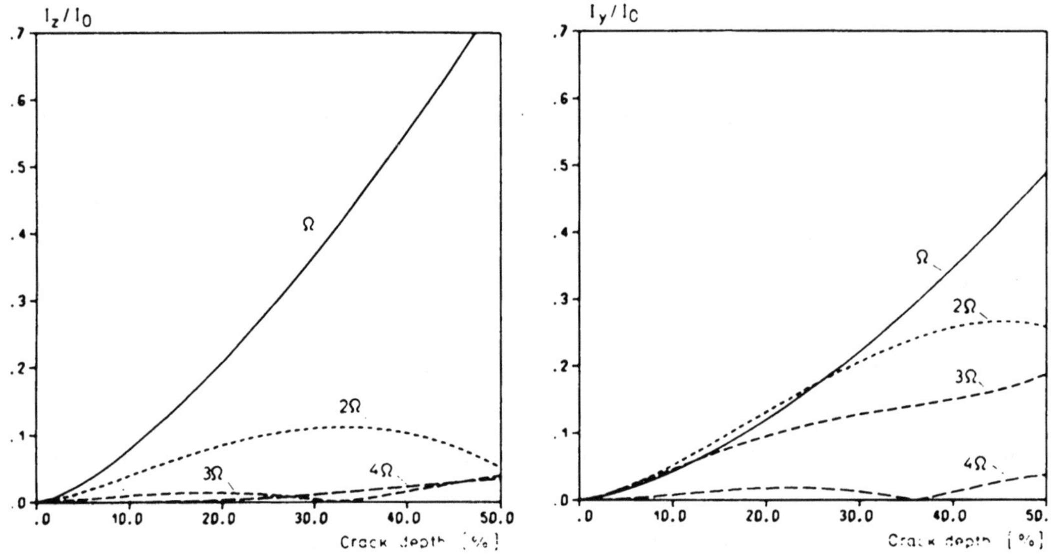

Fig. 19 First harmonic components of the normalized second moments of area l_z/l_0 and l_y/l_0 for the crack model (l_0: uncracked shaft). After Grabowski [15]

In other words, a factor of approx. 20 is achieved. In this case, surely no influence of the vibration on the opening and closing of the crack will occur. Even at the critical speed, when amplitudes up to approx. 250 µm are assumed, such an influence will remain small. That means, however, that normally no relation exists between size and position of the out-of-balance and the vibration caused by a crack.

Only at large crack depth, if due to the crack itself resonance amplitudes in the magnitude of the static deflection appear - as a rule the blades then touch the casing - the presented crack model will be no more valid.

But in theory certainly this case with large amplitudes is interesting too. Above all it can occur, that the crack is permanent open or closed. For such a model some results of Meyer [12] shall be presented here. Similar investigations on a massless shaft with single disk are carried out by other authors.

5. Vibration of a Rotor with Single Disk on Massless Shaft Containing a Breathing Crack (hinge). In recent years some authors have published investigations on the vibrational behaviour of cracked shafts [5,6,7,12]. They employ the rotor model with massless shaft and single disk. For modelling a breathing crack a spring-mounted hinge is assumed [Fig. 20]. The opening and closing chracteristic of the hinge depends on the self-weight deflection as well as on the bending vibration amplitude itself. With closed hinge the stiffness is symmetrical. With open hinge the stiffness changes only in one direction.

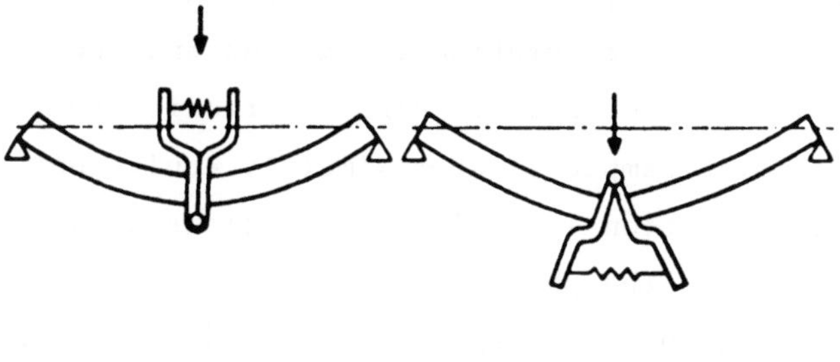

Fig. 20 Crack-model. After Gasch [6]

In rotating system of coordinates ξ and η the crack area is open at positiv and closed at negativ deflection η [Fig. 21]. With small vibration amplitudes during one half revolution we have a symmetrical shaft and during the next half revolution an asymmetrical shaft.

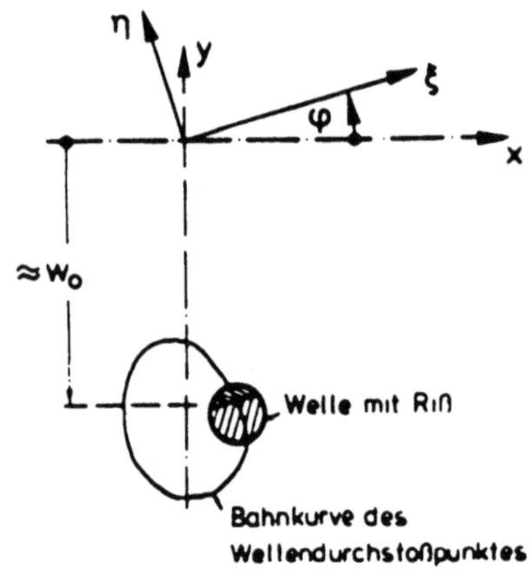

Fig. 21 Vibration orbit and system of coordinates. After Meyer [12]

With large amplitudes the moment of switching at $\eta = 0$ can be shifted, at the limit it is possible that the crack area is constantly open or closed.

With this assumptions for the behaviour of a breathing crack on principle the equations of motion for the shaft with asymmetrical cross section are valid. But the stiffness c_η is now depending on the sign of the deflection η in rotating coordinates. The equations can be solved on analog computer as well as on digital computer.

With regard to the harmonic parts of the moments of inertia in Fig. 19 it is expected that with this crack model in addition to the twice per revolution excitation due to an asymmetrical shaft above all once per revolution vibration and the higher harmonics appear.

Fig. 22 shows one result of stationary-periodical vibrations. In this calculations it is assumed, that the vibration amplitudes are small in relation to the static deflection so that the moments of switching are not depending on the vibration amplitudes. Moreover the calculations are carried out without unbalance, this means, the vibrations are excited only by the crack. There appear once per revolution vibrations and due to the sudden change of stiffness in principle all higher harmonics too.

The stiffness asymmetry in this example is approximately equivalent to a crack depth of the diameter. Additional unbalance vibrations have to be added or subtracted depending on the phase relationship.

Similar to the asymmetric shaft unstable regions [Fig. 23] exist for this crack model. In Fig. 23 it is also supposed that the opening and closing of the crack area does not depend on vibration amplitudes. Beside the three plotted unstable regions there exist additional regions at smaller speeds. However, these regions are very narrow.

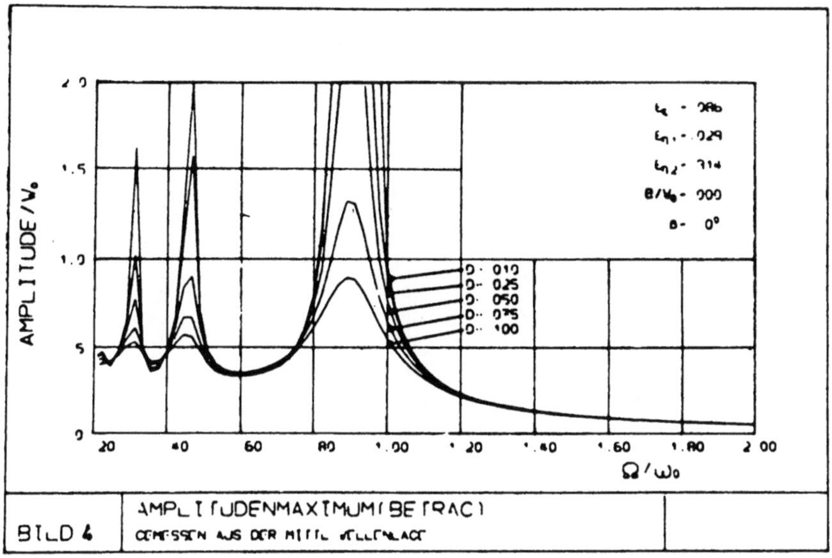

Fig. 22 Vibration amplitudes, breathing crack (hinge). After Meyer [12]

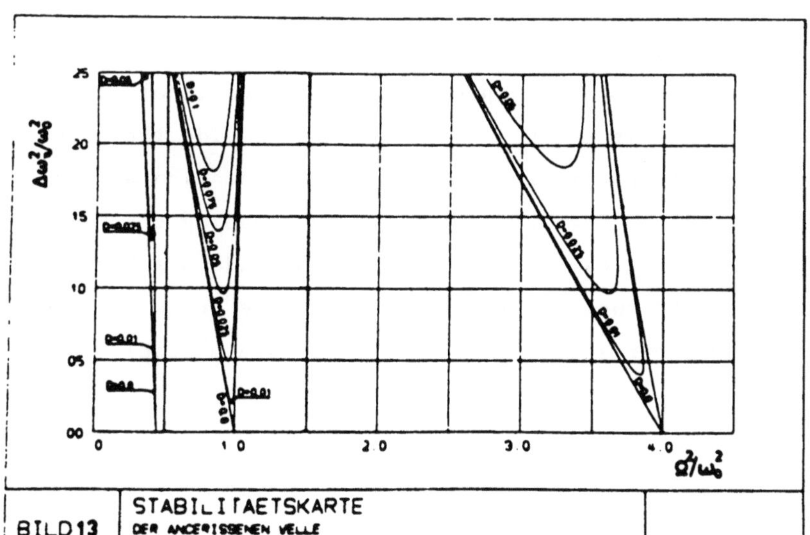

Fig. 23 Stability map. After Meyer [12]

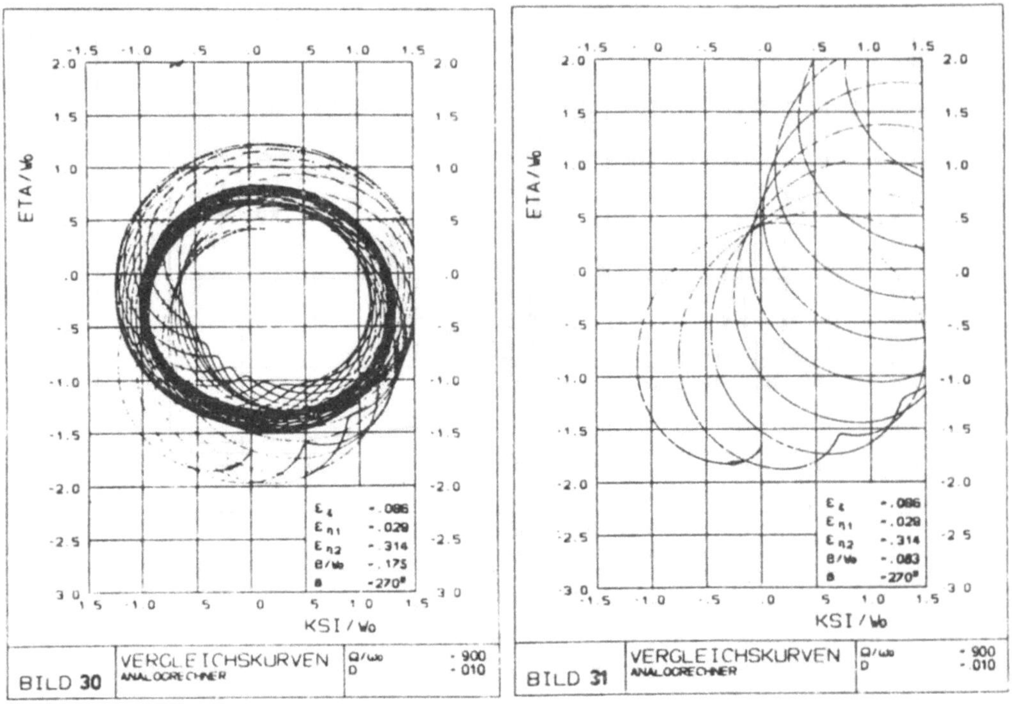

Fig. 24 Vibration orbits. After Meyer [12]

To calculate the vibration orbit in Fig. 24 the static deflection was taken as an initial value. The equations of motion were solved on an analog computer. The rotating speed is near the critical speed of the once per revolution vibration. On the right hand side the crack is constantly gaping as a result of the size of the vibration amplitudes. And for a damping ratio of D = 0.01 a shaft with a gaping crack is unstable.

The result in the figure on the left hand side has been achieved by doubling the unbalance. This effects, that the crack is no more constantly gaping but partially closed. In this case a stable limited amplitude results.

Part 2

Turbine Rotors with Breathing Crack

1. Introduction

In the first part of this paper crack models are presented and the effect of cracks on the vibrational behaviour of simple rotors with single disk on massless shaft is discussed. A gaping crack excites only twice per revolution vibrations. A breathing crack excites mainly once per revolution as well as twice per revolution vibrations; the vibration amplitudes of the higher harmonics are small.

The aim of the investigations in this second part of the paper is to determine whether and in what manner a crack in a turbine rotor can be recognized by measuring the shaft vibration so that large consecutive damage can be prevented. The results of the vibrational analysis of simple rotors are not or only inaccurate transferable to real turbine rotors. Above all they do not allow to estimate the vibration amplitudes at normal speed. However, in order to recognize a crack, first of all the change of the vibrational behaviour at normal speed has to be taken into account. The drive-up and drive-down offer an additional criterion.

Because the analysis of the vibration of a real turbine rotor must include the non-conservative characteristics of the journal bearings, the formulation of the equations of motion in rotating system of coordinates is no longer an advantage. So, in spacefixed coordinates the change in stiffness depending on angle of rotation leads to time dependent coefficients in the equations of motion. These equations cannot be solved exactly. Numerical integration as well as iterative methods (as used by Mayes and Davies [17]) can be employed.

The application of a rotor model (Fig. 1) which is more realistic than the simple model with single disk leads to a large numer of degrees of freedom. This in connection with the numerical solution method creates a considerable need of memory and computing time. Therefore in these investigations the concept of modal analysis was chosen [11] to reduce the numer of degrees of freedom. In a similar way Mayes [8] investigates the crack propagation in rotating shafts.

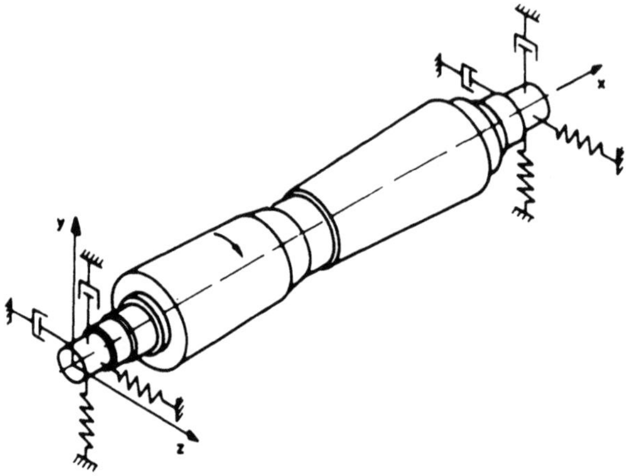

Fig. 1
Rotor model and coordinate system

2. Analysis There are applied approximately the eigenfunctions V(x) and W(x) of the conservative system in the vertical and the horizontal plane.

The range of the local change of stiffness due to a crack covers the length L. A medium stiffness between open and closed crack will be taken approximately for the calculation of the eigenfunctions in this region. This is necessary for the consideration of the major curvature at the crack position, compared with the uncracked shaft.

The deflections in the vertical direction y(x,t) and in the horizontal direction z(x,t) will be composed by the first few eigenfunctions weighted with the generalized time dependent coordinates q(t), as follows,

$$y(x,t) = \sum_{k=1}^{K} Y_{ok}(x)\, q_{yk}(t) \tag{2-1}$$

$$z(x,t) = \sum_{\ell=1}^{L} Z_{o\ell}(x)\, q_{z\ell}(t) \tag{2-2}$$

The transformation yields a system of K + L coupled equations of motion for the generalized coordinates q(t) which is

$$\underline{M}\, \underline{\ddot{q}} + \underline{C}\, \underline{\dot{q}} + \underline{K}(t)\, \underline{q} = \underline{F}(t) + \underline{g}. \tag{2-3}$$

$\underline{M}$, $\underline{C}$ and $\underline{K}$ can be considered as mass matrix, damping matrix and stiffness matrix. On the right-hand side the function F(t) includes the out-of-balance distribution, while $\underline{g}$ contains the self-weight load.

The time dependent elements in the sitffness matrix $\underline{K}(t)$ can be separated and we obtain

$$\underline{M}\,\ddot{\underline{q}} + \underline{C}\,\dot{\underline{q}} + [\underline{K}_m + \Delta\underline{K}(t)]\,\underline{q} = \underline{F}(t) + \underline{g}\ . \quad (2\text{-}4)$$

A Runge-Kutta-Fehlberg procedure of 5th order with constant step-size [14] which proved to be very exact for the necessary computing time, will be used by the numerical integration for the calculation of the instationary vibrational behaviour. The necessary initial values will be taken from the stationary solution of the equation (4) without the term $\Delta\underline{K}(t)$. The time variable system has a stationary periodic solution. After the retransformation according to equations (1) and (2), the deflection of the entire rotor which depends on the angle of rotation is obtained.

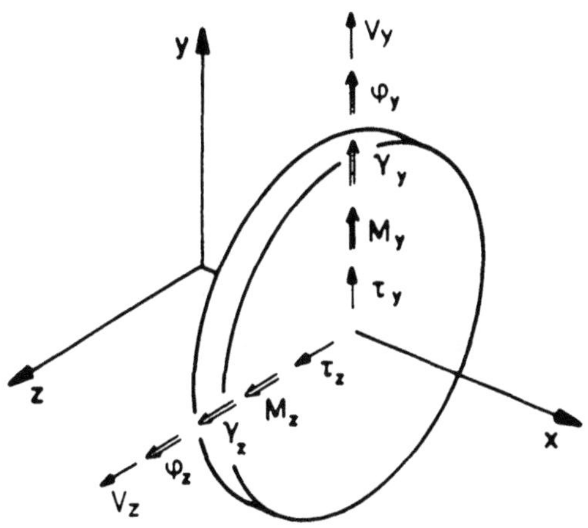

Fig. 2 Sign convention for space fixed coordinate system

The rotor displacements are described in two mutually perpendicular directions y and z with corresponding angular displacements φ_y and φ_z (Fig. 2). They are functions of the coordinate x along the rotor axis. The rotor is supported at N stations by journal bearings. Each bearing is represented by a set of stiffness and damping coefficients such that the bearing reaction can written as

$$\begin{bmatrix} R_y \\ R_z \end{bmatrix}_n = - \begin{bmatrix} k_{yy} & k_{yz} \\ k_{zy} & k_{zz} \end{bmatrix}_n \cdot \begin{bmatrix} y \\ z \end{bmatrix}_n - \begin{bmatrix} c_{yy} & c_{yz} \\ c_{zy} & c_{zz} \end{bmatrix}_n \cdot \begin{bmatrix} \dot{y} \\ \dot{z} \end{bmatrix}_n \quad (2\text{-}6)$$

$$\underline{R}_n = - \underline{K}_n \underline{u}_n - \underline{c}_n \dot{\underline{u}}_n \quad (2\text{-}7)$$

The equations of motion for a differential element of the rotor are

$$\left.\begin{matrix}\mu\ddot{y} - V'_y \\ \\ \mu\ddot{z} - V'_z\end{matrix}\right\} + \sum_{n=1}^{N} \delta(x-x_n)(\underline{K} + \underline{C}\frac{\partial}{\partial t})_n \begin{bmatrix} y \\ z \end{bmatrix} \left\{\begin{matrix} = f_y(t) - \mu g \\ \\ = f_z(t)\end{matrix}\right. \tag{2-8}$$

$$\mu i^2 \ddot{\varphi}_y + 2\Omega\mu i^2 \dot{\varphi}_z + V_z - M'_y = 0$$

$$\mu i^2 \ddot{\varphi}_z - 2\Omega\mu i^2 \dot{\varphi}_y - V_y - M'_z = 0$$

where $f_y(t)$ and $f_z(t)$ are external forces due to the unbalance and μg the shaft weight per unit length.

The deformation equations for the shaft are the conentional beam equations,

$$y' = \varphi_z + \frac{V_y}{kGA}, \quad M_y = EI\varphi'_y , \tag{2-9}$$

$$z' = -\varphi_y + \frac{V_z}{kGA}, \quad M_z = EI\varphi'_z . \tag{2-10}$$

Calculating the eigenfunctions of equation (2-8), the damping matrix $\underline{C}$ and the coupling elements of the stiffness matrix $\underline{K}$, k_{yz} and k_{zy}, are neglected. In this calculation a lumped-parameter approach is employed. Near the crack a symmetric average stiffness is assumed. Therefore each crack depth and each crack positon requires a new calculation of eigenfunctions.

The solution provides two sets of R real eigenfunctions for the rotor displacements,

$$\underline{Y}_o = [Y_{o1}, Y_{o2} \dots Y_{or} \dots Y_{oR}] ,$$

$$\underline{Z}_o = [Z_{o1}, Z_{o2} \dots Z_{or} \dots Z_{oR}] , \tag{2-11}$$

and the angular displacements

$$\underline{\Phi}_{y_o} = [\Phi_{y_{o1}}, \Phi_{y_{o2}} \dots \Phi_{y_{or}} \dots \Phi_{y_{oR}}] ,$$

$$\underline{\Phi}_{z_o} = [\Phi_{z_{o1}}, \Phi_{z_{o2}} \dots \Phi_{z_{or}} \dots \Phi_{z_{oR}}] . \tag{2-12}$$

The modal solution of the transient response of equation (2-8) is of the form

$$\begin{aligned} y(x,t) &= \underline{Y}_o(x)\,\underline{q}_y(t), \\ z(x,t) &= \underline{Z}_o(x)\,\underline{q}_z(t). \end{aligned} \tag{2-13}$$

The components are rearranged in matrix form

$$\underline{u}(x,t) = \begin{bmatrix} y(x,t) \\ z(x,t) \end{bmatrix} \tag{2-14}$$

$$\underline{q}(t) = \begin{bmatrix} \underline{q}_y(t) \\ \underline{q}_z(t) \end{bmatrix} \tag{2-15}$$

and

$$\underline{U}_o(x) = \begin{bmatrix} \underline{Y}_o(x) & \underline{0} \\ \underline{0} & \underline{Z}_o(x) \end{bmatrix} \tag{2-16}$$

Then equation (2-13) becomes

$$\underline{u}(x,t) = \underline{U}_o(x)\,\underline{q}(t) \tag{2-17}$$

and it follows similarly for the angular displacements

$$\underline{\varphi}(x,t) = \underline{\Phi}_o(x)\,\underline{q}(t). \tag{2-18}$$

Additionally, the shear deformation $\underline{\gamma}(x,t)$ is required,

$$\underline{\gamma}(x,t) = \underline{\Gamma}_o(x)\,\underline{q}(t) = [\underline{S}\,\underline{U}_o'(x) - \underline{\Phi}_o(x)]\,\underline{q}(t) \tag{2-19}$$

with

$$\underline{\Gamma}_o = \begin{bmatrix} \underline{\Gamma}_{yo} & \underline{0} \\ \underline{0} & \underline{\Gamma}_{zo} \end{bmatrix} \tag{2-20}$$

and

$$\underline{S} = \begin{bmatrix} 0 & -1 \\ 1 & 0 \end{bmatrix} . \tag{2-21}$$

For the modal transformation of equation (2-8) the complete bearing matrices $\underline{K}$ and $\underline{C}$ are used. External and internal damping is included as well. The result are 2R completely coupled ordinary differential equations in terms of the modal coefficients $\underline{q}_y(t)$ an $\underline{q}_z(t)$,

$$\underline{M}\ddot{q} + \underline{C}\dot{q} + \underline{K}_m q = \underline{F}(t) + \underline{g} \; . \qquad (2\text{-}22)$$

In calculations only the first K eigenfunctions in the y,x-plane and L eigenfunctions in the z,y-plane are used. So, the mass matrix, the damping matrix and the stiffness matrix have the size (K+L)(K+L).

The matrix are defined by

$$\underline{M} = \int_0^{\ell} [\mu \, \underline{U}_o^T \underline{U}_o + \mu i^2 \, \underline{\Phi}_o^T \underline{\Phi}_o] \, dx$$

$$\underline{C} = \int_0^{\ell} [c_e \, \underline{U}_o^T \underline{U}_o - \mu i^2 \, 2\Omega \, \underline{\Phi}_o^T \underline{S} \, \underline{\Phi}_o +$$

$$+ G\varkappa A \, d_i \, \underline{\Gamma}_o^T \, \underline{\Gamma}_o + EI_m d_i \, \underline{\Phi}_o'^T \, \underline{\Phi}_o'] \; dx +$$

$$+ \sum_N \underline{U}_o^T(x_n) \, \underline{B}_n \, \underline{U}_o(x_n) \qquad (2\text{-}23)$$

$$\underline{K}_m = \int_0^{\ell} [EI_m (\underline{\Phi}_o'^T \, \underline{\Phi}_o' - c_i \, \Omega \underline{\Phi}_o'^T \, \underline{S} \, \underline{\Phi}_o') +$$

$$+ G\varkappa A \, (\underline{\Gamma}_o^T \, \underline{\Gamma}_o - c_i \, \Omega \, \underline{\Gamma}_o^T \, \underline{S} \, \underline{\Gamma}_o) \; dx +$$

$$+ \sum_N \underline{U}_o^T(x_n) \, \underline{K}_n \, \underline{U}_o(x_n)$$

$$\underline{F} = \int_0^{\ell} \mu \, \underline{U}_o^T \; [\Omega^2 \, \underline{T} + \dot{\Omega} \, \underline{S} \, \underline{T}]^T \; \underline{e} \, dx$$

$$\underline{g} = \int_0^{\ell} -\mu g \begin{bmatrix} \underline{Y}_o^T \\ \underline{0} \end{bmatrix} dx$$

here $\underline{e}$ is the body fixed radius of the unbalance,

$$\underline{e} = \begin{bmatrix} e_\eta \\ e_\zeta \end{bmatrix} \tag{2-24}$$

and $\underline{T}$ contains the harmonic functions of the angle of rotation $\varphi = \Omega t$,

$$\underline{T} = \begin{bmatrix} \cos \Omega t & \sin \Omega t \\ -\sin \Omega t & \cos \Omega t \end{bmatrix} . \tag{2-25}$$

In addition to the constant stiffness the time dependent stiffness at the crack has to be considered. In the range L the second moments of area are

$$I_m + \Delta I_y(\varphi),\; I_m + \Delta I_z(\varphi),\; I_{yz}(\varphi) = I_{zy}(\varphi). \tag{2-26}$$

These functions are directly stored and during the numerical integration the time dependent part $\Delta\underline{K}(t)$ of the stiffness matrix is recalculated for every time step as follows:

$$\Delta\underline{K}(t) = \int_{x_{c\ell}}^{x_{cr}} E\, \Phi_o'^T\, \Delta\underline{I}(\varphi)\, \Phi_o'\, dx \tag{2-27}$$

with

$$\Delta\underline{I}(\varphi) = \begin{bmatrix} \Delta I_y & I_{yz} \\ I_{zy} & \Delta I_z \end{bmatrix} . \tag{2-28}$$

The final result are K + L linear differential equations given by equation (2-4)

$$\underline{M}\ddot{\underline{q}} + \underline{C}\dot{\underline{q}} + [\underline{K}_m + \Delta\underline{K}(t)]\,\underline{q} = \underline{F}(t) + \underline{g} .$$

3. Numerical Results. According to calculations of the author Fig. 3 illustrates the principle form of the curve of the vibration amplitude close to a bearing of a large LP turbine. As a result of the crack, the critical speed of the once/revolution vibrations is reduced to smaller values and as a rule the amplitudes increase. The original amplitudes

can, however, also become smaller in the case of an opposite phase position between the out-of-balance vibration and the vibration caused by the crack.

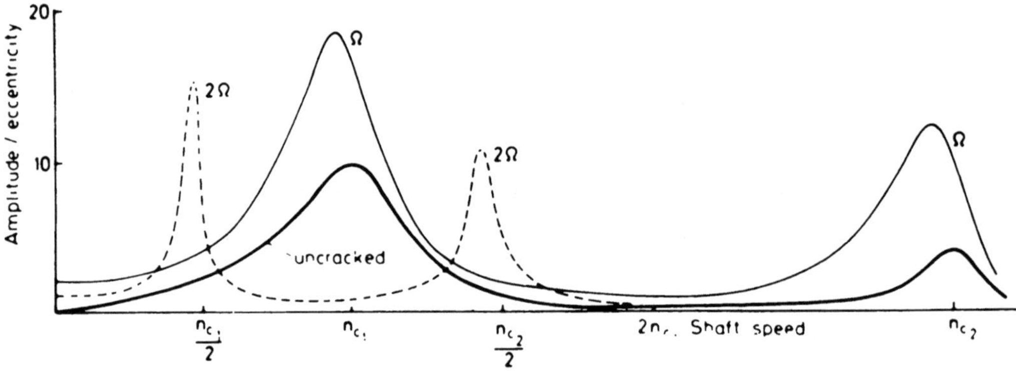

Fig. 3 Amplitude of vibration for a turbine rotor (uncracked and cracked conditions). After Grabowski [15]

In each of the half speeds of the resonance points of the once/revolution vibrations, new resonance points with twice/revolution vibrations arises. This effect is known from generators in which their running speed is near the half of the second critical speed. As a result of the rotor which is not round, the twice/revolution vibrations can cause a non-smooth run.

Generally, the size of the vibration caused by the crack depends on a lot on the crack position, but most of all, it depends on the crack position which is relative to the excited eigenfunction. Where a symmetrical rotor is concerned with a crack in its centre, the second twice/revolution resonance point would completely disappear because the second eigenfunction in the shaft centre would show no curvature. The vibration amplitudes are at the largest when the shaft is cracked on a position which shows the largest curvature of an eigenfunction. This is the same for the once/revolution as well as for the twice/revolution vibrations.

The simple model of a massless shaft with disk is suitable approximately for the range up to almost the n_{c1} in Fig. 3, however, only for a crack located about in the centre of the shaft.

The influence of the crack on vibrational behaviour of a real turbine rotor can vary extensively. The size of the excited vibration amplitudes depends mainly on the following four factors:

- size of the weight influence g_k in equation (2-4); for a single eigenfunction we obtain

$$g_k = \int_0^{\ell} g\ \mu(x)\ Y_{ok}(x)\ dx \qquad (3\text{-}1)$$

  with the shaft weight/unit length $\mu(x)$. The value g_k is practically identical to the corresponding out-of-balance value of a constant eccentricity along the length of the rotor.
- size of the term $\Delta K(t)$ which depends on the angle of rotation in equation (2-4). For a single eigenfunction the corresponding component is determined not only by the depth of a crack, but also by the curvature of an eigenfunction. The position of the crack plays the most important part here.
- for the once/revolution vibration, the distance between the critical speeds and the running speed.
- twice/revolution vibrations are important at the running speeds that are approximately one half the critical speeds.

3.1 Vibration Amplitudes Depending on Crack Depth. An example for the vibration amplitudes of a LP turbine rotor at running speed is shown in Fig. 4. The crack was assumed to be located near the rotor centre. If the once/revolution vibration should have almost the same phase relationship due to out-of-balance and crack, then the amplitudes add up as is the case for example on the right bearing. If the vibrations are oppositely phased, then the original amplitude as a result of the out-of-balance will be smaller at first. At larger crack depths, finally the vibrations caused by the crack will be dominant. Thus, the vibration amplitude can also be smaller at first, along with increasing depth of the crack.

The twice/revolution excitation concerns above all the second eigenfunction. Therefore only relatively small amplitudes occur in the rotor centre. On the other hand, near the bearings the amplitudes are up to crack depths of 40 % larger than those of the once/revolution vibration.

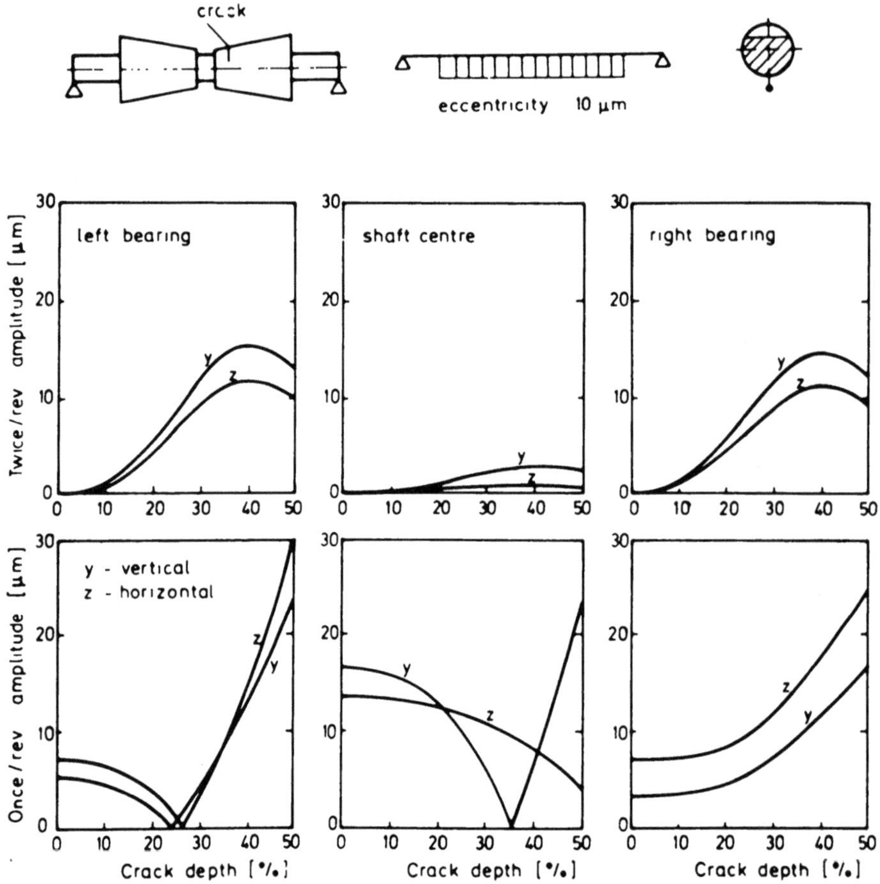

Fig. 4 Vibration amplitudes at normal speed of a cracked LP turbine rotor supported on oil film bearings. After Grabowski [15]

The maximum at crack depth of 40 % is caused by the change of the moments of inertia (Fig. 19 in Part 1). However, this effect is overlapped by the curvature change of the eigenfunctions at the crack position (Fig. 5).

3.2 Vibration Amplitudes Depending on Rotating Speed. Since the number of degrees of freedom can be severely reduced by means of the modal analysis, it is possible to even calculate the vibrational behaviour of trains of rotors. Fig. 6 shows an example for a LP rotor which is coupled with a generator. In this case eight eigenfunctions were needed, four in each plane.

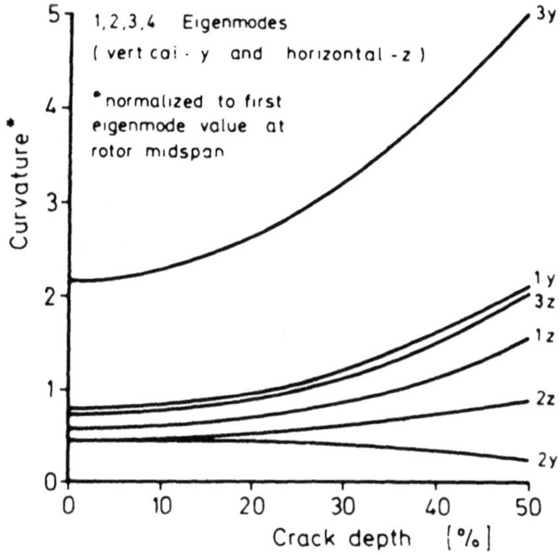

Fig. 5 Curvature change at the crack location for the LP turbine rotor. After Grabowski [15]

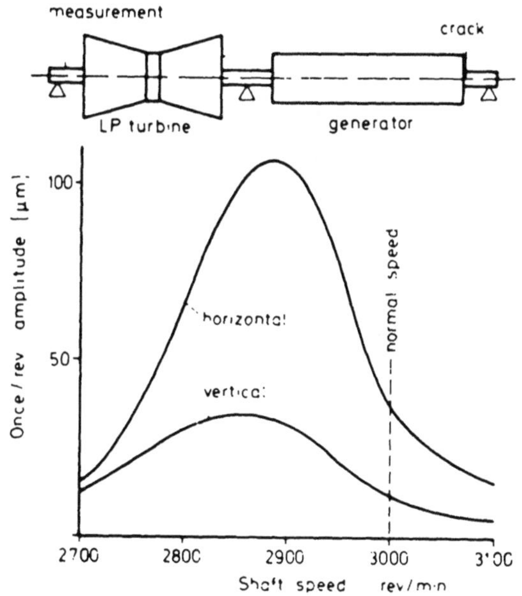

Fig. 6 Vibration amplitudes of a coupled two-rotor system with a 30 percent depth crack. After Grabowski [15]

The plotted once/revolution vibration amplitudes are positioned near the left bearing of the turbine, while the crack is located near the right end of the generator shaft. Such a behaviour - large amplitudes on positions which are far from the crack - has even been observed by cracks on existing turbine plants. Although at the calculation only a crack depth of 30 % was assumed, the excited amplitudes are very large. The calculation occurs without-balance.

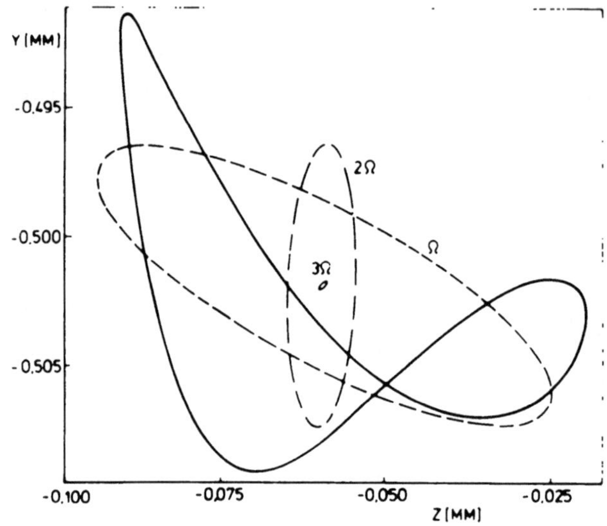

As a result of the asymmetric stiffness and damping of the bearing the harmonic components of the vibration orbits are ellipses; the shape of the complete orbit can vary considerably. An example is shown in Fig.7.

Fig. 7 Shaft vibration orbit and the first harmonic components near an end bearing, LP turbine rotor crack depth 40 percent, rotational speed $n = 2n_{cl}$. After Grabowski [15]

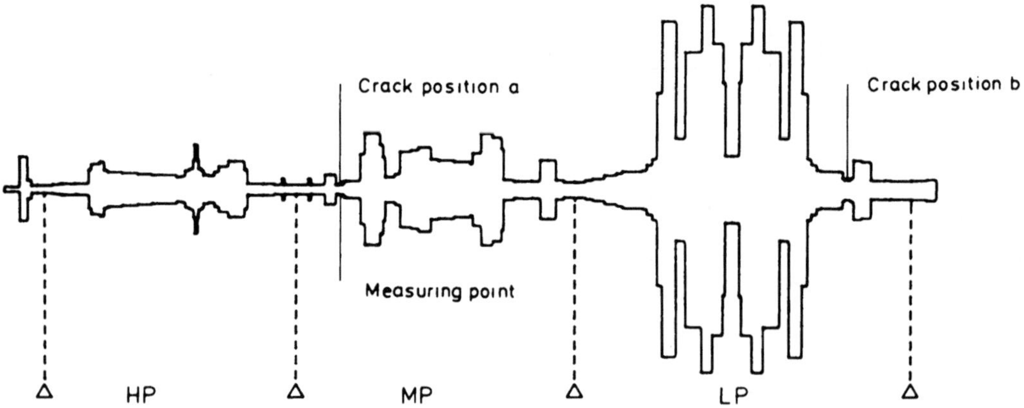

Fig. 8 Mass coverage of a rotor system of a turbine plant, crack positions and measuring point

Fig. 8 shows three coupled turbine rotors of a 300 MW power plant. The vibration amplitudes near the left bearing of the MP-turbine have been calculated for both crack positions a and b.

The once per revolution vibration amplitudes due to the crack as well as the vibration amplitudes due to unbalance are plotted in Fig. 9. The crack depth amounts to 40 % of the diameter, the radius of unbalance is assumed to be constant along the three rotors. At the resonance speed of approximately 1800 rpm both amplitudes have nearly the same size, the major semiaxis amounts to 200 µm. These resonance amplitudes are caused by an eigenmode of the MP-rotor. The resonance peak at approx. 1300 rpm is caused by the LP-rotor. At this speed the crack excites very small vibrations, although the measuring point is located at the crack position. Depending on the phase relationship the vibration amplitudes due to crack and unbalance can be added or subtracted.

The twice per revolution resonance speed of approx. 900 rpm in Fig. 10 is half the critical speed of the MP-rotor. In this case the crack would excite a distinct twice per revolution resonance peak during the drive-up and drive-down which would give a clear indication of a crack. The two lower peaks belong to critical speeds above 3000 rpm.

Fig. 11 shows for crack position b another example with large amplitudes at positions which are far from the crack, in this case up to 300 µm at approx. 1400 rpm. The amplitudes are caused by an eigenmode of the LP-rotor.

Below the speed of 750 rpm the results become instable. Because of the employed method of numerical integration it cannot be decided, whether the mathematical procedure or the physical system is unstable. At this speed one would expect a twice per revolution resonance peak.

The level of the twice per revolution resonance amplitudes at approx. 900 rpm in Fig. 12 is smaller than for crack position a; excited is an eigenmode of the MP-rotor, but at running speed the vibration level is higher than for crack position a.

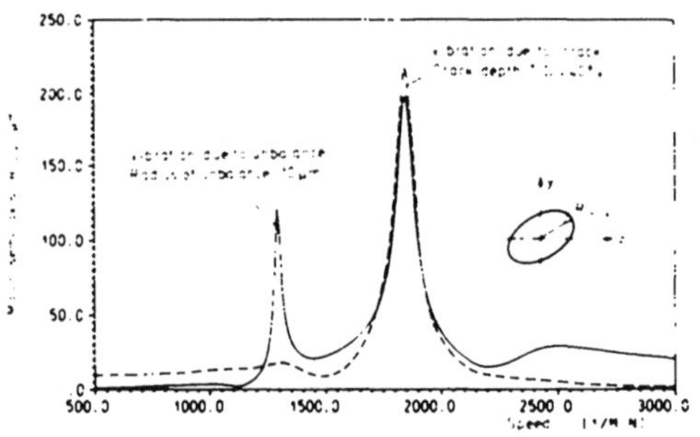

Fig. 9 Once/revolution vibrations for crack position a

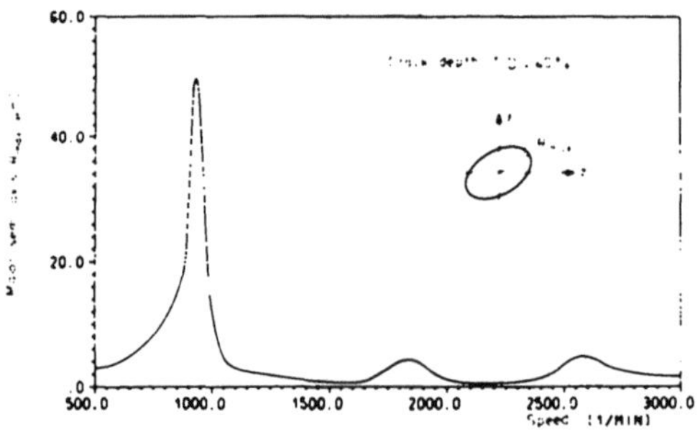

Fig. 10 Twice/revolution vibrations for crack position a

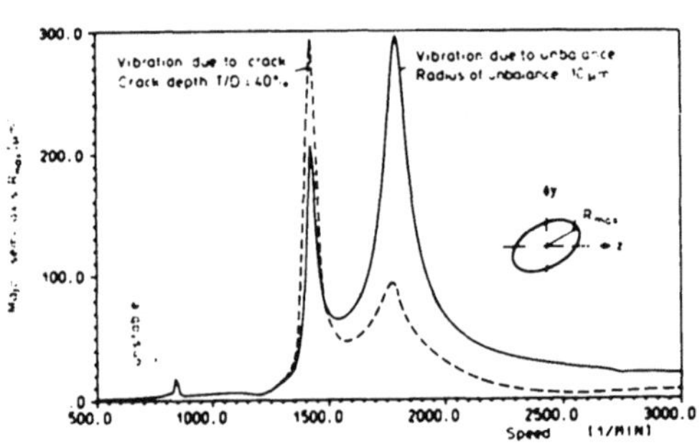

Fig. 11 Once/revolution vibrations for crack postion b

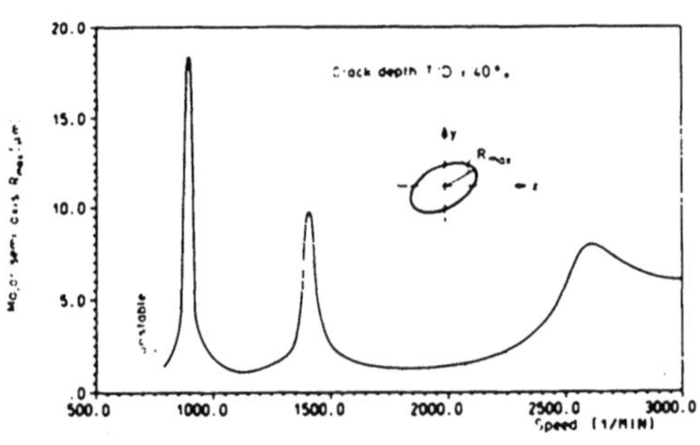

Fig. 12 Twice/revolution vibrations for crack position b

4. EXPERIMENTAL AND THEORETICAL INVESTIGATIONS

4.1 Experimental Rotor and Crack Propagation. To confirm the theoretical results of the vibrational behaviour of cracked shafts experimental investigations were carried out. Figure13 shows the experimental rotor, which is supported by journal bearings in an experimental Helium-compressor housing. At first it was intended to produce the crack by an oscillating load. However, other experiments with this kind of load have shown that a crack surface due to such a treatment is different from a crack surface which is produced by an alternating load. Therefore, we decided to produce the crack during the rotation of the shaft in the rig itself by the application of an external force.

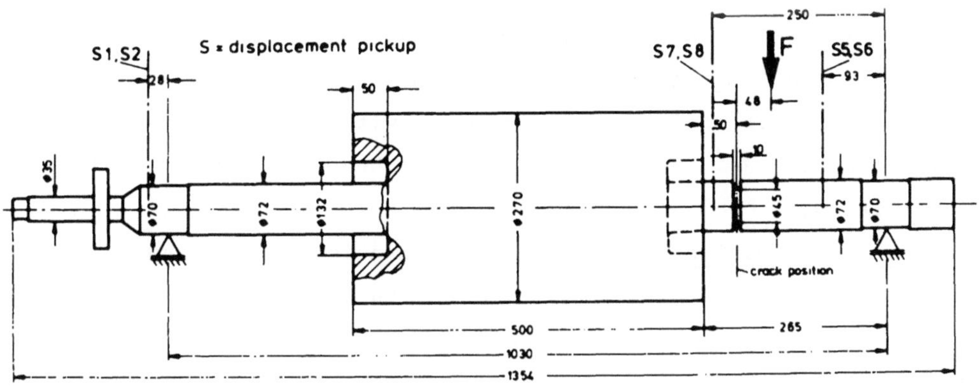

Fig. 13 Experimental rotor

Due to the high bearing load as a result of the external force F (Fig.13) the journal bearing at the crack side of the rotor was replaced by a ball bearing. The load was afforded by a ball bearing too. To reduce the necessary amount of the external force the diameter of the shaft at the crack position was reduced to a diameter of 45 mm with a cut radius of 5 mm. For crack initiation, a 4 mm deep cut was sawed with a thin wire of 70 μm diameter (Fig. 14).

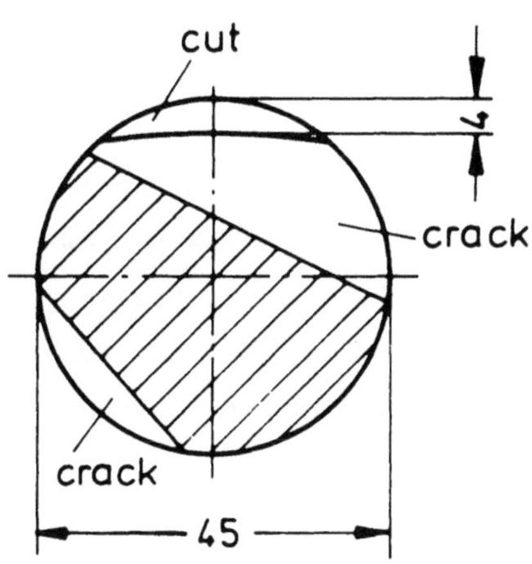

Fig. 14 Cross-section at crack position

For crack propagation external forces of 5000 N and 3000 N, depending on the crack depth, were applied for about 3 hours. The crack propagation was controlled by observing the twice per revolution resonance amplitude near the crack position using a 2-channel Fast-Fourier-Analyser. The rotating speed was equal to the corresponding resonance frequency. Unfortunatley, two cracks came into existence. Probably on the opposite side of the cut, a small groove was the cause of the initiation of the second crack. Therefore, the comparison of the theoretical and experimental results is uncertain.

For the theoretical calculations we have assumed a linear shape of the crack ground (Fig. 14) and a crack depth of 45 % of the diameter. We intend to carry out additional experiments with this rotor, e. g. with different unbalances. Therefore, we did not break the shaft at the crack position.

4.2 Theoretical Model of the Rotor. The vibrational behaviour of the rotor has been calculated with the FEM and the transfer matrix method. To make use of the crack model with L = T, the geometry in the neighbourhood of the crack had to be modified. We have taken one element of diameter of 64 mm and length of 32 mm (Fig. 15). This gives the same flexural shape as the real geometry.

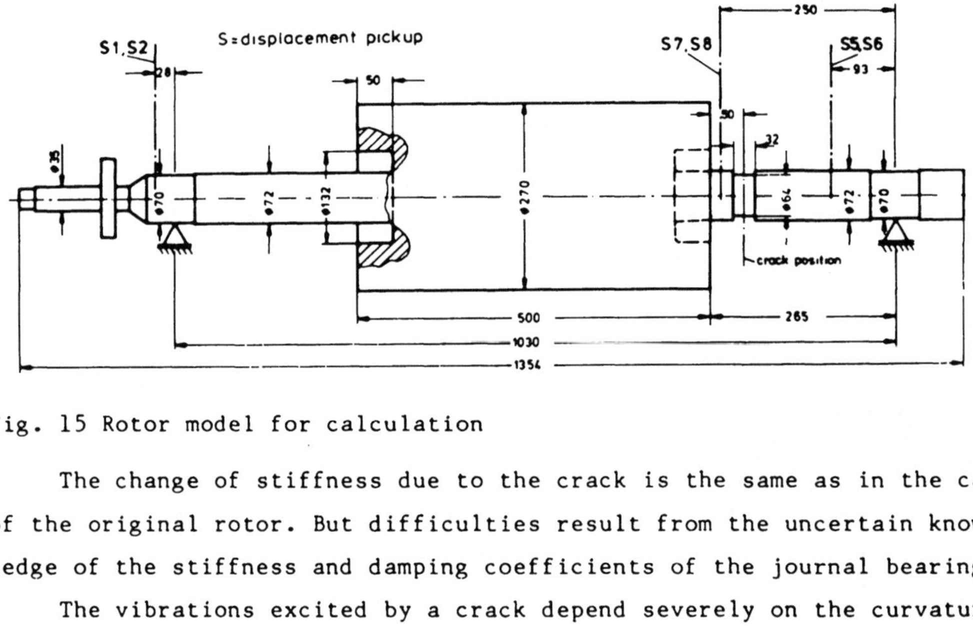

Fig. 15 Rotor model for calculation

The change of stiffness due to the crack is the same as in the case of the original rotor. But difficulties result from the uncertain knowledge of the stiffness and damping coefficients of the journal bearings.

The vibrations excited by a crack depend severely on the curvature of the eigenmodes at the crack position and on the weight influence (see Ref. 15). If

$$\int_{o}^{\ell} g\ \mu(x)\ Y_{ok}(x)\ dx \approx 0, \tag{3-1}$$

in which g is the gravitational constant, $\mu(x)$ the shaft weight/unit length and $Y_{ok}(x)$ the eigenmode, the crack does not excite any vibration. This is approximately the case for the second eigenmodes in the horizontal and the vertical plance (see e.g. Fig. 16) and accordingly for the third and fourth complex eigenmode of the complete system (Fig. 18), too.

Thus from the eigenmodes, one may have an idea about the speed ranges in which crack-excited vibrations can occur.

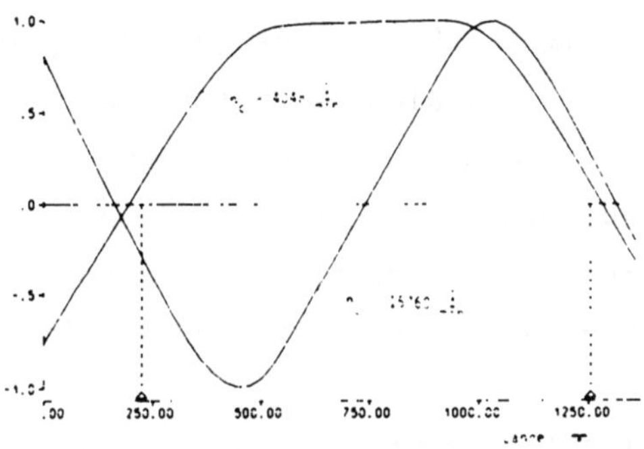

Fig. 16 Vertical eigenmodes of the undamped system (uncracked rotor)

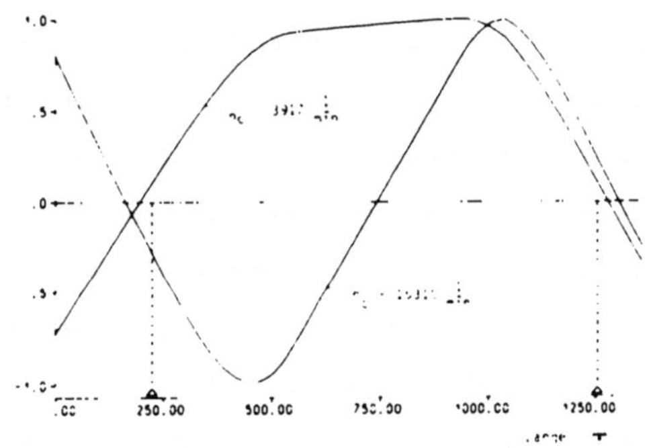

Fig. 17 Vertical eigenmodes of the undamped system (cracked rotor)

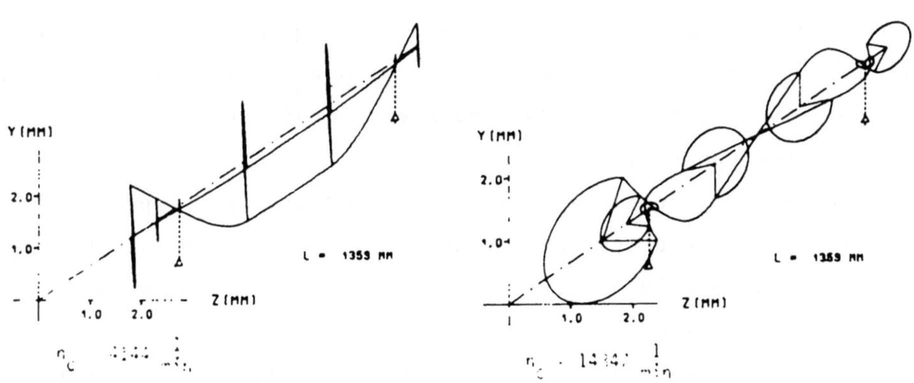

Fig. 18 Second and fourth complex eigenmode of the uncracked rotor, complete damped system

5. THEORETICAL AND EXPERIMENTAL RESULTS

5.1 Eigenfrequencies and Eigenmodes. The results of the numerical investigation of the vibrational behaviour of the cracked rotor are obtained by using the uncoupled eigenmodes of the conservative system (Fig. 17). The modal transformation includes the complete bearing stiffness and damping coefficients.

Figures 16 and 17 show, that the difference between the calculated eigenfrequencies of the uncracked and the cracked rotor amounts to approximately 3 %. The measured frequencies in figure 19 (uncracked) and figure 21 (cracked) differ approximately by 12 %. This indicates that the assumed mean stiffness reduction in the calculation of the eigenfrequencies is too small. In the case of the uncracked rotor the measured and calculated eigenfrequencies are in good agreement. For comparison only the first few eigenmodes of the complete coupled nonconservative rotor (Fig. 18) are calculated with FEM.

5.2 Rotor with Cut. An interesting experimental result is depicted in the figures 19 and 20. The cut with a depth of approximately 10 % of the diameter has no influence on the once per revolution vibration amplitudes. This corresponds to the theory for a gaping crack. (The small difference may be the result of a change in the distortion due to the storage during some days.) The amplitudes at low speeds seem to be due to the runout. The rotor is neither balanced nor equipped with an additional unbalance. The twice per revolution amplitudes may also contain runout, but less than 1 μm. The resonance amplitude increases up to 16 μm. One or two additional resonance frequencies can be observed at higher speeds. The small amplitudes of these resonance vibrations can be explained by the influence of the weight (see Eq. (5)). The rotor with cut has not been theoretically investigated.

5.3 Cracked Rotor. As mentioned, the missing knowledge of the excat bearing stiffness and damping coefficients is a problem when calculating the vibrational behavior of rotating shafts. The theoretical model of this rotor does not include the measured rigid-body eigenmode at 2400 rev/min (Fig. 21 and 23). The resonance frequency at 4000 rev/min is not theoretically determinable, too. But for crack detection

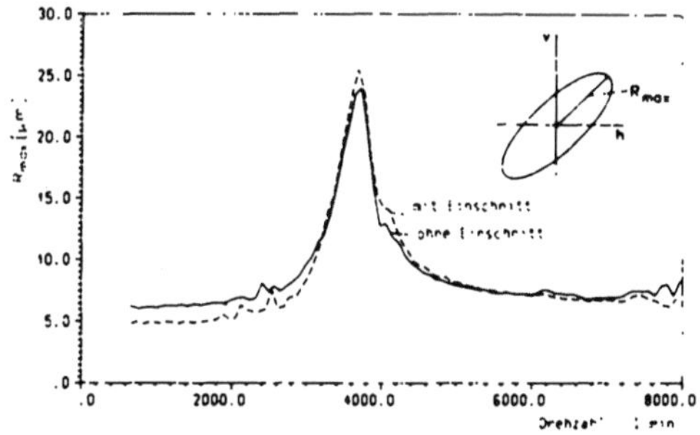

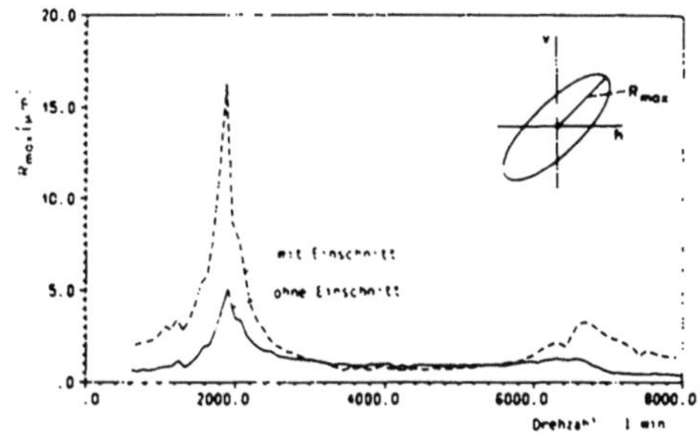

Fig. 19 Once per revolution vibration amplitudes of the rotor without and with cut measured at position S78

Fig. 20 Twice per revolution vibration amplitudes of the rotor without and with cut measured at position S78

these effects are of secondary order. The comparison of the measured (Fig. 21 and 23) and calculated (Figs. 22 and 24) once per revolution vibration amplitude shows a very good agreement at both mesuring planes S56 and S78. We indeed expected a good agreement, but this exact agreement must be an accidental one.

On the other hand, between the measured and calculated twice per revolution resonance amplitudes contain a factor of 2. This is to be explaine with the two cracks in the shaft. The stiffness changes twice per revolution and therefore we have a greater excitation of twice per revolution vibrations than the crack model delivers.

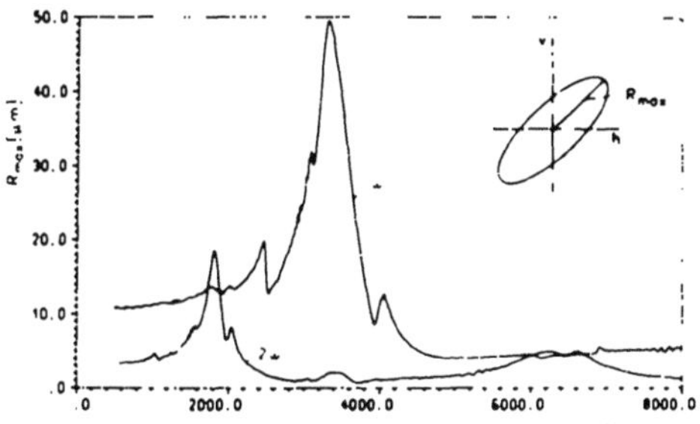

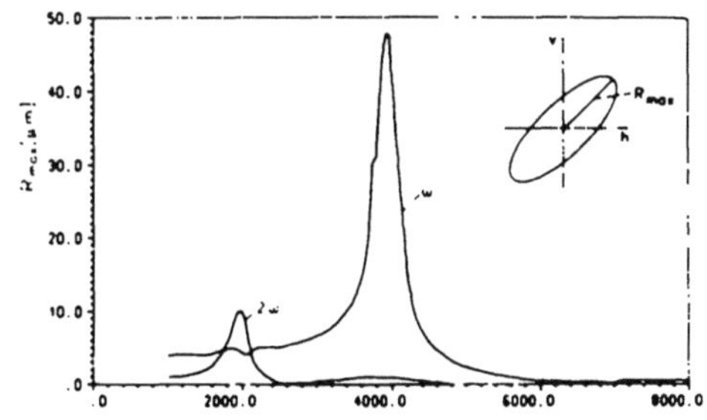

Fig. 21 Measured vibration amplitudes of the cracked rotor at postion S56

Fig. 22 Calculated vibration amplitudes of the cracked rotor at position S56 (crack depth 45 % of the diameter)

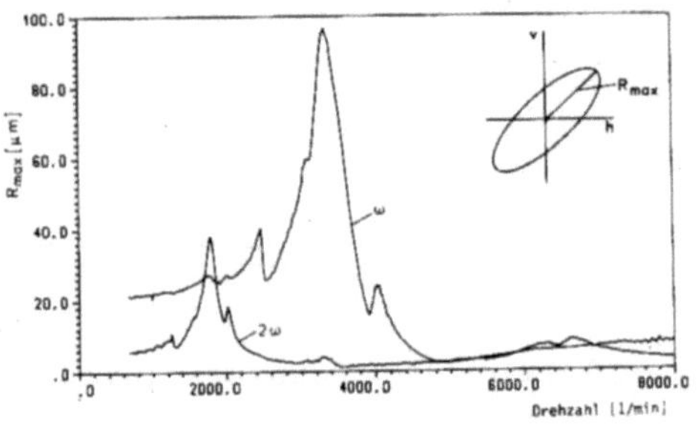

Fig. 23 Measured vibration amplitudes of the cracked rotor at position S78

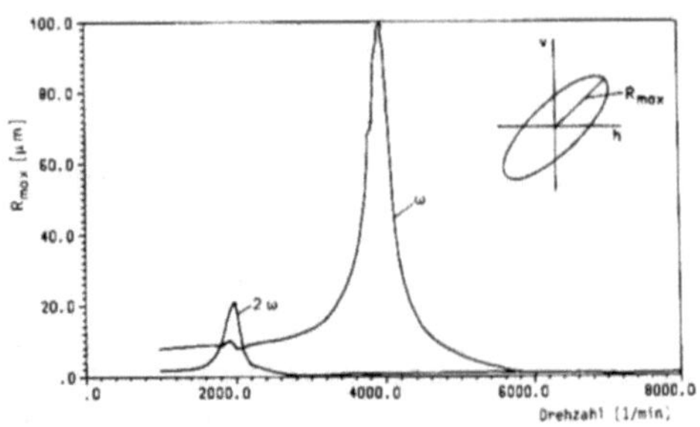

Fig. 24 Calculated vibration amplitudes of the cracked rotor at position S78 (crack depth 45 % of the diameter)

In figures 25 and 26, the complete frequency spectrum of the rotor with cut is compared with the spectrum of the cracked rotor. The scales are the same. This representation gives a good survey of the change of the vibrational behaviour due to the crack.

At the rotational speed of 7000 rev/min a resonance vibration appeared with a frequency of approximately 3500 rev/min for a very short time. Until now we have no explanation for this effect.

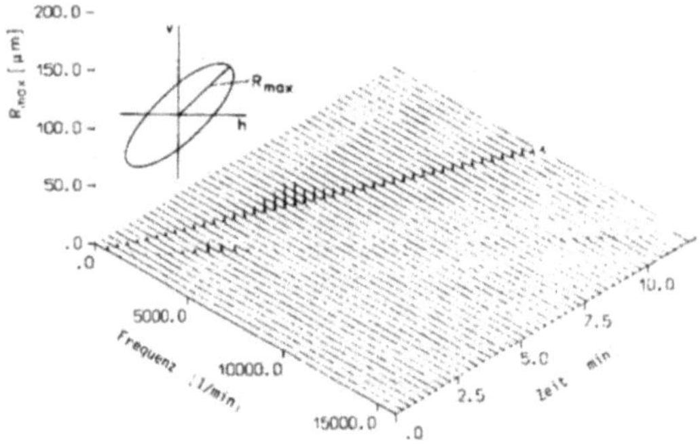

Fig. 25 Frequency spectrum of the rotor with a cut

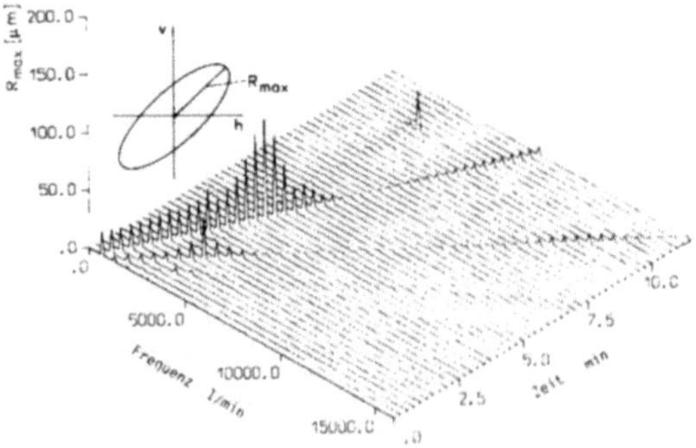

Fig. 26 Frequency spectrum of the cracked rotor

6. Conclusion. To detect a crack in a turbine rotor by means of the shaft vibration monitoring, the influence of the crack on the vibrational behaviour has to be investigated first. For this purpose, a procedure is introduced which uses the modal analysis. The stiffness at the crack position depending on the angle of rotation will be determined according to model. Due to the structure of the computer program, the stiffness can also be obtained from test data.

The numerical results show that in the speed range in which many turbine plants have their normal speed a crack can cause an important change of the shaft vibrations. The crack excites mainly one/revolution and twice/revolution vibrations. Both can be used for crack detection. The size of vibration amplitudes depends mainly on the construction of the rotor and on the crack position.

The comparison of calculation and measurement shows in principle that it is possible to predetermine the vibration amplitudes excited by a crack. But at the same time this investigation shows the problems in modelling the system. The developed crack model seems to be useful.

For crack supervision the phase should also be taken into account, because the amplitude of the sum of the original vibration and the vibration due to crack can become smaller with crack propagation.

In the future the influence of unbalance is to be investigated. Our calculations have shown that this influence is often small for greater turbine rotors because the statical deflection is greater than the vibration amplitudes. But in rotors constructed as in the case of this experimental rotor, an influence can be expected. Here the vibration amplitudes at crack position have the same magnitude as the statical deflection.

Nomenclature

A = cross-sectional area of shaft

b - support damping coefficient

$\underline{C}$ = support damping matrix

c_e = external damping coefficient

c_i = internal damping coefficient

$\underline{C}$ = modal damping matrix

e_η, e_ζ = radius of unbalance in rotating coordinate system

E = Young's modulus

f_y, f_z = external force due to unbalance

$\underline{F}$ = modal matrix of unbalance force

g = gravitational constant

$\underline{g}$ = modal vector of self-weight load

G = shear modulus

i = radius of mass moment of inertia

I = cross-sectional transverse moment of inertia of shaft

$I_m + \Delta I_y$, $I_m + \Delta I_z$, $I_{yz} + I_{zy}$ = second moments of area at crack postion, depending on angle of rotation φ

k = support stiffness coefficient

$\underline{K}$ = support stiffness matrix

$\underline{K}_m$ = constant part of modal stiffness matrix

$\underline{K}(t)$ = time dependent part of modal stiffness matrix

ℓ = length of shaft

L = range of crack extension along the axis of the shaft

M_y, M_z = bending moment

$\underline{M}$ = modal mass matrix, diagonal

q_y, q_z = modal coordinate

R_y, R_z = bearing reaction force

$\underline{S}$ = rectangular rotation matrix

T = crack depth

$\underline{T}$ = transformation matrix

$\underline{u}$ = vector of displacements

$\underline{U}_o$, $\underline{\Phi}_o$, $\underline{\Gamma}_o$ = modal matrix

V_y, V_z = shear force

x = coordinate along rotor axis

x_{c_ℓ}, x_{c_r} = left and right boundary of the crack length L

y,z = radial displacement

Y_o, Z_o = eigenfunction for radial displacement

γ_y, γ_z = angular displacement of shear deformation

Γ_y, Γ_z = eigenfunction for shear deformation

δ = unit delta function

$\varkappa$ = cross-sectional shape factor for shear deformation

μ = shaft mass per unit length

φ = angle of rotation

φ_y, φ_z = angular displacement

Φ_{yo}, Φ_{zo} = eigenfunction for angular displacement

Ω = angular speed of rotation

Indexes and Superscripts

k, K, ℓ, L = number of eigenfunctions

n, N = bearing number

o = eigenfunction

r, R = eigenfunction number

y = y direction

z = z direction

$(\dot{\ })$ = $\frac{\partial}{\partial t}$

$(\ ')$ = $\frac{\partial}{\partial x}$

References

[1] STODOLA,A.: Dampf- und Gasturbinen. Springer, Berlin, 5. Auflage (1922) 931.

[2] KELLENBERGER,W.: Biegeschwingungen einer unrunden, rotierenden Welle in horizontaler Lage. Ingenieur-Archiv. XXVI (1958) 302-318.

[3] TONDL,A.: The Effect of Unequal Moments of Inertia of the Shaft Section on the Motion and Stability of a Rotor. "Some problems of Rotor Dynamics". 1st ed., Champmann & Hall, London.

[4] BISHOP,R.E.D. and PARKINSON, A.G.: Second Order Vibration of Flexible Shafts. Phil. Trans. of Royal Society of London, Vol. 259, No. 1095, (1965).

[5] MAYES,I.W., and DAVIES, W.G.R.: The Vibrational Behaviour of a Rotating Shaft System Containing a Transverse Crack. "Conference on Vibrations in Rotating Machinery", University of Cambridge, Sept. 1976, 53-64.

[6] GASCH,A.: Dynamisches Verhalten des Lavalläufers mit angerissenem Wellenquerschnitt. VDI-Berichte, No. 269 (1976).

[7] HENRY,T.A. and OKAH-AVAE, B.E.: Vibrations in Cracked Shafts. "Conference on Vibrations in Rotating Maschinery", University of Cambridge, Sept. 1976, 15-19.

[8] MAYES,I.W.: Crack Propagation in Rotating Shafts. ASME Paper 77-DET-164, presented at the Design Engineering Technical Conference, Chicago, Illinois, Sept. 26-30, 1977.

[9] Buerhop,H.: Zur Berechnung der Biegesteifigkeit abgesetzter Stäbe und Wellen unter Anwendung von finiten Elementen. VDI-Fortschrittsberichte, Reihe 1, No. 36 (1975).

[10] ZIEBARTH,H., SCHWERDTFEGER,H. and MÜHLE, E.-E.: Auswirkung von Querrissen auf das Schwingungsverhalten von Rotoren. VDI-Berichte 320 (1978) 37-43.

[11] GRABOWSKI,B.: Das Schwingungsverhalten eines angerissenen Turbinenläufers. VDI-Bericht Nr. 320 (1978) 31-36.

[12] MEYER,J.: Zum Schwingungsverhalten einer Lavalwelle mit angerissenem Wellenquerschnitt. Dissertation, HSBW Hamburg 1979.

[13] GRABOWSKI,B.: Zur modalen Simulation des instationären Schwingungsverhaltens von Turboläufern. VDI-Fortschrittsberichte, Reihe 11, No. 25 (1967).

[14] FEHLBERG,E.: Neuere genauere Runge-Kutta-Formeln für Differentialgleichungen zweiter Ordnung bzw. n-ter Ordnung, ZAMM 40 (1960) 449-455.

[15] GRABOWSKI,B.: The Vibrational Behaviour of a Turbine Rotor Containing a Transverse Crack. Trans. ASME, J. Mech Design Vol. 102, Jan. 1980, 140-146.

[16] INAGAKI, T., KAUKI, H. and SHIRAKI,K.: Response Analysis of a General Asymmetric Rotor-Bearing-System. Trans. ASME, J. Mech. Design Vol. 102, Jan. 1980 ,147-157.

[17] MAYES,I.W.and DAVIES, W.G.R.: A Method of Calculating the Vibrational Behaviour of Coupled Rotating Shafts Containing a Transverse Crack. "Second International Conference Vibrations in Rotating Maschinery", University of Cambridge, 2-4 Sept. 1980.

[18] HAAS,H.: Großschäden durch Turbinen- oder Generatorläufer, entstanden im Bereich bis zur Schleuderdrehzahl. Der Maschinenschaden, 50. Jahrgang (1977), Heft 6, 195-204.

CHAPTER 5.2

DYNAMIC BEHAVIOR OF ROTATING TURBINE BLADES

O. Mahrenholtz

1. Introduction

The vibrations of rotating blades are an important subject in turbomachinery. The problem of determining the eigenfrequencies and mode shapes of vibration of pre-twisted, tapered cantilever blades is of importance in the design of turbine and compressor blading. The geometry of a turbine blade is a complex one. The turbine blade can be pre-twisted and tapered both in width and depth.

The blades are to be described by a beam model which includes the following influences:

- the stiffening effect of the centrifugal force field
- tapered blades
- a bladed disk with stagger angle α between 0 and 90 deg
- pre-twisted blades
- shear deformation and rotary inertia
- support elasticity
- coupled vibrations of the disk and its blades
- shrouded blades, banded groups of turbine buckets (packet vibrations)

2. The purpose of calculating vibrations

One basic requirement for the design of reliable turbomachines is the avoidance of resonance situations in the blading. By resonance is meant the coincidence of the eigenfrequencies of the blades with the frequencies of the excitation forces of aerodynamic origin. If this situation cannot be prevented, stresses are increased in the blades amounting to many times the steady values due to gas pressure and centrifugal force. Should these stresses exceed the reserved fatigue strength of the materials used, blade failure will ensue in a short time.

The blading design is particularly difficult in the case of turbomachines, where continuous operation is required not only at a fixed speed but within a speed range. This means that frequency ranges must be taken into account also for the aerodynamic excitation forces acting on the blades. When blades are designed the eigenfrequencies must be kept out of the excitation range of the aerodynamic forces. The design of turbomachine blading from the vibration aspect must satisfy two requirements:

- the amplitude and frequency of the aerodynamic forces acting on the blades must be known
- the eigenfrequencies and the modes of vibration must be found out.

The present work is devoted exclusively to the calculation of the eigenfrequencies of turbine blades and bladed disks.

3. Bending vibration of a rotating untwisted slender beam

Fig. 1 shows the bladed disk considered. The differential equation developed in this chapter note the following assumptions:

- The beam is fixed rigidly to a rotating, axi-symmetric disk

- The disk rotates with a const. angular velocity Ω

- $A(x) = A(x=0) = \text{const.}$

- The beam is untwisted

- $L \gg i_y$; that means, the beam is to be described by the classical beam theory

- $\alpha = 90°$

- The natural vibrations considered are assumed to be of small amplitude, and the material homogeneous and linear elastic.

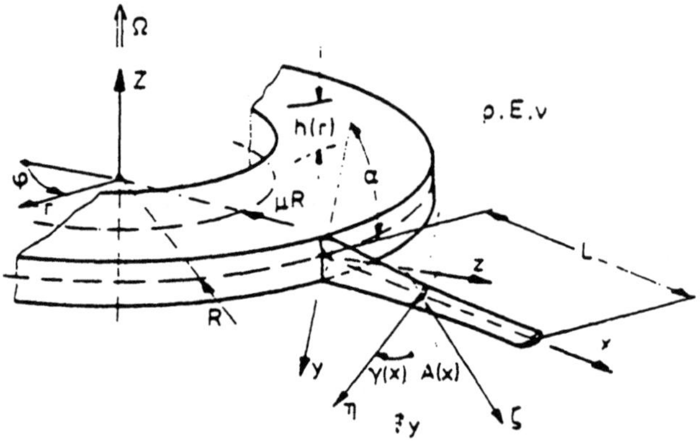

Fig. 1 Bladed disk, notation

Equation of motion:

For a differential element:

$$d\vec{f} - dm\,[\vec{a}_f + \vec{a}_{cor}] = dm\,\vec{a}_{rel} \qquad (5.2\text{-}1)$$

$$d\vec{M} = \vec{0} \qquad (5.2\text{-}2)$$

The equation of motion are formulated on a deformed element (Fig. 2):

- external force:

$$d\vec{f} = (N'dx + n_0 dx)\,\vec{i} + (Q'dx + q_0 dx)\,\vec{k}$$

- acceleration:

$$\vec{a}_f = \vec{a}_B + \dot{\vec{\omega}}\times\vec{p} - \omega^2\vec{p}$$

where

$$\vec{a}_B = -\Omega^2 R\vec{i};\quad \dot{\vec{\omega}} = 0;\quad \omega=\Omega$$

$$\vec{p} = (x+u)\vec{i}+w\vec{k}$$

$$\vec{a}_{cor} = 2\vec{\omega}\times\vec{v}_{rel}$$

$$\vec{\omega} = \Omega\vec{j};\vec{v}_{rel} = \dot{u}\vec{i}+\dot{w}\vec{k}$$

$$\vec{a}_{cor} = 2\Omega(\dot{w}\vec{i}-\dot{u}\vec{k})$$

$$\vec{a}_{rel} = \ddot{u}\vec{i}+\ddot{w}\vec{k}$$

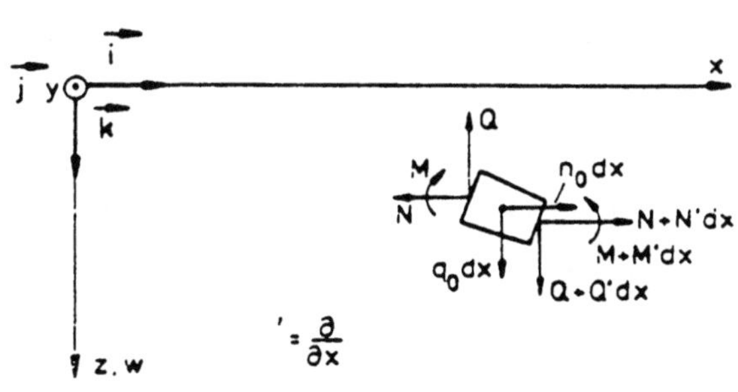

Fig. 2 Internal forces and displacements of a deformed element

Substituting these expressions into equation (5.2-1), we obtain (5.2-3) and (5.2-4)

$$N'+n_0-\mu[-\Omega^2R-\Omega^2(x+u)+2\Omega\dot{w}] = \mu\ddot{u}$$

with $u \ll (R+x)$ follows

$$N'+n_0+\mu[\Omega^2(R+x)-2\Omega\dot{w}] = \mu\ddot{u} \qquad (5.2\text{-}3)$$

$$Q'+q_0-\mu[-\Omega^2 w-2\Omega\dot{u}] = \mu\ddot{w}$$

$$Q'+q_0+\mu[\Omega^2 w+2\Omega\dot{u}] = \mu\ddot{w} \qquad (5.2\text{-}4)$$

Equation (5.2-2) applied to the deformed element yields

$$M'-Q+Nw' = 0 \qquad (5.2\text{-}5)$$

Physical interpretation of the equations (5.2-3) - (5.2-5):
we recognize

- a coupled motion in x- and z-direction by Coriolis-force in equation (5.2-3) and (5.2-4)

- a component of the centrifugal force in x-direction ($\mu\Omega^2(R+x)$ in equation (5.2-3))(Fig. 2a)

- a component of the centrifugal force in z-direction ($\mu\Omega^2 W$ in equation (5.2-4))(Fig. 2a)

- an influence of the centrifugal force N on the moment - shearing force relation in equation (5.2-5)

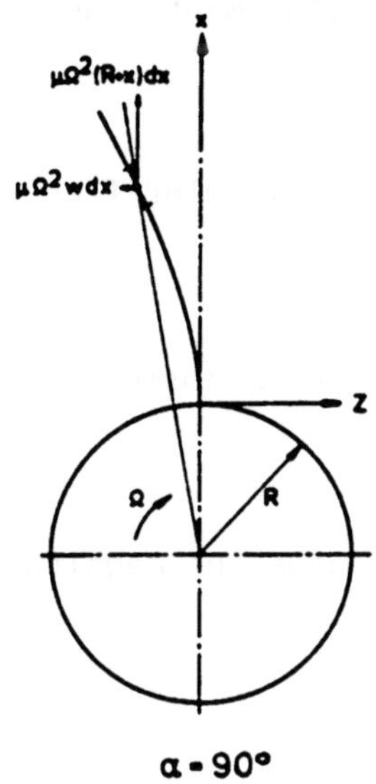

Fig. 2a Components of the differential centrifugal force

Kinematical equations and Hooke's law

Relations between displacements, deformations and internal forces yield:

$$u' = \frac{N}{EA} \tag{5.2-6}$$

$$w'' = -\frac{M}{EI} \tag{5.2-7}$$

If the internal forces are eliminated and we assume that $n_o = 0$ and $q_o = 0$ we obtain from equation (5.2-3) - (5.2-7):

$$(EAu')' + \mu[\Omega^2(R+x) - 2\Omega\,\dot{w}] = \mu\ddot{u} \tag{5.2-8}$$

$$-(EIw'')'' + (EAu'w')' + \mu[\Omega^2 w + 2\Omega\,\dot{u}] = \mu\ddot{w} \tag{5.2-9}$$

For the natural vibrations the displacements can be represented as:

$$u(x,t) = U(x)\cdot \sin \omega t$$

ω - circular frequency

$$w(x,t) = W(x)\cdot \sin \omega t$$

The equations (5.2-8) and (5.2-9) can be simplified by considering the following estimations:

ω and Ω are of the same order of magnitude

$-\Omega^2 (R+x) >> 2\Omega\omega W \qquad (R+x) >> W \qquad$ in Eq. (5.2-8)

$-\Omega^2 W >> 2\Omega\omega U \qquad W >> U \qquad$ in Eq. (5.2-9)

$-|\mu \ddot{u}|_{max} = \mu\omega^2 U << \mu\Omega^2 (R+x) \qquad$ in Eq. (5.2-8)

Therefore, the inertia force equation (5.2-8) and the Coriolis-force in eqs. (5.2-8) and (5.2-9) are negligible. Then eqs. (5.2-8) and (5.2-9) reduce to eqs. (5.2-10) and (5.2-11):

$$N' + \mu\Omega^2 (R+x) = 0 \tag{5.2-10}$$

$$-(EIW'')'' + (NW')' + \mu\Omega^2 W = -\mu\omega^2 W \tag{5.2-11}$$

The boundary conditions for a cantilever beam are:

$$W(o) = W'(o) = 0$$

$$W''(l) = W'''(l) = 0$$

The solution of equation (5.2-10) by integration results

$$\int_x^L N' \, d\xi = -\int_x^L \mu\Omega^2 \, (R+\xi) \, d\xi$$

$$\mu(x) = \rho \, A(x)$$ μ - mass per unit length

$$N(L) = 0$$

$$\Rightarrow N(x) = \rho\Omega^2 \int_x^L A(\xi) \, (R+\xi) \, d\xi \qquad (5.2\text{-}12)$$

Equation (5.2-12) can be calculated for a non-uniform bar by numerical integration. N(x) in equation (5.2-11) can be replaced by the solution of equation (5.2-12). Then, the solution of equation (5.2-11) can be found by various approximation methods, like

- Rayleigh - Ritz method
- Galerkin procedure
- transfer matrix method
- numerical integration techniques (Runge - Kutta - Fehlberg)
- finite element method

Results are shown in Fig. 3 for a rotating beam with const. cross section. The curves increase parabolically with rotating speed; and the eigenfrequencies increase with ε = R/l respectively.

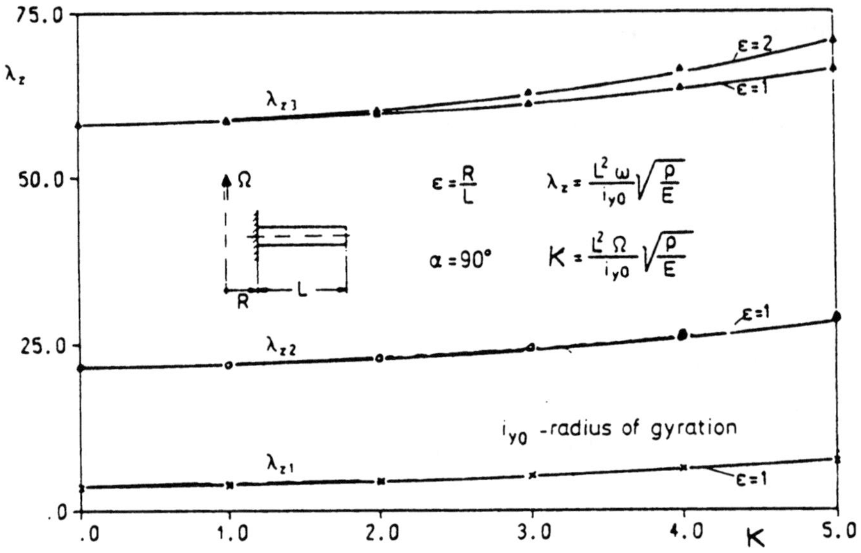

Fig. 5 Eigenfrequencies for bending vibration of a rotating 'Euler-Bernoulli Beam' with const. cross section

An approximation of the stiffening effect of the centrifugal force field is given by SOUTHWELL's equation:

$$\omega^2(\Omega) = \omega_0^2 + \varphi\Omega^2 \qquad (5.2\text{-}13)$$

Ω - rotating speed

ω_0 - circular eigenfrequency at $\Omega = 0$

φ - centrifugal force factor

The factor φ was measured by BÖHM [1] for twisted and untwisted beams:

$\varphi = 1.55 \frac{R}{L} + 1.17 - \sin^2 a$ untwisted beams

$\varphi = 1.53 \frac{R}{L} + 1.19 - \sin^2 a$ twisted beams

Equation (5.2-15) is valid only for the first eigenfrequency. SOUTHWELL's equation is controversial for higher eigenfrequencies. Nevertheless, in [2] values for φ are given.

For the experimental verification of the theoretical results a lot of special test techniques have been developed. The aim of carried out measurements is (1) to prove the basic assumptions which are introduced to describe the blade and (2) to test the accuracy of the numerical approximation. Some results considering both aspects are given in [1] and later in [9] where experimental investigations of mode shapes with help of a holographic technique are described. Typical holograms [9] of two bending modes and - not discussed here in detail - torsional modes of a twisted blade are shown in Fig. 3a.

Mode of vibration 3T f = 4027 Hz

Mode of vibration 2T f = 2385 Hz

Mode of vibration 2T f = 2385 Hz

Mode of vibration 4B f = 1632 Hz

Fig. 3a
Holograms of modes of vibrations of an twisted slender beam with rectangular cross section

4. Consideration of further influences

The equations of motion of a rotating pre-twisted cantilever beam with variable cross section, including shear deformation and rotary inertia are given by BÖHM [1]:

$$\frac{\partial q_y}{\partial x} + \frac{\partial}{\partial x}(n_x^* \frac{\partial v}{\partial x}) + \rho A \Omega^2 \; (v\cos\alpha - w\sin\alpha)\cos\alpha = \rho A \frac{\partial^2 v}{\partial t^2} \tag{5.2-14}$$

$$q_y + \frac{\partial m_z}{\partial x} = \rho \; (I_{yz} \frac{\partial^2 \gamma_y}{\partial t^2} - I_z \frac{\partial^2 \gamma_z}{\partial t^2}) \tag{5.2-15}$$

$$\frac{\partial q_z}{\partial x} + \frac{\partial}{\partial x}(n_x^* \frac{\partial w}{\partial x}) - \rho A \Omega^2 \; (v\cos\alpha - w\sin\alpha)\sin\alpha = \rho A \frac{\partial^2 w}{\partial t^2} \tag{5.2-16}$$

$$q_z - \frac{\partial m_y}{\partial x} = \rho \; (- I_y \frac{\partial^2 \gamma_y}{\partial t^2} + I_{yz} \frac{\partial^2 \gamma_z}{\partial t^2}) \tag{5.2-17}$$

The stress resultant n_x caused by rotation in the equations (5.2-14) and (5.2-16) is obtained from

$$n_x^* = \rho \; \Omega^2 \int_x^L A \; (R + \xi) d\xi \tag{5.2-18}$$

Constitutive equations of a pre-twisted blade described by TIMOSHENKO's beam theory:

$$q_y = k \; GA \; (\frac{\partial v}{\partial x} + \gamma_z) \tag{5.2-19}$$

$$m_z = - E \; (I_z \frac{\partial \gamma_z}{\partial x} + I_{yz} \frac{\partial \gamma_y}{\partial x}) \tag{5.2-20}$$

$$q_z = k \; GA \; (\frac{\partial w}{\partial x} + \gamma_y) \tag{5.2-21}$$

$$m_y = E \; (I_{yz} \frac{\partial \gamma_z}{\partial x} + I_y \frac{\partial \gamma_y}{\partial x}) \tag{5.2-22}$$

The influences of the parameters listed above on the eigenfrequencies are described in the following paragraphs.

4.1. Tapered beams

A tapered beam can be considered in eq. (5.2-11) and (5.2-12) by replacing

$$\mu \rightarrow \mu(x) = \rho\, A(x) \qquad \text{and}$$
$$I \rightarrow I(x) = A(x)\, i^2(x)$$

For a tapered non-rotating beam with rectangular cross section results are shown in Fig. 4. The first eigenfrequency increases with tapering while the second and third eigenfrequencies decrease. Under rotation the eigenfrequency of a tapered beam increases less than for beam with uniform cross section and the same root cross section.

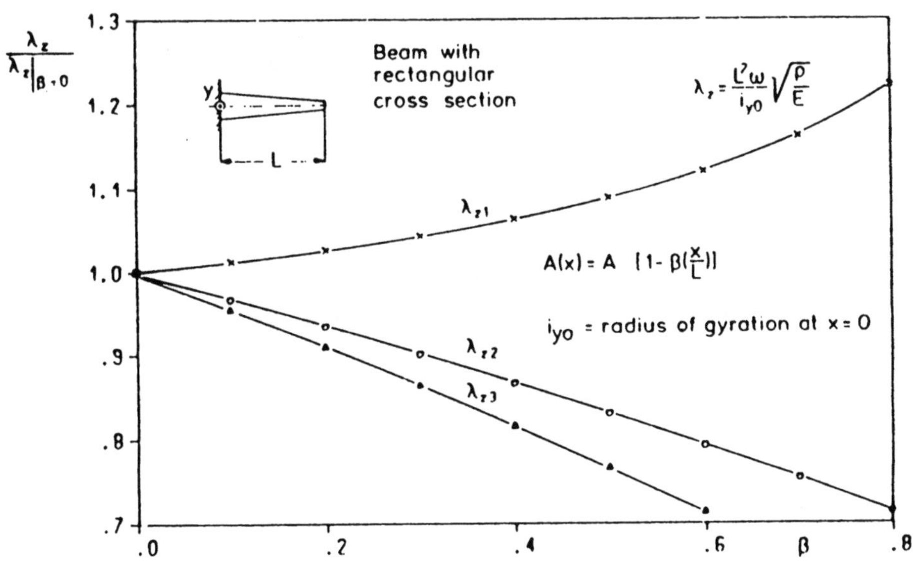

Fig. 4 Eigenfrequencies for bending vibration of a tapered beam

4.2. Stagger angle α

The stagger angle is defined in Fig. 1. There are two different influences on the dynamical behavior of turbine blades.

a) Two extreme cases of the stagger angle α are recognizable (Fig. 5):

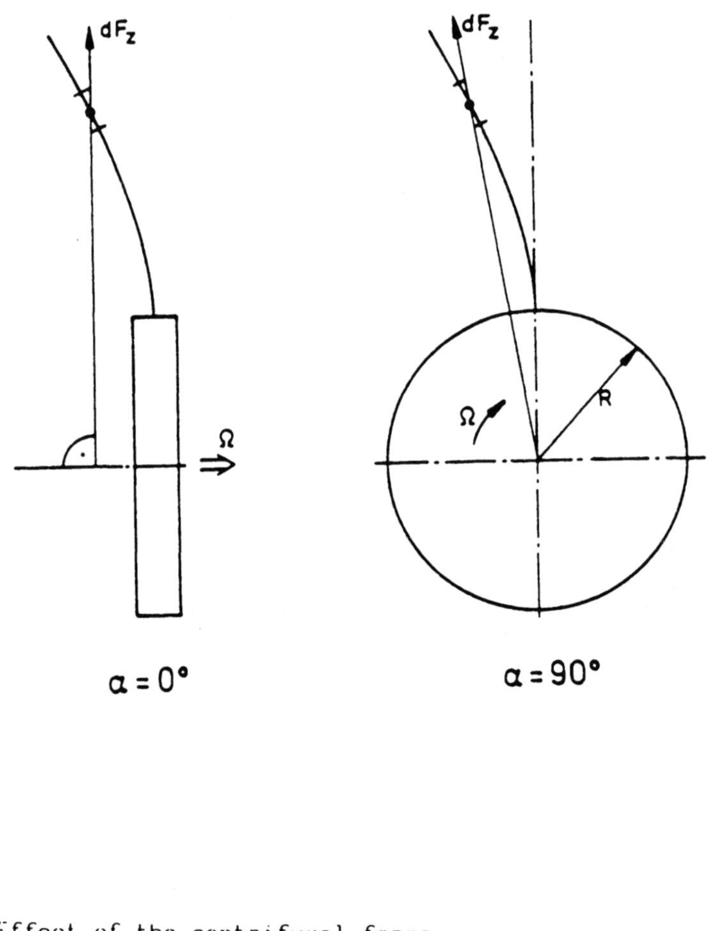

Fig. 5 Effect of the centrifugal force

α = 0 deg: The centrifugal force component $\mu\Omega^2 W$ in eq.(5.2-11) vanishes

α = 90deg: The centrifugal force component $\mu\Omega^2 W$ in eq.(5.2-11) increases the bending. So the centrifugal force component $\mu\Omega^2 W$ reduces the 'stiffness' of the blade.
The influence of the stagger angle α on a rotating beam with constant cross section is shown in Fig. 6.

b) A disk with blades of stagger angle α between 0 and 90deg is considered. In this case, there is a negligible coupling between the flapwise and the chordwise bending of the blade (see eqs. (5.2-14)-(5.2-22)).
The influence of the stagger angle α on the first eigenfrequeny can be taken into account by SOUTHWELL's equation (5.2-13).

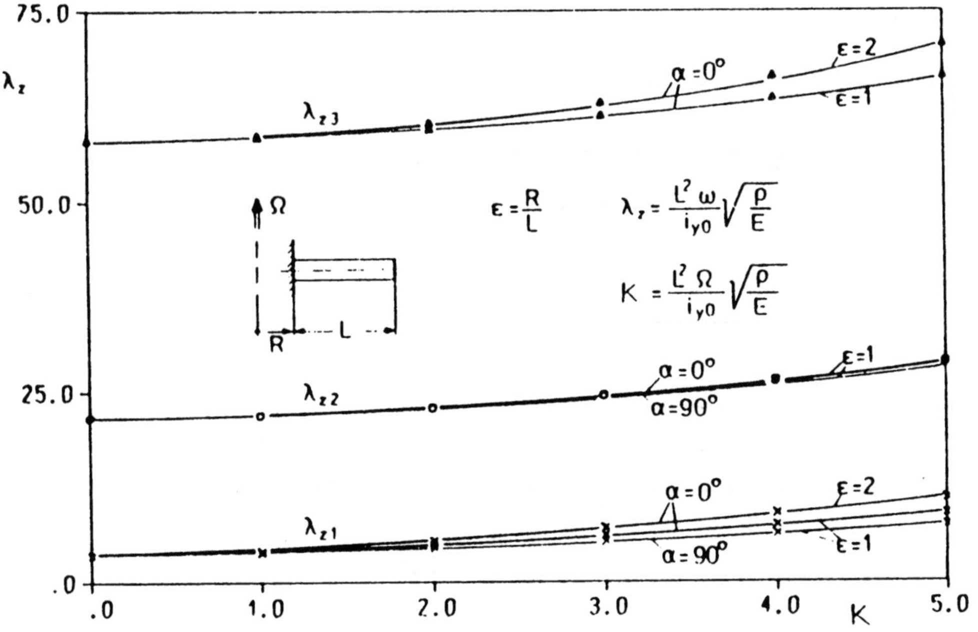

Fig. 6 Eigenfrequencies for bending vibration of a rotating beam with const. cross section and various stagger angle

4.3. Pre-twisted beams

A pre-twisted beam with constant cross section is considered. In this case there is a strong coupling between the flapwise and the chordwise bending of the blade. The influence of the total angle of twist on the first and second eigenfrequency is shown in Fig. 7 and Fig. 8. The curves in Fig. 7 and Fig. 8 are measured by BÖHM [1].

The first eigenfrequency is nearly constant with increasing total angle of twist while the second eigenfrequency is decreasing.

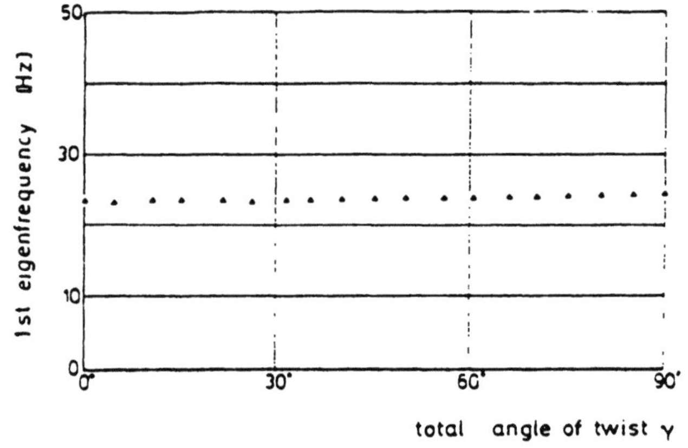

Fig. 7 Measured fundamental frequency of a pre-twisted beam (H:B:L = 3:28:328 MM)

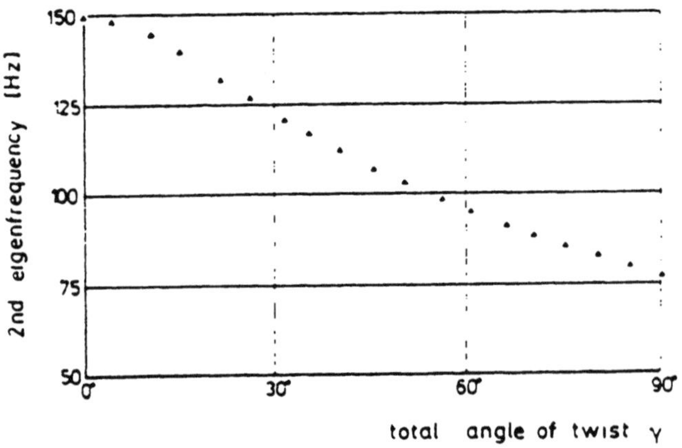

Fig. 8 Measured second mode frequency of a pre-twisted beam (H:B:L = 3:28:328 MM)

Fig. 9 shows measured and calculated eigenfrequencies for bending vibrations of pre-twisted beams. Differential equations for bending vibrations of pre-twisted beams are given by MONTOYA [4] and BÖHM [1] (see eqs. (5.2-14)-(5.2-22)). The influence of the total angle of twist on the first eigenfrequency can be taken into account by SOUTHWELL's equation (5.2-13).

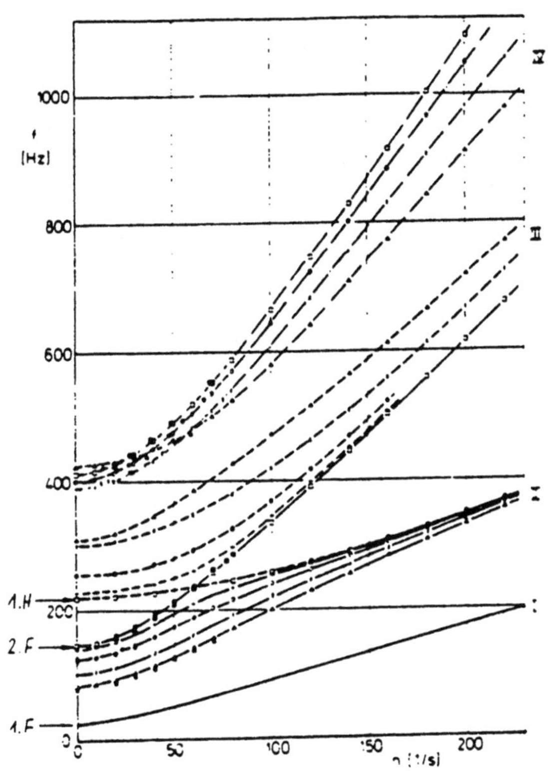

Fig. 9 Measured and calculated eigenfrequencies for bending vibration of a rotating pre-twisted beam

Total angle of twist	0°	32,1°	57,8°	85,8°	15°
Experimental	■	●	/	▲	/
Theoretical	□		x	∴	+

F - Flapwise bending H - Chordwise bending

4.4. Shear deformation and rotary inertia

For a reliable theoretical prediction of the eigenfrequencies of short blades or higher modes, it is necessary to include the effects of shear and rotary inertia. The shear deformation, which is ignored in the Euler - Bernoulli - beam theory, constitutes an additional deformation to the bending deformation. By taking this into account the blade under consideration is assumed to be more elastic, from which it follows directly that the eigenfrequencies are reduced by the influence of the shear deformation. If not only the translation energy but also the rotation energy of an element of length dx is included in the calculation, the kinetic energy of the blade is increased. Since the expression for the kinetic energy contains ω^2 and the elastic energy does not change, taking the rotation energy into account also has the effect of lowering the frequency. In the following differential equations an untwisted, non-rotating blade is described by TIMOSHENKO's beam theory [5].

Fig. 10 shows the influence of shear deformation and rotary inertia on the first three eigenfrequencies.

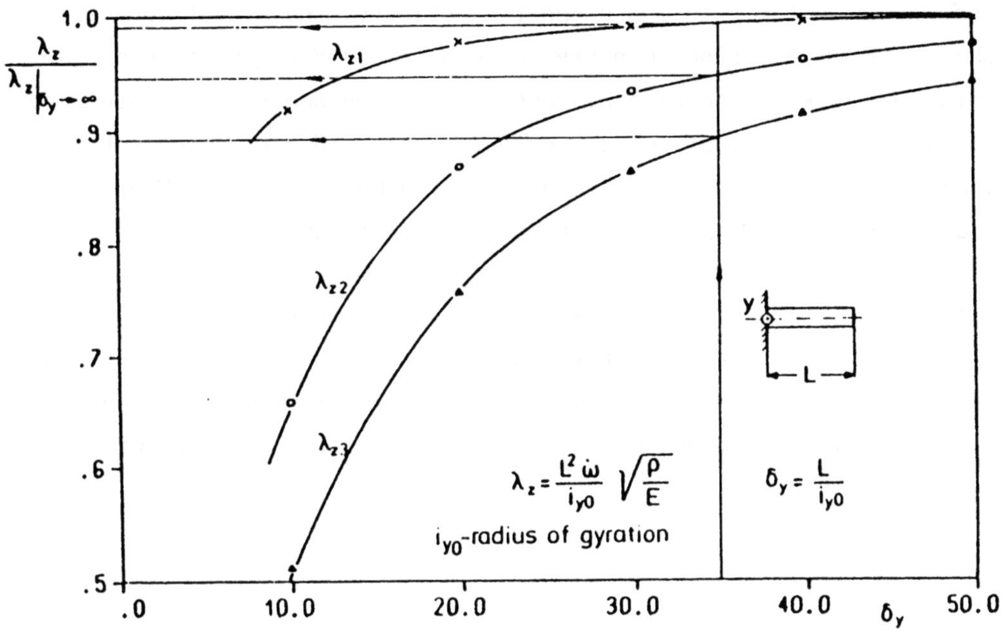

Fig. 10 Influence of shear deformation and rotary inertia on the lateral frequencies of cantilever beams

For example:

A beam with rectangular cross section (30 mm x 10 mm, l = 100 mm) is considered.

$$\rightarrow \delta_y = \frac{L}{i_y} = \frac{L}{(h/\sqrt{12})} = \frac{100}{2.887} = 34.6 \ .$$

The third eigenfrequency is lowered by nearly 11%.

4.5. Support elasticity

Absolutely rigid support implies that no displacement or slope occurs. Owing to the existing support elasticity the eigenfrequencies of the blades fixed at one end are reduced compared to the case of rigid clamping. Even in an assembly where the blade and the disk are manufactured in one piece, at the blade root with x = 0 a displacement or a slope will be induced under the action of a bending moment M or a lateral force Q. BEGLINGER takes these facts into account [6].

The elastic support of the blades is difficult to define, with no general validity in the case of turbomachines, since different constructions of root fixings on disk are used. Often the influence of the elastic support of the blades is described for the first eigenfrequency by the following relation:

$$\omega_{el} = \varphi \cdot \omega_{rigid}$$

ω_{rigid} - eigenfrequency for rigid supported blades

ω_{el} - eigenfrequency for elastic supported blades

φ - measured factor

Fig. 11 shows the factor ψ for different shapes of roots [2].

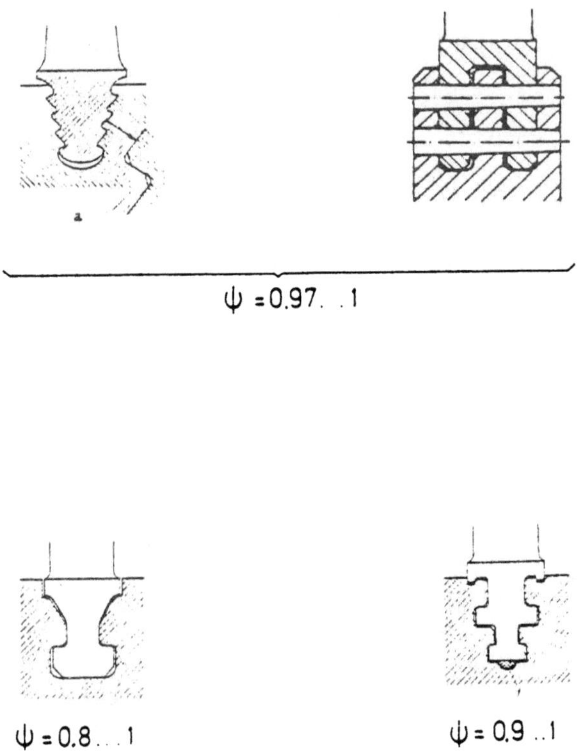

4.6 Coupled vibrations of the disk and its blades

It is important to consider the combined vibrations of the blades mounted on slender disks. For all possible vibrations of the disk and the blade, the differential equations are given in [8]. Each of them can be considered isolated from the other, if well defined boundary conditions can be found either for the disk at the bore and the rim or the blades at their roots or ends. However, to calculate the natural vibrations of the whole system of bladed disk, it is necessary to find forces and displacement relationships at the disk rim and the blade roots (Fig. 12). This will be done by assuming a large number of identical blades, so that the displacements and inner forces of all blades vary harmonically in φ around the disk. Moreover a rigid connection between disk rim and blade root is supposed.

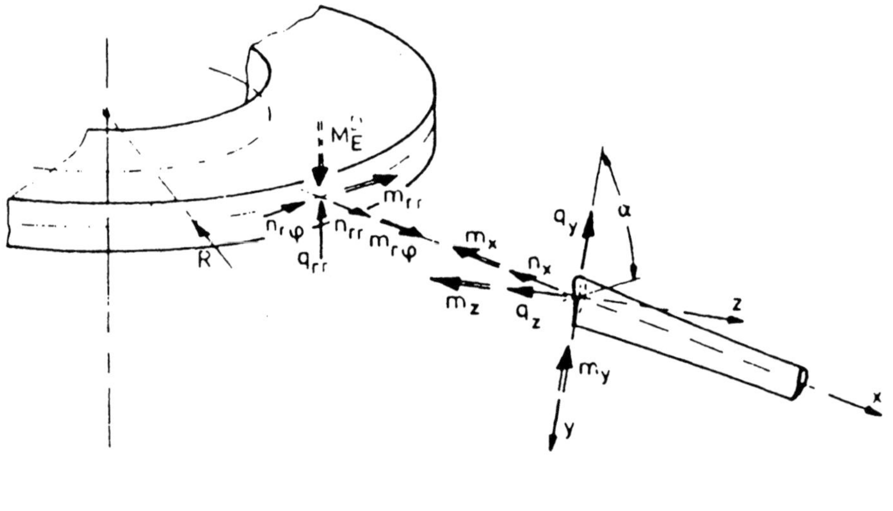

Fig. 12 Connectivity of the disk rim to the blade roots

In Fig. 13, the dimensionless eigenfrequencies, λ_z, for the tangential vibrations of non-rotating blades of constant thickness and with stagger angle α = 90 deg are plotted taking into account the torsional vibrations of the disk in its plane. The first natural mode of vibration is shown in Fig. 13, too. The dashed lines show the first three eigenfrequencies λ_z^B of the blades given by Eulers beam theory when rigid clamping is assumed at the blade root. Fig. 13 shows that there is no effect of disk elasticity on the blade frequencies for small values of $\varepsilon = R/L$. However, for $\varepsilon > 1$ the eigenfrequencies of the system will be considerably below those of the rigid clamped blade, specifically for the higher eigenfrequencies and small values of μ.

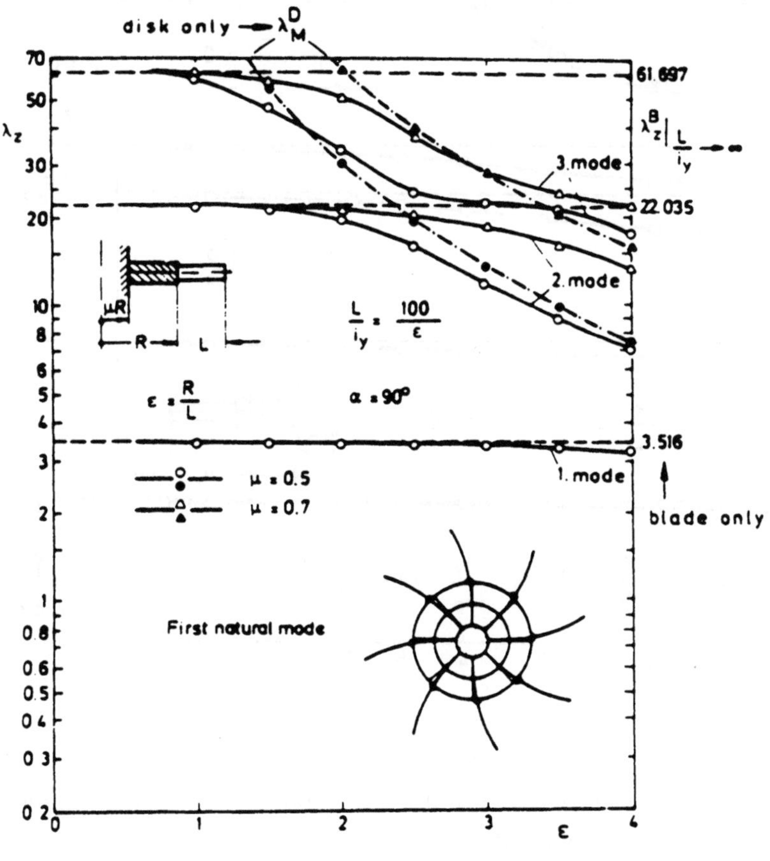

Fig. 13 Eigenfrequencies for torsional vibration of a bladed disk

For stagger angle α = 0 deg. the natural motion of the system consists of bending vibration of the disk and bending and torsional vibration of the blades. Fig. 14 shows the eigenfrequencies over the K nodal diameter for a rotating bladed disk. The horizontal lines represent the first and second bending eigenfrequencies of the blade with rigid clamped support. The dash-dotted lines represent the first eigenfrequencies of bending vibration of the unbladed disk, both for the non-rotating and the rotating case, respectively. The continuous curves show the eigenfrequencies for the bladed disk.

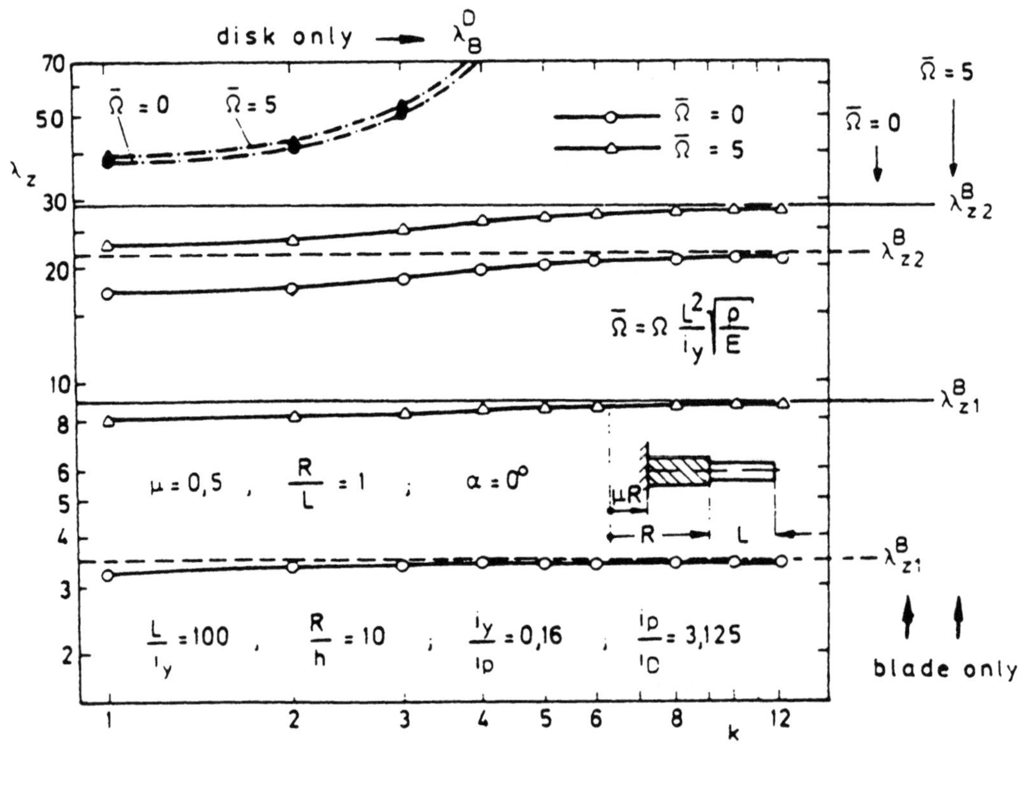

Fig. 14 Eigenfrequencies for bending vibrations of a rotating and a non-rotating bladed disk

Usually the stagger angle α of the blades is between 0 and 90 deg. In this range of α the vibrations of the disk (in-plane and out of-plane) and the torsional and bending vibrations of the blades are completely coupled. A typical mode is shown in Fig. 15. The continuous lines show the eigenfrequencies λ_z of the bladed disk for the two cases α = 50 deg. and α = 80 deg..

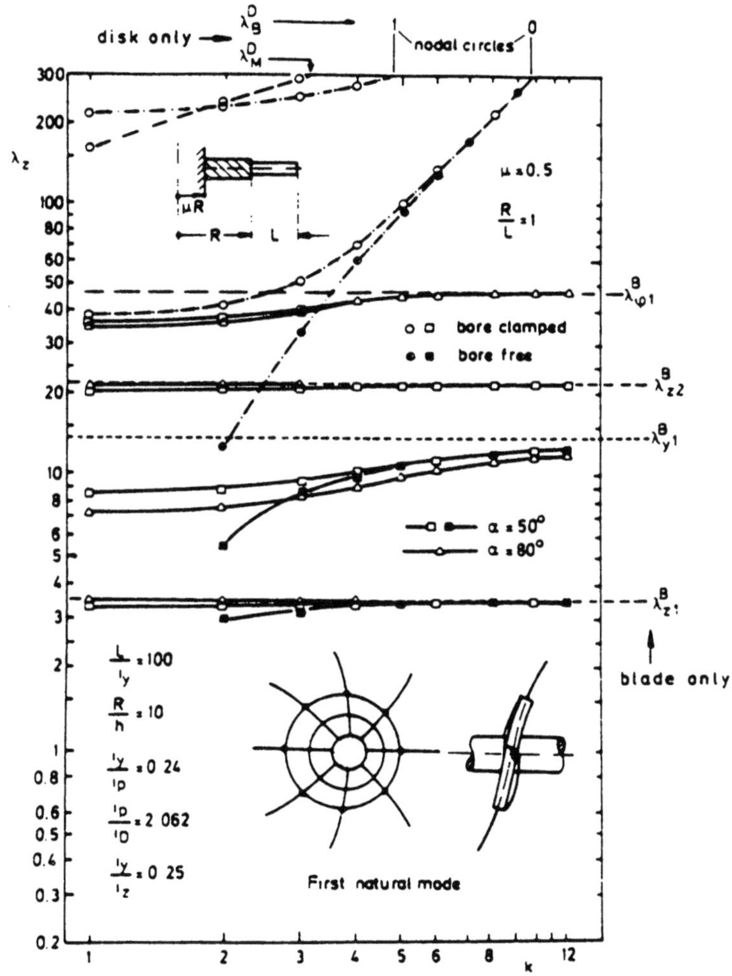

Fig. 15 Eigenfrequencies of the in-plane and out-of-plane vibrations of a bladed disk

An important effect in bladed disk vibrations arises by mistuning the assembly. As Ewins [10] pointed out frequency splitting and shifting occur, and both cause significant change in the dynamical behavior of the system when the response to harmonic excitation is considered. Some results on this topic were recently reported by IRRETIER and SCHMIDT in [11]. As an example, Fig. 16 shows the blade tip deflections of a 24-bladed disk assembly of the first row of 24 modes for both a tuned and a statistically mistuned system.

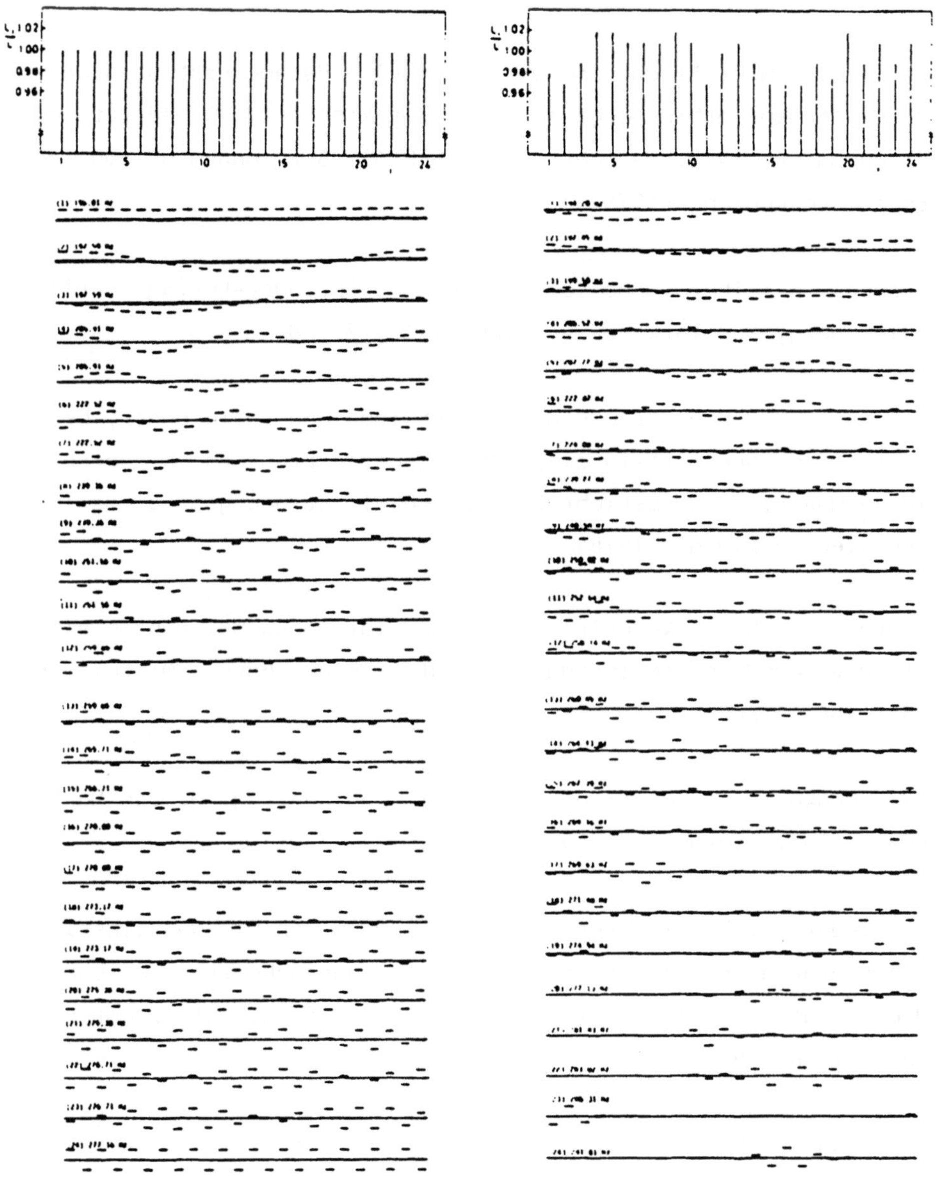

Fig. 16 Mode shapes of a tuned and a statistically mistuned bladed disk

References

[1] Böhm, J.: Theoretische und experimentelle Parameterstudien an schwingenden Turbinenschaufeln im Fliehkraftfeld. Dissertation, TU Hannover, 1978. Bericht 27 des SFB 61 'Strömungsprobleme in der Energieumwandlung (1978)

[2] Traupel, W.: Thermische Turbomaschinen, Springer-Verlag, Berlin, Heidelberg, New York, 2. Auflage, 1968, 2. Bd.

[3] Irretier, H.: Die Berechnung der Schwingungen rotierender, beschaufelter Scheiben mittels eines Anfangswertverfahrens. Dissertation, TH Hannover, 1978. Bericht 28 des SFB 61 'Strömungsprobleme in der Energieumwandlung' (1978)

[4] Montoya, J. G.: Gekoppelte Biege- und Torsionsschwingungen einer stark verwundenen rotierenden Schaufel, Brown Boveri Mitteilungen 53 (1966) 216-230

[5] Timoshenko, S.: Vibration Problems in Engineering. D. von Nostrand Comp., Inc., Toronto, 1955

[6] Beglinger, V., Schlachter, W.: Der Einfluß von Einspannelastizität, Schubdeformation, Rotationsträgheit und Querschnittsverwölbung auf die Eigenfrequenzen von Turbomaschinenschaufeln. Sulzer-Forschungsheft, 1978, 19-26

[7] Belohradsky, H.-J.: Beitrag zur Berechnung der Eigenfrequenzen und Eigenformen von Schaufelpaketen in Dampfturbinen mit variabler Drehzahl. Dissertation, TU Hannover, 1978

[8] Irretier, H.: Coupled Vibrations of Blades in Bending-Bending-Torsion and Disks in Out-of-Plane and In-Plane Motion. ASME-Paper 79-DET-90 (1979) 1-12

[9] Vogt, E.: Untersuchung der Schwingungsformen von Turbinenschaufeln. Seminarbericht des SFB 61 'Strömungsprobleme in der Energieumwandlung (1982) 4-22

[10] Ewins, D.J.: Vibration Characteristics of Bladed Disk Assemblies. J. Mech. Eng. Sci. 15 (1973) 165-186

[11] Irretier, H., Schmidt, K.J.: Mistuned Bladed Disks - Dynamical Behavior and Computation, Proc. IFToMM Conf. on Rotordynamics Problems in Power Plants, Rome (Italy) (1982) 215-226

CHAPTER 5.3

TORSIONAL VIBRATIONS IN LARGE TURBINE GENERATOR SETS

J. Drechsler

1. Excitation Sources for Torsional Vibrations in Turbo Generators

The long turbogenerator shaft trains in modern power plants are loaded by considerable torsional moments, which have to be transmitted between the turbine and the generator. While the torsional moments under stationary full load conditions are well within tolerable limits from a material strength point of view, the oscillating torsional moments imposed on the shaft under transient conditions are large enough to eventually cause a material fatigue failure in the shaft. Thus it is necessary to both control excessive torsional vibration and to check the number and magnitude of excessive torsional vibration incidents, in order to perform life time calculations.

Torsional vibrations will be excited by any sudden load change. Such transient operating conditions can give rise to both sudden changes in the electrical moment and to sinusoidal electrical moments in the generator which can excite dynamic torsional moments exceeding the stationary values by factors up to five.

Among exceptional operating conditions which initiate torsional vibrations in the turbogenerator shaft train are

- short circuit across the generator terminals following full load operation, a classical design case;
- faulty synchronizing;
- transmission line faults (short circuits);
- transmission line switching operations.

Moreover it has been discovered, that when series capacitors are used in transmission lines electrically close to the generator, steady state and transient currents can be generated at frequencies below the normal power system frequency. The existence of such currents causes alternating torques in the generator which can reinforce natural vibration modes of the shaft and result in significant dynamic torques at or near the shaft couplings.

For numerical calculations it has to be observed, that these above mentioned exceptional operating conditions typically occur in sequence:

The first impact on the generator might be a sudden short circuit in a transmission line close to the generator, which in electrical engineering is frequently called fault application.

The second impact results from an electrical disconnnection of the faulty transmission line, which is called fault clearing or fault removal. Usually this is all that has to be done to extinguish the electrical arch, which has been initiated for some reason (typically lightning) and short circuits the line.

After a brief disconnection the third impact is applied, when the line is electrically reconnected. This is also done automatically, the technical term for this operation is high speed reclosing.

High speed reclosing can be successful if the fault is effectively cleared by the disconnection of the line or unsuccessful, if the fault persists after reclosing.

In this case a fourth impact is applied by a second clearing.

As these consecutive shocks are applied, while the shaft still oscillates due to the previous shock, the different shocks can add up to very high dynamic shaft torques, depending on the timing interval between the shocks. These torques can be much higher than in the terminal short circuit case, which has been considered the worst design case until just a few years ago.

1.1. Subsynchronous Resonance

In the subsynchronous resonance case electrical net oscillations are in resonance with some torsional mode in the shaft system. As the load on the electrical power system is inductive to a large extent, it is common practice to compensate the system by series capacitors. Thus the system has the possibility to oscillate with a natural frequency, which according to figure 1 can be calculated to

$$f_c = \sqrt{X_C/X_E} \tag{1}$$

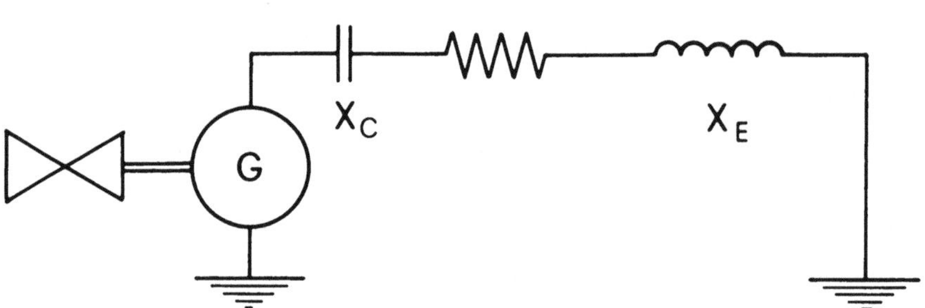

Fig. 1: Turbogenerator with compensated transmission

From a physical point of view the fundamental phenomenon of electrical self excitation of a rotating electrical machine may be regarded as operation as an induction generator. For a constant rotor speed and linear system characteristics the self excited currents may be initiated and grow quite independently of any applied armature voltage or field excitation. If there are no applied voltages, the machine will develop a torque which would in practice grow to some constant value. However if there are applied voltages, the self excited currents will interact with the normal rated frequency currents to generate a pulsating torque, which would be primarily at slip frequency $f_s = f_o - f_e$.

It is this pulsating torque, which may excite torsional movement of the rotor, if the slip frequency is near torsional resonance. Then, in turn, the now appreciable voltages generated by the rotor oscillations will themselves produce relatively large currents which may be said to reinforce the self excited currents.

For the convenience of those primarily concerned with mechanical vibration, the analogy between the mechanical and the electrical system dynamics is summarized below

mass	m	reactance	L
flexibility	$1/k$	capacitance	C
damping	c	resistance	R
deflection	x	charge	$q = \int i\, dt$
velocity	$\dot{x}$	current	$\dot{q} = i$
force	p	voltage	u

$$m\ddot{x} + c\dot{x} + kx = p(t) \qquad L\, di/dt + Ri + 1/C \int i\, dt = u(t)$$

2. Measurement of Torsional Vibrations

Torsional vibrations of rotating shafts cannot be perceived as easily as lateral vibrations and their measurements presents some problems. In principle they can be detected as frequency modulations according to figure 2.

The magnetic pick ups detect the instantaneous frequency of the passing teeth, which is composed of the rotational speed plus or minus the vibration velocity transversal to the sensors. The effects of vibration can be accounted for by measuring on both sides of the shaft and taking the sum of both signals.

But as shaft speed variations are small compared to the rated speed, even a very simple digital time measuring method can be used instead of the frequency demodulation method. The signal can be derived in this case from proximity probes measuring on metal pieces, which can easily be glued to any accessible shaft section, so that they form a tooth wheel like pattern. The two proximity probes will produce modulated sinusoidal signals, which can

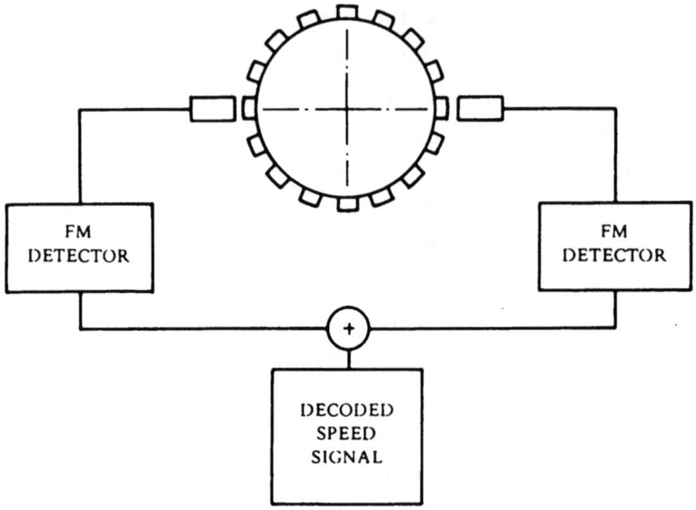

Fig. 2: Instantaneous speed measuring system

be added analogously. The resulting signal will contain frequency modulations due to torsional vibration and partition errors, see figure 3. This signal can be pulse shaped and the period time between pulses can be measured digitally with a cristal clock.

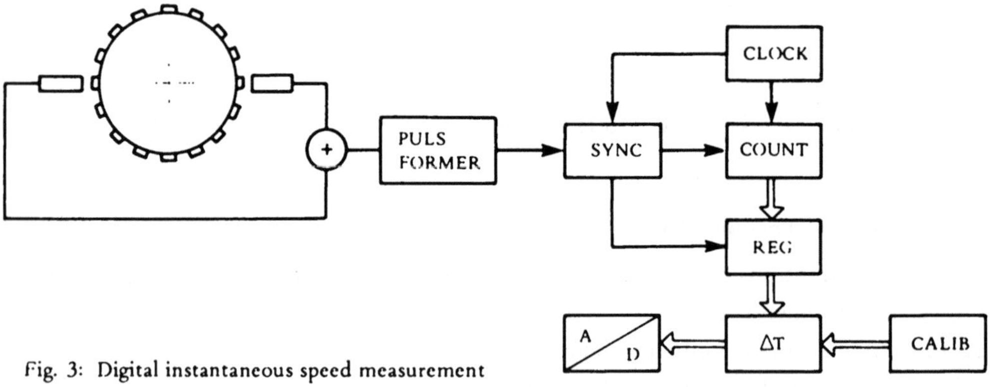

Fig. 3: Digital instantaneous speed measurement

The partition errors can be accounted for by a simple and obvious calibration procedure performed during steady state conditions with no torsional vibration present.

The period time for an ideal signal without partition errors T_i is related to the raw period time between two pulses by the relationship

$$T_i = T_{ri} \, \phi_i / (\phi_i + \Delta\phi_i) \tag{2}$$

with ϕ_i being the partition angle and $\Delta\phi_i$ the partition error.

At constant speed the angles ϕ_i and $\Delta\phi_i$ can be expressed by their time equivalents, which yields

$$\phi_i = \omega T_i$$

$$\Delta T_{cal,i} = T_{ri} - T_i \tag{3}$$

All additional variations in period time are then due to variations in the instantaneous rotor speed.

Again because of small variations in speed, we can calculate

$$\omega_i = 1/T_i = 1/T_o \ (1 - \Delta T_i/T_o) = \omega_o + \Delta\omega \tag{4}$$

$$\Delta\omega = - \Delta T_i/T_o^2 \tag{5}$$

where ΔT_i can be calculated to

$$\Delta T_i = T_{ri} - \Delta T_{cal,i} - T_o \tag{6}$$

The advantage with these calculations is that they can easily be performed in real time.

3. Reconstruction of Torsional Shaft Dynamics by Means of an Observer

In order to get a complete picture of the torsional vibrations in a turbo-generator shaft train, several measuring stations have to be installed. Because this can be practically very difficult, the most advanced torsional monitor equipment determines torsional shaft vibrations and associated shaft torques by means of an observer. Such an observer consists of a torsional analog model of the rotor, which is supplied with information on the instantaneous generator moment and the turbine moments (u) and the instantaneous rotor speed (y) in at least one measuring point.

Figure 4 shows the principal structure of such an observer. The matrix A contains a modal shaft model in state space representation. The matrix B is the input matrix for the control variables u, in this case the external moments on the shaft, C is the output matrix which represents a model of the measuring system and K is the feedback matrix, which supplies the model with information on the behaviour of the real system.

For the determination of the matrix A a modal analysis of the torsional dynamics of the shaft system has to be performed. This the most effective approach, because only

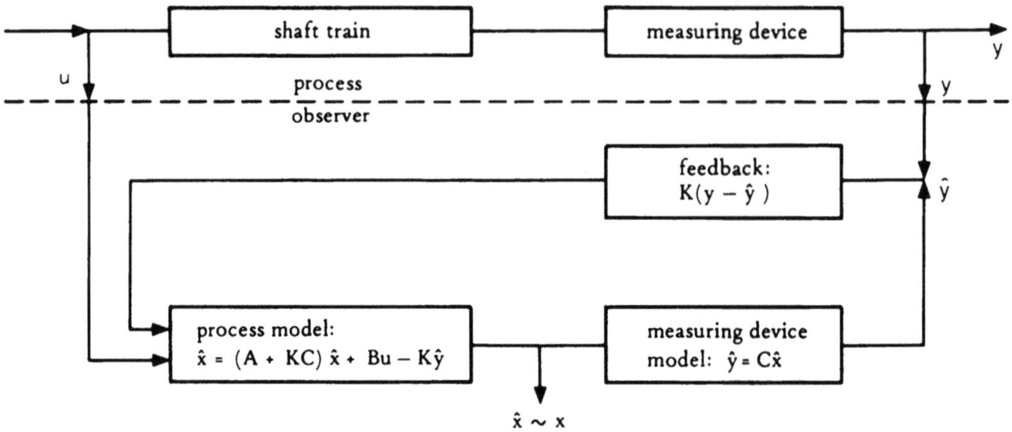

Fig. 4: Structure of an observer

relevant torsional modes have to be considered in the model. Thus the number of unknowns is reduced.

The matrix B is determined by relating the measurable terminal voltages and phase currents at the generator to the air gap moment and the stream pressure at the turbines to the turbine moments. This moment distribution has to be analysed into modal components, which distributes the torsional excitation to the individual modes.

The output matrix C relates the chosen state space variables, that is the amplitudes of the torsional modal components to the resulting instantaneous speed at the chosen measuring points, by means of the performed modal analysis of the system.

The feedback matrix K can be calculated by preassigning the roots of the observer matrix (A + KC) to a suitable set Λ_n. These roots should be chosen considerably better damped and larger than the eigenvalues of the shaft train.

3.1. Governing Equations for Torsional Vibrations

Consider a rotor element of the length dx according to the figure 5. Newtons law applied to this rotor element states that the resulting torque on the element must be equal to the rotational inertia J(x) dx multiplied by the angular acceleration.

$$J(x)\ dx\ d^2\varphi/dt^2 - m_T(x)\ dx - dM_T/dx\ dx = 0 \tag{7}$$

Hooke's law relating the torsional deformation $d\varphi/dx$ to the torsional moment $M_T(x)$ acting on the element, reads

$$M_T(x) = K(x)\ d\varphi/dx \tag{8}$$

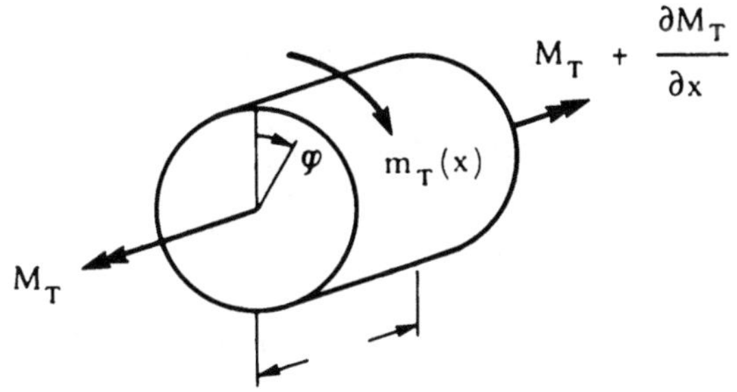

Fig. 5: Equilibrium for shaft element

The torsional stiffness K(x) is related to the material property and shaft geometry by the relationship

$$K(x) = G\ I_T(x) \tag{9}$$

where G represents the shear elastcity of the material and I_T the cross sectional torsional resistance.

Differentiating eq. (8) and substituting into eq. (7) we obtain

$$J(x)\ \ddot{\varphi}(x) = (K(x)\ \varphi'(x))' + m_T(x) \tag{10}$$

The boundary condition for a usual shaft is that there is no internal torque acting at the ends of the shaft, that is

$$\begin{aligned} M_T(0) &= 0 \quad \varphi'(0) = 0 \\ M_T(1) &= 0 \quad \varphi'(1) = 0 \end{aligned} \tag{11}$$

The solution to these equations is

$$\varphi(x, t) = \varphi_0 + \omega t + \sum_{n=1}^{\infty} \phi_n(x)(a_n \sin \lambda_n t + b_n \cos \lambda_n t) \tag{12}$$

where the λ_n represent the as yet unknown eigenvalues and the ϕ_n the torsional mode shapes of the shaft.

3.2. Numerical Solution of the Torsional Eigenvalue Problem

For the actual determination of the mode shapes and the resonant frequencies we can use a very simple numerical procedure, which has been described some 60 years ago by

Holzer and Tolle.

Substituting the solution ϕ_n eq. (12) into eq. (10) and (11) we obtain

$$d/dx(K(x) \cdot d/dx\phi_n(x)) + \lambda_n^2 J(x)\phi_n(x) = 0 \tag{13}$$

$$d/dx\,\phi_{n,x=o,L} = 0 \tag{14}$$

Rewriting eq. (13) as two first order equations by reintroducing the internal torque, we obtain

$$d/dx\,\phi(x) = M_T(x)/K(x) \tag{15}$$

$$d/dx\,M_T(x) = -\lambda_n^2 J(x)\phi(x) \tag{16}$$

These equations can easily be integrated for an assumed λ_n, by f.i. rewriting them as finite difference formulas

$$\phi_{i+1} = \phi_i + M_i/K_i\,\Delta x \tag{17}$$

$$M_{i+1} = M_i - \lambda_n^2 J_i\,\phi_i\,\Delta x \tag{18}$$

Starting with the appropriate boundary conditions $\phi_o = 1$ and $M_o = 0$ and integrating over the whole shaft length, we get a residual moment M_r at the last shaft section, which is a function of the selected value for λ. Plotting this function we will get a graph similar to figure 6. Wherever the residual moment is zero we have found an eigenvalue λ_n, the associated mode shape ϕ_n and the modal torque distribution M_{Tn}.

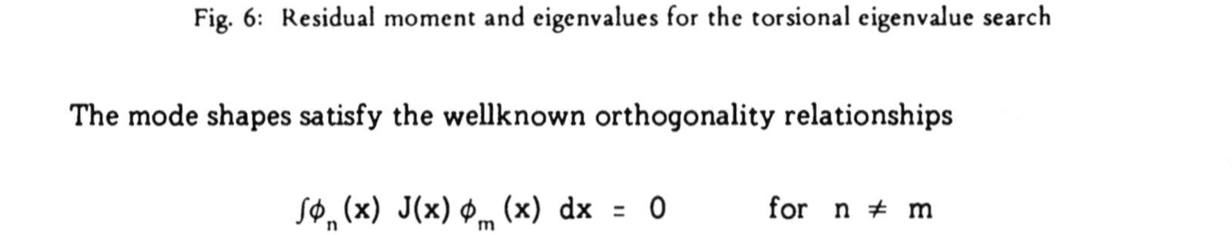

Fig. 6: Residual moment and eigenvalues for the torsional eigenvalue search

The mode shapes satisfy the wellknown orthogonality relationships

$$\int\phi_n(x)\,J(x)\,\phi_m(x)\,dx = 0 \quad \text{for } n \neq m$$

$$= J_n \quad \text{for } n = m \tag{19}$$

which yield the modal inertia J_n associated with the n-th mode shape, and the generalized stiffness, which is

$$K_n = \lambda_n^2 \; J_n \tag{20}$$

The influence of the exciting torsional moments on the individual modes, that is the generalized torsional moments can be calculated from the relationship

$$m_{Tn}(t) = \int m_T(x, t)\, \phi_n(x) \; dx \tag{21}$$

The instantaneous speed response in the selected measuring point x_i is

$$\varphi(x_i, t) = \sum_{n=1}^{N} \phi_n(x_i) \; q_n(t) \tag{22}$$

The $q_n(t)$ are the generalized coordinates, which are equal to the selected state space variables denoted x in figure 4 according to the terminology of the control theory. The instantaneous torque distribution is obtained by substituting the modeshape function in eq. (22) by the modal torque distribution $M_{Tn}(x)$.

4. Torsional Damping Characteristics of Large Turbo Generator Units

While the exact determination of the damping characteristics is not necessary for the construction of the observer, the assessment of realistic damping factors is indispensable for the estimation of the influence of transient loads on shaft fatigue and related life expectancy calculations.

The effective torsional damping consists of a superposition of material damping, oil film damping, steam friction damping and electrical damping, each of which is a complicated function depending on many different parameters, as there are the magnitude of torsional stress, the load on the machine and oil viscosity.

Thus the damping characteristics can only be assessed by measurements. In (15) the results of comprehensive measurements on four different shaft trains are reported. The authors used a modal approach and determined the modal damping associated with the first three torsional modes. Typical for all the measured damping curves is a decrease of the damping coefficient during the decaying time, as to be seen in figure 7. This is due to the non linear material damping which depends on the magnitude of the torsional stress and is increasing for higher stresses.

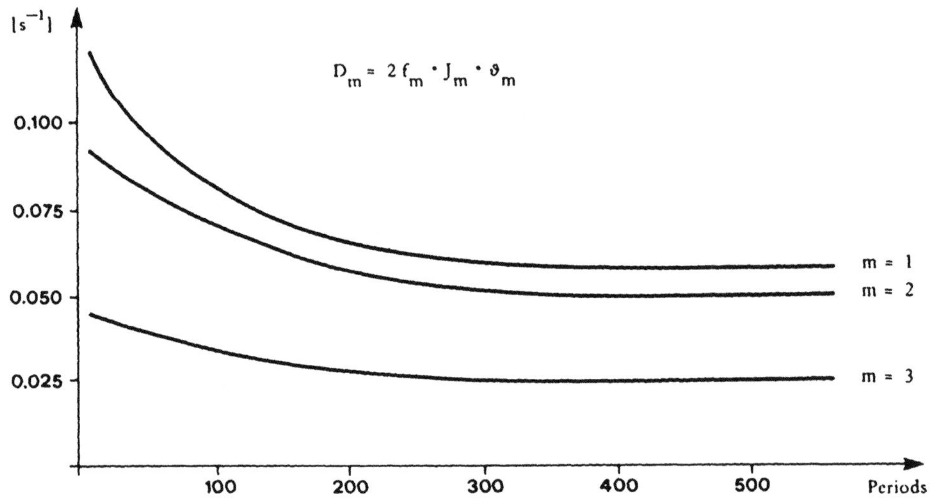

Fig. 7: Modal damping coefficients for mode 1 to 3 from experiments by BERGER and KUHLIG with a 1000 MW, 3000 RPM turbo generator unit

5. Control of Torsional Vibrations in Turbo Generator Units

Due to the nature of the problem most of the countermeasures to control vibrations in turbo generator units aim to reduce the excitation sources on the electrical side of the system. They can be classified as follows:

5.1. Design of the Turbo Generator Unit

- use of low resistance pole face damper windings to reduce the net negative resistance of the generator at subsynchronous frequencies:
- design of the shaft train so that the lowest torsional natural frequency is greater than the subfrequency ($f_0 - f_e$) of the system;
- use of series reactances between the generator and the transmission system, to detune the resonant net work as seen by the generator.

5.2. Design of the Transmission System

- layout of the series compensation such that the subharmonic frequency of the net work as seen from the generator is lower than the lowest torsional resonant frequency;
- connection of an appropriate sized reactor parallel to the series capacitor so that they will block the flow of subsynchronous currents;
- connection of a bypass damping filter for the subsynchronous oscillations parallel with the series compensation consisting of a damping resistor in series with a parallel combination of a reactor and a capacitor tuned to the system frequency.

5.3. Preventive Clearing

- fast fault clearing within a few cycles to reduce larger phase angle miss matches and consequent closing shocks;
- unit tripping, as back up protection in case of long duration short circuit;
- blocking relays for reclosing or synchronisation sensitive to the difference between the generator and the system angles.

5.4. Filtering and Damping Close to the Generator

- use of blocking filters in series with the generator tuned to the subsynchronous frequencies consisting of reactors and capacitors in parallel;
- use of bypass damping filters in series with the generator tuned to the system frequency consisting of a reactor and capacitor in series with a damping resistor in parallel;
- use of an active filter in series with the generator to cancel the subsynchronous voltage generated by the rotor oscillations, thus preventing self excitation due to the interaction of the electrical and the mechanical system;
- use of a tyristor controlled reactor in parallel with the generator modulated in response to measured oscillations of the generator;
- use of additional signals derived from rotor oscillations in the generator exciter system. By properly phased shifting and amplifying these signals damping to subsynchronous oscillations is provided.

5.5. Operation of the Power System

- scheduling of system configuration and degree of series compensation as to avoid high transient torques or subsynchronous resonance.

Unfortunately restrictions exist for practically all the above mentioned counter measures..

In the turbogenerator shaft train there are constraints on shaft and bearing size, which practically determine the range of the torsional natural frequencies.

Reducing damper winding resistance lowers the main damping effects.

Inserting series reactances between the generator and the system always reduces the system stability margin.

Only in radial transmission systems a direct relationship exists between the degree of series compensation and the subsynchronous resonance frequency to a reduction of the stability margin.

In an interconnected system the resonance frequencies are resulting from the layout. Avoiding ali subsynchronous resonance problems would result in an uncompensated system.

Excitation systems with additional signals derived from rotor oscillations will provide damping only to small steady-state oscillations.

REFERENCES

[1] Hammons, T.J.: Shaft torsional phenomena in governing large turbo generators with non linear steam valve stroking constraints. IEEE/P.E.S. Winter meeting, New York, Febr. 1980, Paper 80-053.

[2] Raczkowski, C., Kung, G.C.,: Testing T/Gs – torsional vibration. Electrical world, April 1980.

[3] Kazovskii, Rubisov, Askenova and Livshits: Transients in shaft trains of large turbine sets. Eletrotekhnika, vol. 50, no. 1(1979).

[4] Rodgers, G. and Smith, J.: Mathematical models of synchronous electrical machines. Int. J. for numerical methods in Engineering, vol. 6, (1973).

[5] Triezenberg, D.M.: An efficient state variable transmission line model. IEEE Trans. on Power Apparatus and Systems, Vol. PAS-98 No. 2 March/April 1979.

[6] Ahlgren, L., Johanson, K.E., Gadhammar, A.: Estimated life expenditure of turbine-generator shafts at network faults and risk for subsynchronous resonance in the swedish 400 kV system. IEEE Trans. on Power Apparatus and Systems, vol. PAS-97, no. 6, Nov/Dec 1978.

[7] Ballance, J., Goldberg, S.: Subsynchronous resonance in series compensated transmission lines. Paper T 73 167-4, IEEE, PES Winter Meeting, New York, Jan/Febr 1973.

[8] Bowler, C., Ewart, D.: Self excited torsional frequency oscillations with series capacitors. Paper T 73 218-5, IEEE, PES Winter Meeting, New York, Jan/Febr 1973.

[9] Goldberg, S., Schmus, W.: Subsynchronous resonance and torsional stresses in turbine-generator shafts. IEEE Trans. on Power Apparatus and Systems, vol. PAS-98, no. 4, July/Aug. 1979.

[10] Triezenberg, D.: Characteristic frequencies and mode shapes for turbogenerator shaft torsional vibrations. IEEE Trans. on Power Apparatus and Systems, vol. PAS-99, no. 1, Jan/Febr. 1980.

[11] Gilles, E.: Modelling – a measurement aid. Process Automation 1980.

[12] Stein, J., Fick, H.: The torsional stress analyser for continuously monitoring turbine generators. IEEE Trans. on Power Apparatus and Systems, vol. PAS-99, no. 2, March/April 1980.

[13] Wilharm, H., Fick, H.: Einsatz eines Beobachters zur genauen Bestimmung von Torsionschwingungen. VDI Berichte Nr. 320, 1978.

[14] Hurley, J., South, W.: Torsional monitor equipment for turbine generator units. The American Power Conference, Illinois Inst. of Technology, Chicago, III, April 1979.

[15] Berger, H., Kuhlig, T.: Erfassung der Dämpfung bei Torsionsschwingungen grosser Turbosätze. VDI Berichte Nr. 320, 1978.

[16] Barnett, S.: Introduction to mathematical control theory. Clarendon Press, Oxford 1975.

LIST OF CONTRIBUTORS

Jürgen DRECHSLER, Gardeschützenweg 133, D-1000 Berlin 45, Federal Republic of Germany.

Lothar GAUL, Hochschule der Bundeswehr, 2000 Hamburg 70, Federal Republic of Germany.

Burkhard GRABOWSKI, Institut für Mechanik, Universität Hannover, Appelstrasse 11, 3 Hannover 1, Federal Republic of Germany.

Oskar MAHRENHOLTZ, Institut für Meerestechnik, Technische Universität Hamburg-Harburg, Eissendorferstrasse 38 (Technikum), 2100 Hamburg 90, Federal Republic of Germany.

Rainer NORDMANN, University of Kaiserlautern, Pfaffenbergstrasse, Kaiserlautern, Federal Republic of Germany.

Zdsislaw A. PARSZEWSKI, The University of Melbourne, Parkville, Victoria, 3052, Australia.

Neville F. RIEGER, Stress Technology Incorporated, 1800 Brighton-Henrietta Town Line Road, Rochester, New York 14623, USA.

Volker SCHLEGEL, Institut für Mechanik, Universität Hannover, Federal Republic of Germany.

Helmut SPRINGER, Institut for Dynamics of Machinery, Technical University of Vienna, Karlsplatz 13, 1040 Vienna, Austria.

MIX
Papier aus verantwortungsvollen Quellen
Paper from responsible sources
FSC® C105338

If you have any concerns about our products,
you can contact us on
ProductSafety@springernature.com

In case Publisher is established outside the EU,
the EU authorized representative is:
Springer Nature Customer Service Center GmbH
Europaplatz 3, 69115 Heidelberg, Germany

Printed by Libri Plureos GmbH
in Hamburg, Germany